Horizons in World Physics, Volume 263

NEW TRENDS IN QUANTUM COHERENCE AND NONLINEAR OPTICS

Horizons in World Physics Series

Volume 263	**New Trends in Quantum Coherence and Nonlinear Optics** Edited by Rafael Drampyan
Volume 262	**Astrophysics and Condensed Matter** Edited Thomas G. Hardwell
Volume 261	**Superconductivity Research Advances** Edited by James E. Nolan
Volume 260	**Instabilities of Relativistic Electron Beam in Plasma,** Edited by Valery B. Krasovitskii
Volume 259	**Self-Focusing of Relativistic Electron Bunches in Plasma,** Edited by Valery B. Krasovitskii
Volume 258	**New Topics in Theoretical Physics,** Edited by Henk F. Arnoldus and Thomas George
Volume 257	**Boson's Ferromagnetism and Crystal Growth Research,** Edited by Emerson D. Seifer
Volume 256	**Horizons in World Physics,** Edited by Victor H. Marselle
Volume 255	**Frontiers in General Relativity and Quantum Cosmology Research,** Edited by Victor H. Marselle
Volume 254	**Optics and Electro-Optics Research,** Edited by Albert V. Berzilla
Volume 253	**Trends in General Relativity and Quantum Cosmology,** Edited by Albert Reimer
Volume 252	**Chemical Physics Research Trends,** Edited by Benjamin V. Arnold
Volume 251	**Quantum Dots: Research Developments,** Edited by Peter A. Ling
Volume 250	**Quantum Gravity Research Trends,** Edited by Albert Reimer
Volume 249	**General Relativity Research Trends,** Edited by Albert Reimer
Volume 248	**Spacetime Physics Research Trends,** Edited by Albert Reimer
Volume 247	**New Developments in Quantum Cosmology Research,** Edited by Albert Reimer
Volume 246	**Quantum Cosmology Research Trends,** Edited by Albert Reimer
Volume 245	**Horizons in World Physics,** Edited by Tori V. Lynch
Volume 244	**Horizons in World Physics,** Edited by Albert Reimer
Volume 243	**Theoretical Physics 2002, Part 2,** Edited by Henk F. Arnoldus and Thomas F. George
Volume 242	**Models and Methods of High-T_c Superconductivity: Some Frontal Aspects, Volume 2,** Edited by J. K. Srivastava and S. M. Rao
Volume 241	**Models and Methods of High-T_c Superconductivity: Some Frontal Aspects, Volume 1,** Edited by J. K. Srivastava and S. M. Rao
Volume 240	**Horizons in World Physics,** Edited by Albert Reimer
Volume 239	**Theoretical Physics 2002, Part 1,** Edited by Henk F. Arnoldus and Thomas F. George
Volume 238	**Theoretical Physics 2001,** Edited by Henk F. Arnoldus and Thomas F. George
Volume 237	**Generalized functions in mathematical physics: Main ideas and concepts,** Edited by A.S. Demidov
Volume 236	**Edge Excitations of Low-Dimensional Charged Systems,** Edited by Oleg Kirichek
Volume 235	**On the Structure of the Physical Vacuum and a New Interaction in Nature (Theory, Experiment, Applications),** Edited by Yu. A. Baurov
Volume 234	**Electron Scattering on Complex Atoms (Ions),** Edited by V. Lengyel, O. Zatsarinny and E. Remeta
Volume 233	**An Introduction to Hot Laser Plasma Physics,** Edited by A.A. Andreev, A.A. Mak and N.A. Solovyev
Volume 232	**Electron-Atom Scattering: An Introduction,** Maurizio Dapor
Volume 231	**Theory of Structure Transformations in Non-equilibrium Condensed Matter,** Edited by Alexander L. Olemskoi
Volume 230	**Topics in Cosmic-Ray Astrophysics**, Edited by M. A. DuVernois
Volume 229	**Interactions of Ions in Condensed Matter**, Edited by A. Galdikas and L. Pranevièius
Volume 228	**Optical Vortices**, Edited by M. Vasnetsov and K. Staliunas
Volume 227	**Frontiers in Field Theory**, Edited by Quantum Gravity and Strings, R.K. Kaul, J. Maharana, S. Mukhi and S. K. Rama
Volume 226	**Quantum Groups, Noncommutative Geometry and Fundamental Physical Interactions**, Edited by D. Kastler, M. Russo and T. Schücker
Volume 225	**Aspects of Gravitational Interactions**, Edited by S.K. Srivastava and K.P. Sinha
Volume 224	**Dynamics of Transition Metals and Alloys**, Edited by S. Prakash
Volume 223	**Fields and Transients in Superhigh Pulse Current Devices** Edited by G.A. Shneerson
Volume 222	**Gaussian Beams and Optical Resonators,** Edited by A. N. Oraevsky
Volume 221	**Laser Cathode-Ray Tubes,** Edited by Yu. M. Popov
Volume 220	**Quantum Electrodynamics with Unstable Vacuum,** Edited by V.L. Ginzburg
Volume 219	**The Muon Method in Science,** Edited by V.P. Smilga and Yu.M. Belousov
Volume 218	**Radiative and Collisional Characteristics of Ions in Hot Plasmas,** Edited by I.I. Sobel'man
Volume 217	**Nonlinear and Quantum Optical Phenomena in Nonequilibrium Media** , Edited by V.A. Shcheglov

Volume 216	**Semiconductors and Insulators: Optical and Spectroscopic ...,** Edited by Yu.I. Koptev
Volume 215	**Relativistic Theory of Gravity,** Edited by A.A. Logunov
Volume 214	**Research in Quantum Field Theory**, Edited by V.I. Man'ko
Volume 213	**Nonlinear Theory of Strong Electromagnetic Wave-plasma...,** Edited by O.N. Krokhin
Volume 212	**Dynamics of Elementary Atomic-Molecular...**, Edited by O. S. Scheglov
Volume 211	**Optics and Lasers,** Edited by G.G. Petrash
Volume 210	**Theory and Technology of High Temperature Superconductors,** Edited by Yu.I. Koptev
Volume 209	**Theory of Interaction of Multilevel Systems...**, Edited by V.I. Man'ko
Volume 208	**Physical Properties of High Temperature Superconductors,** Edited by Yu.I. Koptev
Volume 207	**Generalized Mayer Series in Classical Statistical Mechanics**, Edited by Igor J. Ivanchik
Volume 206	**Mathematical Physics, Applied Mathematics and Informatics,** Edited by Yu.I. Koptev
Volume 205	**Squeezed and Correlated States of Quantum Systems**, Edited by M.A. Markov
Volume 204	**Sakharov Memorial Lectures in Physics**, Edited by L.V. Keldysh and V.Ya. Fainberg
Volume 203	**Fiber Optics: Research and Development,** Edited by Ye.M. Dianov
Volume 202	**Gas Dynamics**, Edited by Yu.I. Koptev
Volume 201	**Generation of Nonlinear Waves and Quasistationary ...**, Edited by L.M. Kovrizhnykh
Volume 200	**Laser Thermonuclear Targets and Superdurable Microballoons**, Edited by A.I. Isakov
Volume 199	**Spatiotemporal Characteristics of Laser Emission,** Edited by M.V. Pyatakhin and A.F. Suchkov
Volume 198	**Theory of Nonstationary Quantum Oscillators**, Edited by M.A. Markov
Volume 197	**Reliability Problems of Semiconductor Lasers**, P.G. Eliseev
Volume 196	**Pulsars,** Edited by A.D. Kuzmin
Volume 195	**Physical Characteristics and Critical Temperature of High Temperature Superconductors**, Edited by M.M. Sushchinskiy
Volume 194	**Biophysical Approach to Complex Biological Phenomena,** Edited by E. Volkov
Volume 193	**Research on Chemical Lasers** Edited by A.N. Orayevskiy
Volume 192	**Atomic and Ionic Spectra and Elementary Processes in Plasma** , Edited by I.I. Sobelman
Volume 191	**Polarization of the Vacuum and a Quantum Relativistic Gas ...**, A.Ye. Shabad
Volume 190, (Supp. Vol.)	**Superconductors with A15 Lattice and Bridge Contacts Based ...**, Edited by M. M. Sushchinskiy
Volume 190	**Metal-Optics and Superconductivity**, Edited by A.I. Golovashkin
Volume 189	**Clusters of Galaxies and Extragalactic Radio Sources**, Edited by A.D. Kuz'min
Volume 188	**Integral Systems, Solid State Physics and Theory of Phase Transitions. Pt. 2. Symm. and Alg. Structures in Physics**, Edited by V.V. Dodonov and V.I. Man'ko
Volume 187	**Quantum Field Theory, Quantum Mechanics and Quantum Optics: Pt. 1. Symm. and Alg. Structures in Physics,** Edited by V.V. Dodonov and V.I. Man'ko
Volume 186	**Photoproduction of Pions on Nucleons and Nuclei**, Edited by A.A. Komar
Volume 185	**Injection Lasers in Optical Communication and Information Processing Systems**, Edited by Yu. M. Popov
Volume 184	**Electron Processes in MIS-Structure Memories**, Edited by A.F. Plotnikov
Volume 183	**Invariants and the Evolution of Nonstationary Quantum Systems**, Edited by M.A. Markov
Volume 182	**Luminescence of Wideband Semiconductors**, Edited by M.D. Galanin
Volume 181	**Pulsed Metal and Metal Halogenide Vapor Lasers**, Edited by G.G. Petrash
Volume 180	**Inelastic Light Scattering in Crystals**, Edited by M.M. Sushchinskiy
Volume 179, (Supp. Vol. 2)	**Electron-Excited Molecules in Nonequilibrium Plasma**, Edited by N.N. Sobolev
Volume 179, (Supp. Vol. 1)	**Interaction of Ultrashort Pulses with Matter** , Edited by M.D. Galanin
Volume 179	**X-Ray Plasma Spectroscopy and the Properties of Multiply-Charged Ions**, Edited by I.I. Sobel'man
Volume 178	**The Delfin Laser Thermonuclear Installation: Operational Complex ...**, Edited by G.V. Sklizkov
Volume 177	**Stoichiometry in Crystal Compounds and Its Influence on Their Physical Properties**, Edited by Yu. M. Popov
Volume 176, (Supp. Vol)	**Quantization, Gravitation and Group Methods in Physics** Edited by A.A. Komar
Volume 176	**Classical and Quantum Effects in Electrodynamics** Edited by A.A. Komar
Volume 175, (Supp. Vol.)	**Luminescence and Anisotropy of Zinc Sulfide Crystals** Edited by M.D. Galanin
Volume 175	**Luminescence Centers of Rare Earth Ions in Crystal Phosphors**, Edited by M.D. Galanin
Volume 174, (Supp. Vol. 2)	**Electron Liquid Theory of Normal Metals** Edited by V.P. Silin

Volume 174, (Supp. Vol. 1)	**Nonequilibrium Superconductivity**, Edited by V.L. Ginzburg
Volume 174	**Thermodynamics and Electrodynamics of Superconductors,** Edited by V.L. Ginzburg
Volume 173	**Quantum Mechanics and Statistical Methods** , Edited by M.M. Sushchinskiy
Volume 172	**Phase Conjugation of Laser Emission** , Edited by N.G. Basov
Volume 171	**Research on Laser Theory** , Edited by A.N. Orayevskiy
Volume 170	**The Theory of Target Compression by Longwave Laser Emission** , Edited by G.V. Sklizkov
Volume 169	**The Physical Effects in the Gravitational Field Black Holes**, Edited by M.A. Markov
Volume 168	**Issues in Intense-Field Quantum Electrodynamics**, Edited by V.L. Ginzburg
Volume 167	**Group Theory, Gravitation and Elementary Particle Physics**, Edited by A.A. Komar
Volume 166	**The Nonlinear Optics of Semiconductor Lasers**, Edited by N.G. Basov
Volume 165	**Solitons and Instantons, Operator Quantization**, Edited by V.L. Ginzburg

Horizons in World Physics, Volume 263

New Trends in Quantum Coherence and Nonlinear Optics

Rafael Drampyan
Editor

Nova Science Publishers, Inc.
New York

For permission to use material from this book please contact us:
Telephone 631-231-7269; Fax 631-231-8175
Web Site: http://www.novapublishers.com

LIBRARY OF CONGRESS CATALOGING-IN-PUBLICATION DATA
Available upon request.

ISBN 978-1-60741-025-6

Published by Nova Science Publishers, Inc. ✦ New York

CONTENTS

PREFACE

In memory of my mother

The aim of this book is to present new results in rapidly developing fields of nonlinear optics such as atomic coherence, magneto-optics of atomic media, Doppler-free spectroscopy, time-domain spectroscopy of dense atomic vapors, frequency comb spectroscopy and atom-surface interactions.

The near-resonance interaction of monochromatic and tunable radiation sources with matter allows the determination of atomic and molecular levels with high sensitivity and accuracy. In linear spectroscopy the consideration is restricted to two-level systems. Intense laser fields and irradiation of atomic systems by a single field or simultaneously by several fields produce a variety of nonlinear optical phenomena associated with both two-level and multi-level systems. Among different nonlinear processes associated with three-level atomic systems, the application of two laser fields to the so-called Λ configuration leads to the preparation of the atom in a coherent superposition of states, which is a non-absorbing state. The process was termed *coherent population trapping* (CPT) [1] indicating the presence of a coherent superposition of atomic states. An intimately linked phenomenon is *electromagnetically induced transparency* (EIT) [2], where as a result of a coherent superposition of atomic states in Λ, V or cascade configurations an initially highly opaque medium becomes almost transparent. However, under special conditions the interaction of two laser fields with four-level N- or W-type system leads to the opposite effect - enhancement of absorption resulting from atomic coherence induced by optical radiation, which was termed *electromagnetically induced absorption* (EIA) [3]. Spectral features arising from atomic coherence can be much narrower than the atomic natural linewidth, hence they can be useful in high precision spectroscopy. The light-induced atomic coherences govern not only dissipative, but also dispersive and nonlinear properties of the atomic medium, particularly steep dispersion in the medium.

The effects of atomic coherence have a variety of applications. Enhanced optical nonlinearities in three-state system [4], second harmonic generation [5] and four-wave mixing enhanced by EIT [6], ultraslow [7] and fast [8] light-speed propagation and storage of light in an atomic medium [9], lasing without inversion [1] and application of velocity selective CPT to laser cooling of atoms [10] have been studied. The majority of the atomic coherence investigations have been performed in gaseous media.

Nonlinear magneto-optical processes (NMOP) in an atomic medium under resonant laser excitation [11,12] are important tools in laser spectroscopy and precision magnetometry and have been the subject of extensive investigations. The discovery of the coherent effects such as CPT, EIT and EIA opened a new avenue for development of high resolution spectroscopy and high precision magnetometry, quantum information storage and processing [12], atomic frequency standards [13] and searching for parity violation in atoms caused by weak interaction [14].

Saturated absorption, polarization spectroscopy and ultrafast nonlinear spectroscopy are other powerful methods for high resolution spectroscopy. Polarization spectroscopy has application in the control of spectral features for frequency stabilization of diode lasers, which are widely used as light sources for atomic frequency standards and cooling of atoms.

Investigations of the last decade have shown many related new results and approaches in these fields. These are NMOP studies in the context of high precision magnetometry, quantum information storage and processing, magnetic resonance imaging and magnetic particle detection; the use of ultra-thin (nanometric length) atomic vapor cells in magneto-optical experiments, which show new peculiarities of both coherent effects and NMOP; the use of frequency and amplitude modulated laser beams for NMOP studies in atomic vapors with the aim to extend the range of ultra-precise magnetometers to stronger, particularly geomagnetic fields; application of counter-propagating beam technique to reveal the Doppler free nonlinear Faraday rotation signals, as a spectroscopic tool and for applications; enhanced nonlinearity due to atomic coherence in multi-state systems; experimental study of EIT in solids; the use of the phase stabilized frequency domain combs of femtosecond laser pulses for velocity selective optical pumping of atomic hyperfine levels, as a direct frequency comb spectroscopy; time–domain nonlinear spectroscopy of dense atomic vapors; atom-surface interaction and laser induced adsorption and desorption of alkali atoms on the surface of solid materials for nanotechnology applications.

The book is composed of nine chapters concerning coherent effects and magneto-optical processes and high resolution spectroscopy in atomic systems, written by authors from research centers and University research laboratories in the USA, Australia, UK, Poland, Croatia, Russia and Armenia, all having many years of research experience in their particular fields. The book starts with an introductory review to the field of atomic coherence effects such as CPT, EIT and EIA, emphasizing the relatively new effect of EIA and a review of magneto-optical processes presented by leading experts. The reviews are followed by original articles on quantum coherence effects in the atomic ladder system, high resolution spectroscopy using both conventional cm-length cells and extremely thin nano-cells, time-domain spectroscopy of dense atomic vapors, femtosecond laser frequency comb spectroscopy and atom-surface interactions. The original articles, both experimental and theoretical, describe new trends and ideas in the field of atomic coherence, NMOP and high resolution spectroscopy. The book has a special focus on technical approaches, particularly (i) the preparation of nanometric length vapor cells, (ii) fabrication of high-temperature cells to obtain dense atomic vapor, (iii) the use of Doppler-free signals for laser - frequency stabilization, (iv) adsorption and desorption techniques.

Chapter 1 presented by Alexander Akulshin, Russell McLean, Andrei Sidorov and Peter Hannaford (*Swinburne University of Technology, Melbourne, Australia*), is a review of EIT and EIA and nonlinear effects such as negative or "superluminal" group velocity of light, enhanced nonlinearity and four-wave mixing EIT and EIA media and "storage of light", based on their recent experiments.

In the following *chapter 2* the review on magneto-optical processes is done by Wojciech Gawlik and Szymon Pustelny (*Instytut Fizyki, Uniwersytet Jagiellonski, Krakow, Poland*).

In *chapter 3* David Sarkisyan and Aram Papoyan (*Institute for Physical Research, Armenia*) describe peculiarities of absorption, magneto-optical processes and EIT in experiments with unique ultra-narrow nanometric length atomic vapor cells. The principal scheme of the cell and the technique of cell length variation are presented.

In *chapter 4* Virginia Lorenz (*JILA, University of Colorado and National Institute of Standards and Technology, and the University of Colorado, Boulder, USA*) and Steven Cundiff (*JILA, University of Colorado and National Institute of Standards and Technology, Boulder, USA*) describe time domain nonlinear optical spectroscopy in dense atomic vapors based on their recent results.

In *chapter 5* Ifan Hughes (*Durham University,UK*) gives a theoretical treatment of polarization spectroscopy and experimental results for alkali-metal atoms. The application to laser frequency locking is presented.

Chapter 6 is devoted to the study of the accumulation of population and coherence in the rubidium atom due to the femtosecond (fs) pulse train excitation. The authors developed modified direct frequency comb spectroscopy which combines fs frequency comb and weak probe diode laser; this chapter is written by Ticijana Ban, Damir Aumiler and Goran Pichler (*Institute of Physics, Zagreb, Croatia*).

In *chapter 7* application of quantum coherence in the Rb ladder system to frequency up-conversion, state selective cold atom imaging and quantum squeezing is presented by Robert Scholten, Simon Bell, David Sheludko, James White, Lincoln Turner, Thijs Meijer, Bart Smeets, Matt Jeppesen, Christoph Hofmann and Martijn Jasperse (*School of Physics, University of Melbourne, Australia*).

In *chapter 8* Rafael Drampyan (*Institute for Physical Research, Armenia*), Andrew Greentree (*School of Physics, University of Melbourne, Australia*) and Alan Durrant (*The Open University, UK*) present the results of a joint experimental study on nonlinear Faraday rotation (NFR) in Rb vapor with a counter-propagating beam Doppler-free configuration and discuss the application of dispersive NFR signals to laser- frequency stabilization.

In *chapter 9* Tigran Vartanyan, S.G. Przhibelskii and V.V.Khromov from *St. Petersburg State University of Information Technologies, Mechanics and Optics (Russia)* give a review of their theoretical and experimental results of the last years concerning photo-induced absorption and desorption of alkali atoms on the surface of solid materials. Applications to the manufacturing of surface micro- and nano- structures are discussed.

The combination of reviews and original papers in this book makes the Volume useful both for researchers working in the field covered by the topic of the book and for University faculty and advanced students specializing in the field of laser physics.

I express my gratitude to all contributors for supporting the idea of the book and finding time to write the chapters.

I am very grateful to all my friends and colleagues who supported my efforts in preparation of the book.

Rafael Drampyan,
Ashtarak, Armenia, October, 2008

References

[1] E.Arimondo, "*Coherent Population Trapping in Laser Spectroscopy*" in Progress in Optics, XXXV, edited by E.Wolf (Elsevier Science B.V. New-York, 1996), 257-354.

[2] K.-J.Boller, A.Imamoglu, S.E.Harris, *Phys. Rev. Lett.* **66** (1991) 2593-2596; see for review J.P.Marangos, *J.Modern Optics*, **45** (1998) 471-503; M.Fleichhauer, A.Imamoglu, J.Marangos, "Electromagnetically induced transparency: Optics in coherent media", *Reviews of Modern Physics*, **77** (2005) 633-673.

[3] A.M.Akulshin, S.Barreiro, A.Lezama, *Phys. Rev. A*, **57** (1998) 2996-3002.

[4] Harris, S.E.; Field, J.E.; Imamoglu, A. *Phys. Rev. Lett.*, **64** (1990) 1107-1110.

[5] K.Hakuta, L.Marmet, B.P.Stoicheff, *Phys. Rev. Lett.* **66** (1991) 596-599.

[6] G.Z.Zhang, M.Katsuragawa, K.Hakuta, R.I.Thompson, B.P.Stoicheff, *Phys.Rev.* **52** (1995) 1584- 1593; C.Dorman, I. Kucukkara, J.P.Marangos, *Phys.Rev. A* **61**, (1999) 013802; C.Dorman, J.P.Marangos, *Phys.Rev. A* **58**, (1998) 4121-4132; A.Merriam, S,J.Sharpe, H.Xia, D.Manuszak, G.Y.Jin, S.E.Harris, *Optics Letters*, **24** (1999) 625-627.

[7] A.Kasapi, M. Jain, G.Y.Yin, S.E.Harris, *Phys. Rev. Lett.* **74** (1995) 2447-2450; L.V.Hau, S.E.Harris, Z.Dutton, C.H.Behroozi, *Nature (London)*, **397** (1999) 594; J.Marangos, ibid. 559-560.

[8] S.Chu, S.Wong, *Phys.Rev.Lett.* **48** (1982) 738.

[9] D.F.Phillips, A.Fleischhauer, A.Mair, R.L.Walsworth, M.D.Lukin, *Phys.Rev.Lett.* **86** (2001) 783-786.

[10] A. Aspect, E.Arimondo, R.Kaiser, N.Vansteenkiste, C.Cohen-Tannoudji, *JOSA, B*, 6 (1989) 2112; S-Q.Shang, B.Sheehy, H.Metcalf, P.vander Straten, G.Nienhuis, *Phys.Rev.Lett.* **67** (1991) 1094.

[11] W.Gawlik, "Optical nonlinearity and atomic coherences" in *Modern Nonlinear Optics*, M.Evans and S.Kielich, eds.,Vol. LXXXV of Advances in Chemical Physics Series (Wiley, New York, 1994), part 3.

[12] D.Budker, W.Gawlik, D.F.Kimball, S.M.Rochester, V.V.Yashchuk, A.Weis, "Resonant nonlinear magneto-optical effects in atoms", *Rev. Mod. Phys.* **74** (2002) 1153-1201.

[13] S. Knappe, V. Shah, P.D.D. Schwindt, L. Hollberg, J. Kitching, L-A. Liew, J. Moreland, *Appl. Phys. Lett.* **85** (2004) 1460-1462.

[14] M.A. Bouchiat, C. Bouchiat, "Parity violation in atoms", *Rep. Prog. Phys.* **60** (1997) 1351- 1396.

In: New Trends in Quantum Coherence...
Editor: Rafael Drampyan, pp. 1-43
ISBN 978-1-60741-025-6

Chapter 1

ELECTROMAGNETICALLY INDUCED ABSORPTION AND TRANSPARENCY DUE TO LONG-LIVED ZEEMAN COHERENCE

Alexander Akulshin*[*]*, Russell McLean, Andrei Sidorov and Peter Hannaford
Centre for Atom Optics and Ultrafast Spectroscopy,
ARC Centre of Excellence for Quantum-Atom Optics,
Swinburne University of Technology, Melbourne, Australia

Abstract

The rapidly developing field of quantum communication has emerged from the fusion of different branches of physics and computer science. It promises very fast, reliable and secure data transmission between distant sites. The practical implementation of quantum networks will more likely be realized using solid state technology; however, resonant atomic media can provide an ideal testing ground for quickly exploring different ideas before developing expensive and time-consuming technologies.

It is well known that quantum interference effects in resonant atomic media can dramatically change the absorption and dispersion of light propagating in an atomic sample. This forms the basis of a wealth of interesting and sometimes counterintuitive phenomena, which have the demonstrated potential for storage and retrieval of information carried by light.

In this chapter we discuss the two complementary effects of electromagnetically induced transparency (EIT) and electromagnetically induced absorption (EIA) due to long-lived Zeeman coherences in atomic media. Both effects result in very narrow absorption profile resonances and steep sign-reversible dispersion. We show that in addition to ground-level coherence, coherent population oscillations between the ground and excited levels contribute to resonances with sub-natural width. We present a study of the optical properties of coherently prepared atomic media paying particular attention to steep sign-reversible dispersion and enhanced Kerr nonlinearity, which are crucial to such phenomena as few-photon nonlinearity, slow and fast light, and storage and retrieval of light pulses.

[*]E-mail address: aakoulchine@swin.edu.au

PACS 05.45-a, 52.35.Mw, 96.50.Fm.

Keywords: Coherent population trapping, electromagnetically induced transparency, electromagnetically induced absorption, slow light, fast light.

1. Introduction

The interaction of resonant radiation with an atomic medium can generate coherence between atomic states that dramatically changes the optical properties of the medium. Consider an experiment in which a linearly polarized laser beam whose optical frequency is scanned across the transitions $5S_{1/2}(F_g = 2) - 5P_{1/2}(F_e = 1, 2)$ (where F_g and F_e are the total angular momenta of the ground and excited levels, respectively) of the ^{87}Rb D_1 absorption line, propagates through a cell which contains ^{87}Rb vapour (Fig. 1). The light intensity is modest, the Rabi frequency of the laser radiation being $\Omega \sim 0.1\Gamma$, where Γ is the width of the excited level. If background magnetic fields, which are usually of order of 1 G, are reduced to the mG level, the absorption of resonant light, in particular on the transition $F_g = 2 - F_e = 1$, is significantly reduced compared to the non-compensated case. This is the result of a non-absorbing coherence prepared by the monochromatic light via two-photon Raman transitions within the degenerate ground level. This linear superposition of the ground-level Zeeman states is decoupled from the optical field at the Raman resonance and is populated through the spontaneous decay from the upper level. If the degeneracy of the ground level is removed by an external magnetic field, the monochromatic light is no longer able to produce the non-absorbing coherent superposition, leading to lower transmission of the resonant light through the atomic sample.

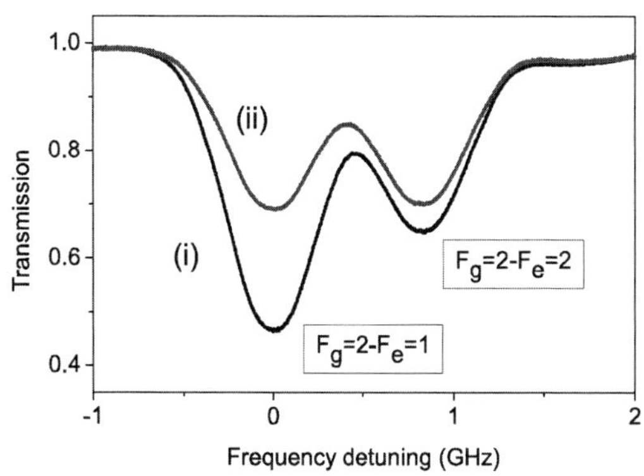

Figure 1. Spectral transmission profiles of linearly polarized resonant laser radiation propagating through a vapour cell that contains an atomic vapour of ^{87}Rb. A background magnetic field of approximately 1 G is (i) present or (ii) compensated. Optical frequency detuning from the transition $5S_{1/2}(F_g = 2) - 5P_{1/2}(F_e = 1)$ varies slowly enough to prepare the non-absorbing coherent state.

This simple illustration demonstrates how readily ground-level coherence can be generated and how strongly it can affect the absorption. Light-induced coherences play an important role and govern not only dissipative, but also dispersive and nonlinear properties of atomic media. The coherence may be in the lower or upper level of an atomic transition, or both. Amongst the best-known examples of coherent quantum-mechanical effects that can occur in atomic media are coherent population trapping (CPT) and electromagnetically induced transparency (EIT).

In CPT and EIT the interference of excitation pathways leads to suppression of the absorption of the resonant radiation fields involved in the interaction. It occurs when electromagnetic radiation couples a pair of ground states via the excited state in a three-state Λ system. In such a system, light will readily produce a redistribution of the atomic population amongst the levels - incoherent optical pumping, but CPT is the result of atoms being pumped into a coherent superposition of the two lower states that is decoupled from the exciting radiation fields, so that absorption is suppressed. Atomic population is trapped in the coherent superposition state.

Although the terms CPT and EIT are often used interchangeably, they describe distinct phenomena. In essence, CPT describes the accumulation of atomic population in the superposition state that is decoupled from resonant light, while EIT is an effect of the high transmission of a weak probe field through an optically thick medium in the presence of and as a result of a strong coupling field applied to a linked transition. Certainly, the two effects are closely related.

1.1. Brief History of Coherent Effects in Atomic Media

There are many papers, reviews and books (for example [1, 2, 3, 4, 5, 6]) devoted to coherent effects in atomic media, which have a variety of manifestations in different areas of modern optics. Here we outline just a few historical milestones and features of the CPT/EIT phenomenon relevant to the sections that follow.

The first observation of CPT was made by Alzetta *et al* in 1976 [7]. They observed the suppression of fluorescence from a sodium vapour excited by a multimode dye laser in a varying magnetic field when the laser mode separation matched the frequency between Zeeman shifted ground-level hyperfine states. In the same year, Whitley and Stroud [8] published a theoretical description of a similar effect in a three state cascade system with two laser fields. Theoretical papers followed, in 1978 by Arimondo and Orriols [9], and in 1979 by Orriols [10], while Gray, Whitley and Stroud [11] demonstrated CPT on the sodium D_1 line with two dye lasers with variable frequency detuning.

The CPT/EIT effect occurs over the spectral range in which the Raman condition for the probe and coupling fields is satisfied. This region can be much narrower than the upper-level natural linewidth if the two lower states of the Λ system have long lifetimes. The width of the CPT/EIT resonance is determined by the rate at which the coherent superposition state decays. It is often the transit time of the atoms through the laser beams that determines the ultimate width of the resonances. Mutual coherence of the radiation fields is also crucial for high spectral resolution.

Most studies of CPT in the 1980s were performed on the sodium D lines with dye lasers. When it was demonstrated that extended cavity diode lasers could have sufficiently high coherence to produce high contrast CPT absorption resonances in Rb and Cs vapour cells with a spectral resolution below the hundred kHz range, the combination of compact and inexpensive diode lasers and atomic media such as Cs and Rb vapour, which can be contained in glass cells and used at room temperature, allowed simpler experimental set-ups to be used for studying coherent effects with higher spectral resolution [12].

The coherences associated with CPT/EIT also strongly affect signals obtained in experiments that use a Faraday effect arrangement. In the nonlinear Faraday effect, the polarization of light propagating through the atomic medium is modified by the refractive index changes associated with the EIT coherences. The effect was first used to detect coherences by Gawlik *et al* [13], and subsequently used by many authors, including Kanorsky *et al* [14], and Budker *et al* [15].

CPT and EIT underpin many optical phenomena, including adiabatic passage, lasing without inversion, slow light, and sub photon recoil laser cooling.

In stimulated Raman adiabatic passage (STIRAP), partially overlapping pulses produce population transfer between two quantum states of an atom or molecule. The process involves the initial creation of a population trapping coherent state and its subsequent adiabatic evolution. In a Λ state system the process can have the remarkable property that there is no population in the excited state, so that no spontaneous emission can occur. Such a scheme was first recognized theoretically by Oreg *et al* in 1984 [16], and first demonstrated experimentally by the transfer of population from one vibrational state to a higher one in a molecular sodium beam [17] before it was applied to atoms.

Agrawal [18] showed that in a Λ state system the nonlinearites associated with laser induced coherent superpositions offered a new mechanism for the phase conjugation that is the basis of degenerate four wave mixing. Furthermore, the process may occur at significantly lower light intensities than in a two-state system. These predictions were confirmed by experiments of Mlynek *et al* [19, 20] and Bloembergen *et al* [21].

Velocity selective coherent population trapping (VSCPT), first demonstrated by Aspect *et al* [22] is a CPT-related technique that permits laser cooling below the photon recoil limit. In VSCPT, atoms, as a result of interacting with counter-propagating beams, are pumped into an uncoupled state having a well-defined momentum, where they no longer interact with the laser radiation.

Scully in 1991 [23] identified a number of applications of EIT where dispersion and enhanced refractive index are important, including lasing without inversion, proposed by Kocharovskaya and Khanin [24], Scully *et al* [25], and Harris [26]. The dispersive properties of EIT were discussed also by Harris *et al* [27, 28] who recognized the significance of the highly dispersive properties of EIT to light propagation through an optically thin medium at speed much less than c. One of the first attempts to experimentally measure the dispersive properties of EIT directly was by Xiao *et al* in 1995 [29] in a cascade state system in rubidium. Schmidt *et al* [30] in 1996 simultaneously measured absorption and dispersion in a Λ system in Cs, and observed much steeper dispersion.

Subluminal light propagation and low values for group velocity are observed under EIT conditions, where high dispersion coexists with low absorption. Harris and co-workers observed a group velocity of $c/165$ in lead vapour in 1995 [31]. Other experiments followed, and the field attracted widespread attention in 1999 when Lene Hau's group demonstrated light slowed to 17 meters per second by exploiting the extremely steep dispersion in an ultracold sodium cloud in which EIT was present [32]. This result also showed that EIT experiments might be a useful tool for studying and manipulating Bose-Einstein condensates.

Other demonstrations of EIT-induced slow light, by Kash *et al* [33] in an optically dense warm rubidium gas at around 360 K, and Budker *et al* [15] in an antirelaxation-coated rubidium vapor cell, quickly followed.

1.2. Applications of Coherent Atomic Media

Besides the natural scientific interest of a quantum interference process producing macroscopically observable effects, CPT and EIT are important because of the many applications that they have led to. One of the first was the proposal by Walls and Zoller in 1980 to use the power broadening of a CPT resonance to induce optical bistability [34]. The effect was observed in 1982 by Mlynek *et al* [35].

Thomas *et al* in 1981 and 1982 [36, 37] discussed and demonstrated Ramsey interferometry in a three-state Λ system in a sodium atomic beam, using bi-chromatic phase coherent laser radiation separated by the ground-level hyperfine splitting (1772 MHz). They pointed out the potential application to metrology. Miniature vapour-cell atomic clocks developed at NIST, Boulder, in which the reference is based on CPT resonances, are being demonstrated to have stability comparable to clocks with conventional microwave-based resonances [36, 37].

Precision magnetometry based on CPT with sensitivity approaching that of SQUID devices has also been demonstrated. These devices measure the shift of CPT resonances in a magnetic field due to the Zeeman effect. This field has been reviewed in [38] and [39].

More recently, three-state Λ systems have been explored in the context of quantum information and storage. An example is the widely recognized protocol of Duan *et al* [40]. EIT has been used to controllably generate, transmit and store single photons [41] which is an essential requirement for creating a deterministic source of single photons needed for quantum networks [42].

2. Complementary Coherent Effects in Two-Level System

2.1. Dark and Bright Coherent Superpositions

Ground-level coherence in an atomic medium with an optical transition between degenerate ground and excited levels leads to significantly different consequences, depending on the degeneracy of the levels. The interaction of monochromatic light and atoms with transitions of the type $F_g \rightarrow F_e = F_g - 1$ (for arbitrary F_g) and $F_g \rightarrow F_e = F_g$ (for integer F_g) can produce a coherent superposition of ground-level Zeeman states that is decoupled from the excited state [43]. Population is trapped in this coherent dark state. This causes absorption and fluorescence to be reduced or even eliminated and is the basis of the effect of Zeeman EIT. It was also rigorously proven that for a $F_g \rightarrow F_e = F_g + 1$ transition in a degenerate two-level atomic system interacting with a classical field, no long-lived coherent non-absorbing state exists. This finding was confirmed by experimental observations performed on the sodium D_1 line using a Hanle-type technique [44].

However, the situation is dramatically different if the $F_g \rightarrow F_e = F_g + 1$ transition is a cycling one. There is still no non-absorbing coherent state in this case; instead, the atomic medium under the action of resonant radiation exhibits enhanced absorption above the linear absorption level [45]. This effect has been named electromagnetically induced absorption (EIA). Because an atomic medium with a $F_g \rightarrow F_e = F_g + 1$ cycling transition subjected to classical resonant light appears brighter due to enhanced fluorescence, this coherent superposition of ground-level Zeeman states is designated a *bright* superposition.

This significant enhancement of absorption of resonant light, and consequently resonant fluorescence, occurs, as was pointed out in [45, 46], under the following conditions: (i) the optical transition is closed; (ii) the degeneracy of the upper level is higher than that of the lower level, i.e., $0 < F_g < F_e$; and (iii) the light polarization allows coupling of different states of the ground level by a Raman transition.

Subnatural-width absorption and fluorescence resonances of opposite polarity, namely EIT and EIA, were observed on the rubidium D_2 line using bi-chromatic laser radiation with a tunable frequency offset between the components [45, 46, 47] and in a Hanle-effect configuration [48].

2.2. Theoretical Outline

The response of a pure two-level system, driven by a monochromatic drive wave and tested by weak probe radiation with a variable frequency offset is well known [49], and predictions based on a two-level model have proven to be powerful for the interpretation of a large number of experimental observations. However, in most cases the atomic levels are degenerate and the vectorial nature of the resonant light plays an essential role. It is obvious that a more general two-level atomic system with degenerate lower and upper levels could be applied to a much wider range of phenomena.

While in the 1990s the conditions for the existence of a superposition state decoupled from resonant light and responsible for CPT and EIT in a degenerate two-level system (DTLS) were well understood [43, 44], an atomic system containing an $F_g \rightarrow F_e = F_g + 1$ cycling transition, where no dark state exists, remained unexplored.

Both the closed and open cases of such a system were analyzed by Lezama *et al* using a density matrix formalism [46, 47]. The numerical calculation of the response of a DTLS to the action of an arbitrarily intense resonant drive field and a weak probe in the presence of a magnetic field was performed. Different manifestations of the coherent nature of the atom-field interaction, such as large absorption variations, steep dispersion and wave mixing were analyzed and the corresponding spectra calculated. Most of these spectra revealed narrow resonances whose width was predominantly determined by the ground-level relaxation rate. It was also confirmed that population distribution among ground-level Zeeman states resulting from incoherent optical pumping by the control beam does not lead to EIA/EIT resonances with sub-natural width.

Good agreement with experimental spectra of EIT and EIA in a DTLS shows that the density matrix formalism is well suited for numerical modeling of the response of the atomic system with arbitrary values of the total angular momentum of the lower and upper levels and for any choice of the polarizations of the optical waves.

Resonances of EIA with sub-natural width are closely related to the anomalous ground-state Hanle effect, which appears in the form of enhanced fluorescence in a narrow magnetic field range which corresponds to the sub-natural width Zeeman shift of the magnetic states of the ground level. Thorough analysis was carried out for a general closed $F_g \rightarrow F_e =$

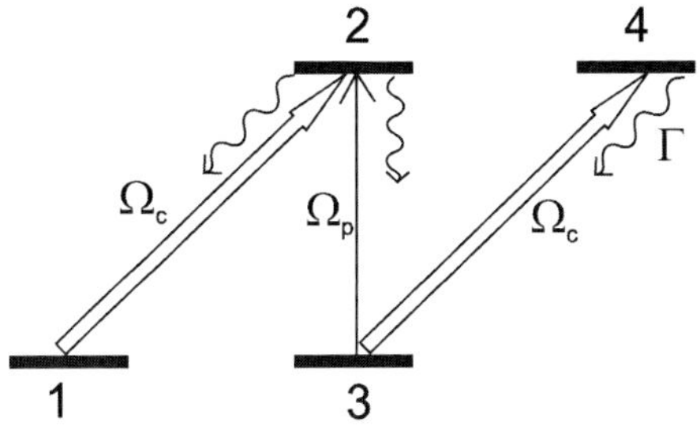

Figure 2. N-type atomic system driven by a control field Ω_c and a probe field Ω_p.

$F_g + 1$ system and analytical expressions for the populations of upper states were obtained for the hypothetical case of a $F_g = 1 \rightarrow F_e = 2$ cycling transition [50]. It was shown that due to the interaction with the light the atoms accumulate in the Zeeman states of the ground level that are most strongly coupled to the excited level.

A serious effort was made by Taichenachev *et al* to find a specific mechanism which is responsible for EIA and to reveal the underlying physics [51]. They considered a four-state N-type configuration with degenerate upper and lower levels (Fig. 2) and explained the appearance of the EIA resonances as a result of spontaneous coherence transfer.

In this model the control or drive field interacts with two transitions ($|1\rangle \rightarrow |2\rangle$ and $|3\rangle \rightarrow |4\rangle$) whereas the probe field is applied to the transition $|3\rangle \rightarrow |2\rangle$. State $|4\rangle$ spontaneously decays only to state $|3\rangle$, while state $|2\rangle$ decays to both ground states $|1\rangle$ and $|3\rangle$. Although this N-type model contains just a fraction of the complete DTLS for the $F_g \rightarrow F_e = F_g + 1$ optical transition, it allows simple analytical expressions for enhanced absorption with sub-natural width to be obtained.

The theoretical treatment involves solving the density matrix equations. The measurable parameter is the absorption of the probe field, which is proportional to the imaginary part of the off-diagonal element ρ_{23}. It is important to emphasize that the equation for the ground-level coherence includes a term describing the spontaneous transfer of the upper-level coherence.

In order to simplify the derivation of the analytical expression, the control field E_c is assumed to be significantly stronger than the probe field E_p but the corresponding Rabi frequencies Ω_c and Ω_p are smaller than the natural linewidth of the transition ($\Omega_p \ll \Omega_c \ll \Gamma$). Using perturbation theory the steady-state solution for $\rho_{23}^{(1)}$ is

$$
\begin{aligned}
\rho_{23}^{(1)} &= \frac{-i\beta\Omega_p}{\Gamma/2 - i\delta_p}\left[\rho_{33}^{(0)} + \frac{\alpha\Omega_c}{\beta\Omega_p}\rho_{13}^{(1)}\right] \\
&= \frac{-i\beta\Omega_p\rho_{33}^{(0)}}{\Gamma/2 - i\delta_p}\left[1 + \frac{\alpha\Omega_c^2}{\beta^2\Omega_c^2 - \delta_c(\delta_c - \delta_p) + i(\delta_c - \delta_p)\Gamma/2}\right],
\end{aligned}
$$

where superscripts (0) and (1) are the zero and first order perturbation solutions for the small parameter Ω_p/Ω_c. The first term in the square bracket is the standard linear absorption while the second term represents the contribution of the ground level coherence $\rho_{13}^{(1)}$. This term leads to the increased absorption in the narrow range of δ, as shown in Fig. 3(b). Thus,

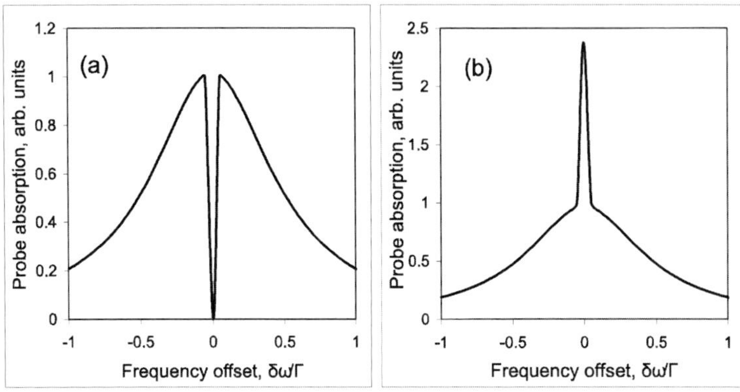

Figure 3. Absorption of a probe beam (a) in a Λ-system in case of electromagnetically induced transparency and (b) in an N-type atom in the case of the electromagnetically induced absorption, as a function of the normalized frequency offset $\delta\omega/\Gamma = (\omega_p - \omega_c)/\Gamma$ for the values of Rabi frequencies $\Omega_c/\Gamma = 0.2$ and $\Omega_p/\Gamma = 0.04$.

coherence established between the two upper states and transferred to the lower states by spontaneous emission can lead to the absorption enhancement of the probe field.

Extensive numerical modelling of a DTLS by Goren *et al* and [52] has helped the understanding of the EIA phenomenon. They found that EIA resonances occur in a degenerate two-level $F_g \to F_e = F_g + 1$ system for the case where both optical fields have the same polarization [52].

The analytical approach and numerical modeling of EIA to date have contributed greatly to our understanding of the processes involved. Even so, the picture we have of EIA is incomplete. More work is needed to produce a general model able to describe the interaction of a quasi-degenerate two-level atomic system with arbitrarily intense bichromatic resonant radiation, which is capable of evaluating optical parameters of coherent media under realistic experimental conditions, and which is universal and applicable to both EIT and EIA. This would be very useful for qualitatively explaining a wide range of phenomena in atomic media with long-lived Zeeman coherences, such as light storage and nonlinear wave mixing effects.

2.3. Observation of Coherent Resonances

We have used a double frequency sweeping technique to demonstrate both the striking difference and the similarity between EIA and EIT, as well as to observe both effects on the same absorption line under the same experimental conditions.

Consider the interaction of an alkali vapour with a bichromatic laser beam comprising orthogonally polarized drive and probe components with tunable frequency offset $\delta = \nu_p - \nu_d$. Since the Doppler broadening of the D_2 lines at room temperature is larger than the hyperfine splitting of the excited $P_{3/2}$ levels for Cs and Rb, the bichromatic radiation interacts with atoms in three different velocity groups independently, each group making a transition to a different hyperfine state in the $P_{3/2}$ level. The velocity component v_i in the direction of the laser beam is determined by the detuning of the laser frequency ν

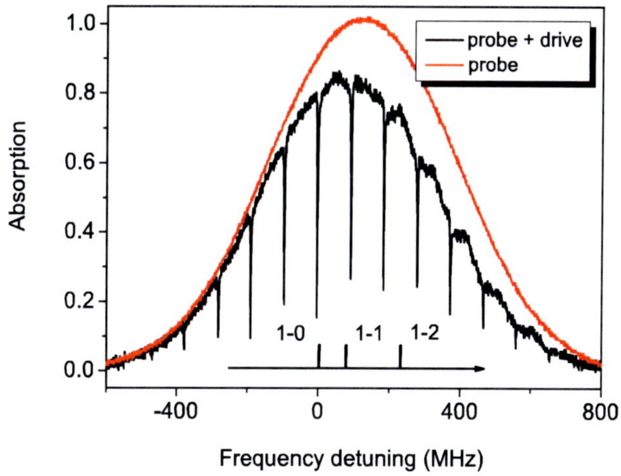

Figure 4. Normalized probe absorption with and without the drive component as a function of the optical frequency ν detuned from the $5S_{1/2}(F_g = 1) - 5P_{3/2}(F_e = 0)$ transition in ^{87}Rb when the frequency offset δ is swept approximately 20 times faster than the optical frequency itself. The positions of hyperfine components from the $F_g = 1$ ground level are indicated.

from the corresponding transition at ν_{i0}: $v_i = 2\pi(\nu - \nu_{i0})/k$, where k is the wave vector. The contribution of each of these groups to the absorption of the probe wave is different not only because of redistribution of population within the level due to incoherent optical pumping, but also as a consequence of coherence that can give rise to absorption features of either polarity.

The laser frequency ν is slowly scanned across the Doppler profile, while the frequency offset δ is swept approximately 20 times faster but over a range comparable to the natural width Γ. Thus, during the scan the probe and drive components always interact with the same atomic velocity group since $\Delta v = 2\pi(\nu_d - \nu_p)/k \leq \Gamma/k$, where Δv is the spread of the atomic velocity components in the direction of the laser beam.

Figures 4 and 5 show probe absorption profiles for the transitions from two different states of the ^{87}Rb ground level ($5S_{1/2}(F_g = 1, 2)$). The bichromatic beam can produce ground-level Zeeman coherence via a two-photon Raman process involving one photon from the drive field and a second one from the probe when δ=0. This happens many times during the slow laser frequency scan, resulting in the sharp absorption resonances.

Non-absorbing superpositions of Zeeman states in the ground level were prepared through the $5S_{1/2}(F_g = 1) - 5P_{3/2}(F_e = 0, 1)$ and $5S_{1/2}(F_g = 2) - 5P_{3/2}(F_e = 1, 2)$ transitions. The maximum contrast of the coherent absorption dips with respect to the linear probe absorption is about 70% (Fig. 4).

Bright superpositions responsible for enhanced absorption were induced in the $F_g = 2$ ground hyperfine level through the closed $5S_{1/2}(F_g = 2) - 5P_{3/2}(F_e = 3)$ transition. The drive component saturates the probe absorption within the natural width except for a narrow region at the two-photon Raman resonance (Fig. 5). The maximum contrast of the absorption peaks due to EIA is less than for the EIT case; however, on the high frequency

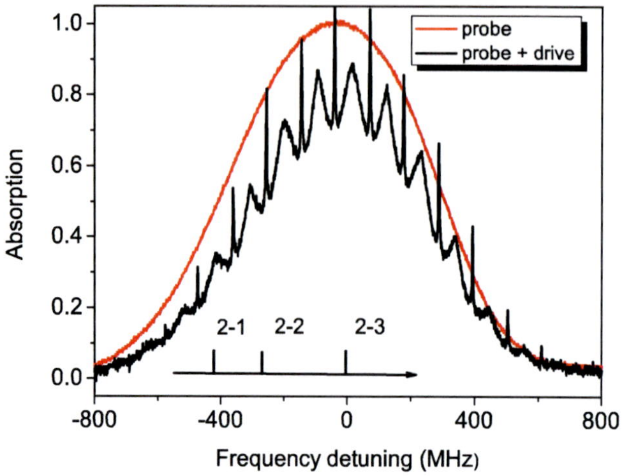

Figure 5. Normalized probe absorption with and without the drive component as a function of the optical frequency detuning from the $5S_{1/2}(F_g = 2) - 5P_{3/2}(F_e = 3)$ transition in ^{87}Rb, when the frequency offset δ is swept more quickly. The positions of hyperfine components from the $F_g = 2$ ground level are indicated.

slope the peaks are well above the linear absorption level. On the low frequency slope EIA is partly compensated by EIT because in this spectral region light interacts with Rb atoms mainly on the open transitions producing non-absorbing dark states.

It is clear that the probe absorption on a degenerate two-level transition is sharply reduced or enhanced by the long-lived coherent superposition of ground-level Zeeman states generated by two-photon Raman transitions.

2.4. Coherent Population Oscillations

Sub-natural absorption resonances observed in atomic media are usually associated with ground level coherence. However, the excitation of an open, purely two-level atomic system by co-propagating drive and probe waves with a tunable frequency offset can also result in a sub-natural resonance, as was predicted by Baklanov and Chebotaev [53, 49]. The periodic modulation of the ground-state population at the beat frequency between the probe and pump fields and multi-photon Raman scattering is responsible for this effect. In the 1970s this effect was studied in connection with parameters of Doppler-free absorption resonances, such as the shape, contrast and width, in the presence of a strong counter- or co-propagating wave [54, 55, 56]. Including the effect, named coherent population oscillation (CPO), improved the description of coherent, nonlinear processes in strongly driven resonant systems and more precise evaluation of nonlinear susceptibility became possible [57, 58]. Recently CPO has been widely employed to produce slow light in solid state media [59]. However, the effect appears to be seldom if ever discussed in connection with sub-natural absorption resonances observed in atomic media. A possible reason is that in a real atom with a complex energy level structure, light-induced coherence between states of the ground level can mask the effects of two-level CPO.

Recently it has been experimentally shown that for a two-level transition a variety of experimental parameters allow sub-natural resonances produced by long-lived Zeeman coherence and CPO to be distinguished [60] .

2.5. Experimental Methods

Two approaches can be employed to generate and analyze a ground-level Zeeman coherence. In the *level crossing* (or *Hanle*-type) method, the energies of the Zeeman states are shifted by an applied magnetic field to be in resonance with the two-photon Raman transition in monochromatic light [61, 48]. This approach is concerned with the intensity of fluorescence or transmitted light, rather than polarization of the resonant light as in the nonlinear Faraday effect [13, 14, 62].

In the *spectral* method, a frequency offset between co-propagating probe and drive fields is scanned to satisfy the resonant condition between different Zeeman states that belong to the same ground level [63].

In this Chapter we present results of our experimental studies of the most striking optical properties of atomic media with a long-lived Zeeman coherent superposition prepared in a degenerate two-level atomic system using the spectral method, paying particular attention to steep sign-reversible dispersion, pulse propagation and enhanced Kerr nonlinearity due to *bright* superpositions.

The experiments discussed below were done in resonant atomic media, where strong optical dipole-allowed atomic transitions allow the generation of long-lived atomic coherences essential for the effects to be seen. Alkali atoms are especially attractive for ground-level coherence studies because of the small depolarizing collision cross sections of their ground levels, their high density at room temperature, and the fact that the resonance transition wavelengths of several of them match available diode lasers.

EIA and related effects due to ground-level Zeeman coherence have been studied mainly with Rb and Cs atoms, which have very similar level structures [64, 65, 66, 67]. The energy-level scheme for the optical transitions in the D_2 Cs line is shown in Fig. 6a. The structure of the D_2 Rb line is similar. Recently EIA studies have been extended to lithium atoms [60]. A diagram of the relevant energy levels of ^{6}Li is shown in Fig. 6b. The isotope ^{6}Li with natural abundance about 7.5 % has a somewhat unusual structure within the D_2 line due to its I=1 nuclear spin: there are only three hyperfine levels in the $2P_{3/2}$ level compared to four levels for other alkali atoms. The hyperfine splitting of the lowest excited level $2P_{1/2}$ is 26.1 MHz, whereas the 4.6 MHz splitting of the excited level $2P_{3/2}$ is smaller than the natural width of the optical transition of about 5.9 MHz [68]. Thus, ^{6}Li atoms can be excited into ground-level coherent superpositions with opposite absorbing properties, resulting in EIT or EIA.

There are common features amongst a wide variety of experiments related to ground-level coherence studies. Ground-level coherence resonances have also been experimentally investigated in buffer gas vapour cells (for example [69, 15]), in collimated atomic beams [37, 44, 66] and laser-cooled atomic media [70].

The double sweeping technique helps in finding spectral regions where coherent effects are most pronounced. At frequency detunings δ in the hundreds of MHz range, high contrast Doppler-free satellites of EIA and EIT resonances can be observed. The origin, peculiarities

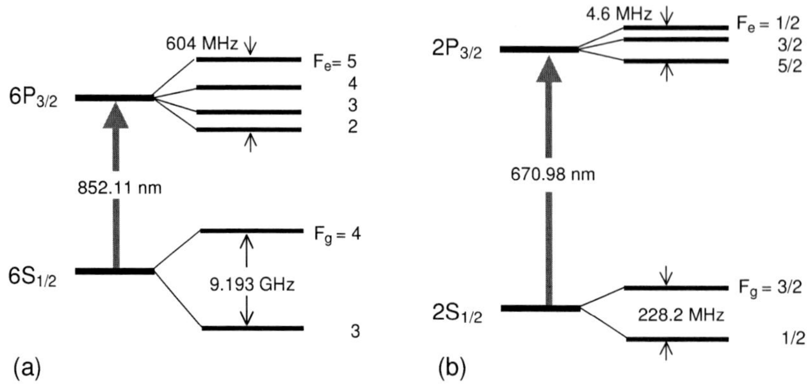

Figure 6. A schematic energy level diagram of the Cs (a) and ^{6}Li (b) D_2 lines.

and possible applications of these absorption features in the D_2 lines of alkali atoms have been discussed in [71]. The parameters of the coherence resonances were analyzed by scanning just the frequency offset δ at a fixed optical frequency ν.

The experiments were performed with cesium and rubidium vapors contained in cells and with a collimated atomic beam of ^{6}Li. Atomic media were excited by the resonant radiation within the D_2 absorption lines at 852, 780 and 670 nm for Cs, Rb and Li atoms, respectively.

To achieve the maximum spectral resolution, mutual coherence of the probe and drive (or coupling) fields is essential while the absolute linewidth is less important. Mutually coherent laser fields with tunable optical frequency offset in the MHz range can be easily prepared by two acousto-optical modulators in series or parallel using radiation from the same laser. This is sufficient offset for studying a quasi-degenerate two-level system. If a larger frequency offset is needed two independent highly coherent or phase-locked lasers should be used.

Homogeneous magnetic fields are essential for ensuring that Zeeman shifts do not depend on the spatial coordinates of atoms along the length of the atom-laser interaction region. Heating of the vapour cell, if needed, must also be done in a way that does not generate magnetic fields.

A simplified scheme of the experimental set-up is shown in Fig. 7. An atomic cell a few cm long with no buffer gas was placed in a μ-metal cylindrical shield to reduce the ambient magnetic field to less than 5 mG. A controllable longitudinal magnetic field is produced by a solenoid mounted inside the magnetic shield. The temperature of the cells was set within the range 20-70 ^{0}C.

Tunable single-mode extended-cavity diode lasers were used as resonant light sources. The linewidth of the lasers was about 1 MHz. Two acousto-optic modulators were used to produce mutually coherent drive (or coupling) and probe beams with tunable small frequency offset ($\delta = |\nu_P - \nu_D| \leq \Gamma/2\pi \simeq 6$ MHz), where Γ is the natural width of the optical transition.

The two beams were then carefully combined into a bichromatic beam and passed through the cell. It is important to minimize the mutual divergence of the probe and cou-

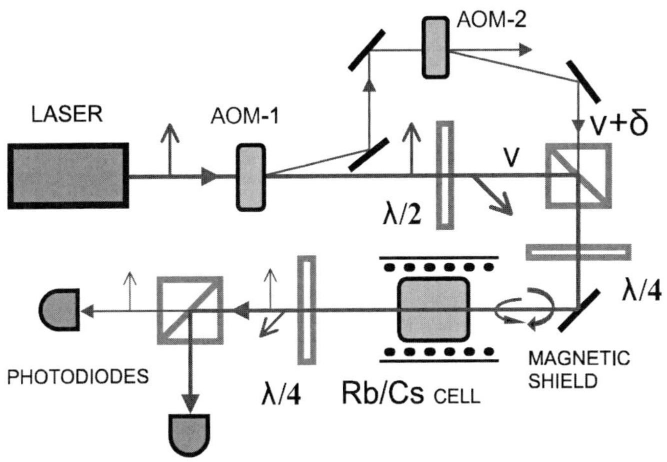

Figure 7. Scheme of experimental set-up for the spectral method of observing EIT/EIA in degenerate two-level atoms.

pling components, although divergence of the bi-chromatic beam as a whole does not affect the width of the subnatural resonances [12].

The polarizations of both components were set with polarizers and quarter-wave plates. The typical diameter of the bichromatic beam in the cell was about 0.6 cm, and the maximum intensities of the drive and probe components were 40 and 10 mW/cm^2. Gaussian-shaped as well as sharp-edge probe pulses and a constant intensity drive component were directed through the atomic vapour to examine light propagation in the coherently prepared medium. The components were separated after the cell by a quarter-wave plate and a Glan-Thomson prism (extinction ratio better than 1000). The intensities of the probe and drive components transmitted through the atomic cell were measured with fast photodiodes. The signals were analyzed and processed with a digital oscilloscope and a RF spectrum analyzer.

3. Experiments on ^{6}Li

3.1. CPT and CPO on the ^{6}Li D_1 Line

The most common way to produce EIT in the D lines of alkali atoms is to couple both ground hyperfine levels to a common excited state [1, 2]. However, the separation of effects due to CPO and ground-level coherence can be realized more clearly in a quasi-degenerate two-level atomic system.

If bichromatic light excites atoms from either of the two hyperfine ground levels, the strongest EIT resonance for the ^{87}Rb D_1 line occurs on the $5S_{1/2}(F_g = 2)\rightarrow 5P_{1/2}(F_e = 1)$ transition. In the case of the ^{6}Li D_1 line it was found that the largest reduction of the resonant fluorescence occurs for the same type of optical transition, i.e., $F_g \rightarrow F_e = F_g - 1$. In this case atoms from all ground-level magnetic states can be connected by Λ-type links, which lead to non-absorbing CPT states [43].

If the atomic two-level system is completely degenerate and the quantization axis has an arbitrary direction, all possible two-photon Raman transitions can contribute to the EIT

(a)

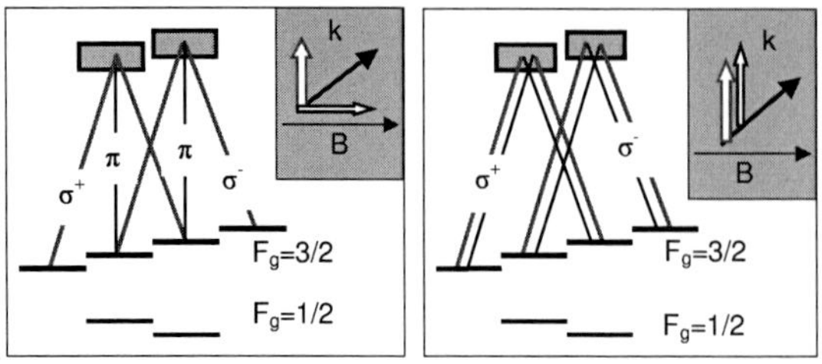

(b)

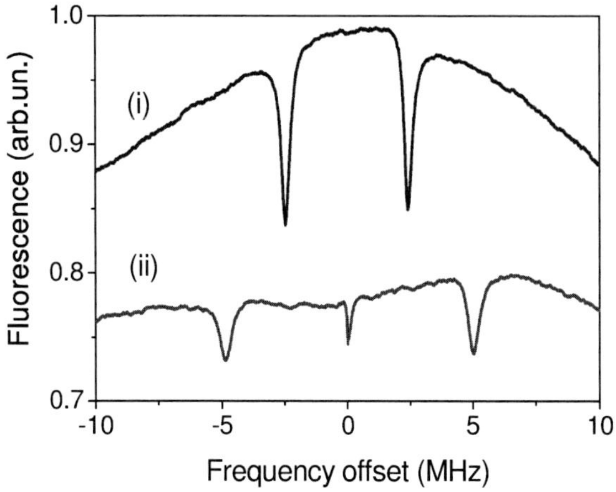

Figure 8. (a). Raman transitions responsible for the dark Zeeman coherence for different polarizations of the drive and probe components. (b). Resonant fluorescence for the ^{6}Li D_1 line as a function of the frequency offset δ. The laser frequency is tuned on the $2S_{1/2}(F_g = 3/2) - 2P_{1/2}(F_e = 1/2)$ transition. The optical components have (i) linear orthogonal polarizations or (ii) parallel polarizations. A magnetic field of approximately 2.5 G is applied orthogonal to the light beam and collinear to the atomic beam.

resonance at zero frequency offset. To separate the contributions from CPT and CPO, an external magnetic field B was applied. This shifts Zeeman states of the ground level by an amount $\Delta_F = g\mu_B m_F B/h$ that depends on the magnetic quantum number m_F of the state and the strength of magnetic field B. g is the ground level gyromagnetic factor, h is Planck's constant and μ_B is the Bohr magneton.

The external field may also define a convenient basis. If the quantization axis is collinear to the wave vector $\overrightarrow{k}$ of the linearly polarized light the interaction with atoms results in σ^+ and σ^- transitions. This is the natural basis for zero magnetic field and for a field which is collinear to the light wave vector, $(\overrightarrow{B} \| \overrightarrow{k})$. If the magnetic field is orthogonal to the wave vector $(\overrightarrow{B} \perp \overrightarrow{k})$ and collinear to the light polarization, π transitions occur. When the magnetic field is orthogonal to both $\overrightarrow{k}$ and the light polarization, σ^+ and σ^- excitations occur.

Thus, once a magnetic field is applied and the quantization axis defined, the optical transitions between the ground and excited levels produced by the probe and drive components can be categorized. We consider two specific cases. A drive component with linear polarization orthogonal to the magnetic field can induce $\sigma^\pm$ transitions, while a probe component with polarization parallel to the magnetic field induces π transitions as shown in

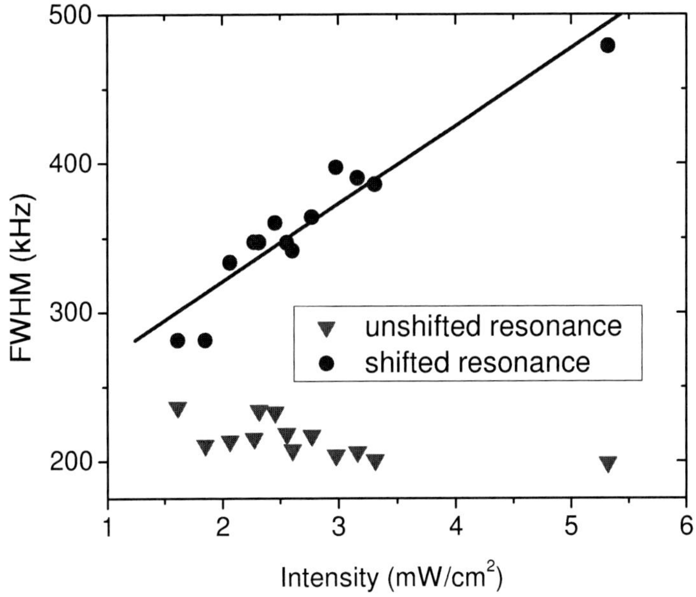

Figure 9. The experimentally measured width of the shifted and unshifted EIT resonances for the $2S_{1/2}(F_g = 3/2) \rightarrow 2P_{1/2}(F_e = 1/2)$ transition within the ^{6}Li D_1 line as a function of the total light intensity. A magnetic field ($B \simeq 1$ G) is applied along the atomic beam.

Fig. 8(a). In the case of identical linear polarizations orthogonal to the magnetic field, both components produce $\sigma^{\pm}$ transitions.

When the degeneracy of the ground hyperfine levels is removed by the applied magnetic field, the EIT resonance splits into several resonances. In order to find resonant conditions for a nearly degenerate two-level system all Raman transitions involving one photon from each optical component should be considered. The Raman transitions responsible for the dark Zeeman coherence in the case of the $2S_{1/2}(F_g = 3/2) \rightarrow 2P_{1/2}(F_e = 1/2)$ optical transition are shown in Fig. 8(a).

There are two high-contrast EIT resonances for orthogonal linear polarizations, curve (i) in Fig. 8(b). As expected both fluorescence dips are shifted from the zero frequency offset by $\Delta = g\mu_B B/h$. The appearance of a narrow unshifted or magnetically insensitive resonance for parallel polarizations, curve (ii) in Fig. 8(b), does not fit into the Λ-type coupling scheme. Different states of the ground level could not be coupled by any Raman transition and ground-level coherence can not be prepared at a zero frequency offset. The unshifted resonance corresponds to a transition from a given Zeeman state to itself and, consequently, the resonance is insensitive to the applied magnetic field, as mentioned in [45, 47]. We attribute this fluorescence dip to CPO.

At the same time two magnetically sensitive dips in curve (ii) of Fig. 8(b) are the result of the light-generated ground-level Zeeman coherence. They are similar in shape to the resonances observed for the light components with orthogonal polarizations, but shifted from zero frequency offset by 2Δ.

It was found that power broadening of the magnetically sensitive and insensitive EIT

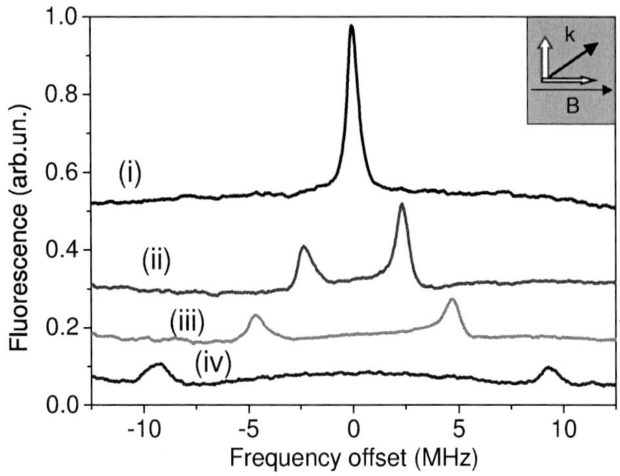

Figure 10. Splitting of EIA resonances in an external magnetic field using the geometry shown in the inset. The frequency of the bichromatic beam is tuned to the $2S_{1/2}(F_g = 3/2) - 2P_{3/2}$ transitions. The optical components have orthogonal linear polarizations. The external magnetic field is (i) 0, (ii) 1.25 G, (iii) 2.5 G and (iv) 5 G, respectively.

resonances is very different. The power broadening of EIT resonances due to hyperfine or Zeeman ground-level coherence has been studied in detail theoretically and experimentally [12, 2, 45, 47, 72]. The width Γ of such EIT resonances at low light intensity grows linearly with the Rabi frequency of the driving field Ω_D ($\Gamma = \Omega_D\Omega_P/\gamma$, where γ is the homogeneous linewidth). If $\Omega_D > \Gamma$ the width becomes a linear function of the drive intensity: $\Gamma = \Omega_D^2/\gamma$. However, within the uncertainty of experimental measurements, the width of the unshifted EIT resonance does not depend on the intensity at all, while the magnetically dependent resonance due to CPT between Zeeman states exhibits the expected power broadening as shown in Fig. 9.

Thus, the high-contrast EIT resonances are due to both CPT amongst the magnetic states from the ground hyperfine level $F_g = 3/2$ and CPO. The contribution from Zeeman coherence is dominant for orthogonal drive and probe polarizations. However, for parallel polarizations both effects make comparable contributions to the overall EIT resonance.

3.2. EIA on the ^{6}Li D_2 Line

To produce an EIA medium on the D_2 line the laser was tuned to the appropriate closed $F_g \rightarrow F_e = F_g + 1$ transition, i.e., $2S_{1/2}(F_g = 3/2) \rightarrow 2P_{3/2}(F_e = 5/2)$. With no applied magnetic field, a sharp enhancement of fluorescence due to EIA occurs at zero frequency offset (δ=0). The maximum contrast of the sub-natural peak relative to the total resonant fluorescence, approximately 40%, was observed with orthogonal linear polarizations.

For linear orthogonal polarizations a magnetic field applied orthogonal to the laser beam and parallel to the polarization of the drive component splits the EIA peak into two resonances, as shown in Fig. 10, in a similar way to EIT resonances for the D_1 line. The positions of the two sub-natural peaks are consistent with Λ-type Raman transitions between magnetic states of the ground level.

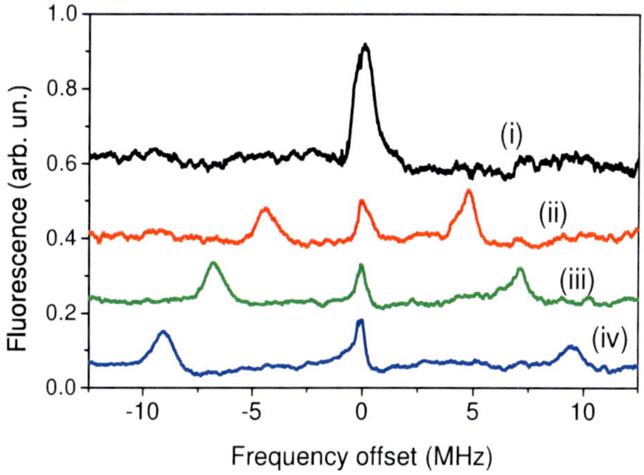

Figure 11. Splitting of EIA resonances in an external magnetic field for the case of orthogonal circular polarizations. The frequency of the bichromatic beam is tuned on the $2S_{1/2}(F_g = 3/2){\rightarrow}2P_{3/2}$ transitions. The optical components have orthogonal circular polarizations. The external magnetic field is (i) 0, (ii) 2.5 G, (iii) 3.75 G and (iv) 5 G, respectively.

In the case of orthogonal circular polarizations, (Fig. 11), the EIA peak splits into three resonances with sub-natural width. Again, the unshifted resonance does not correspond to any Raman transitions between magnetic states of the ground level and, as in the case of unshifted EIT peaks, can be attributed to CPO. This illustrates that CPO can result in enhanced absorption and fluorescence if the magnetic states of the ground level interacting most strongly with the light are more heavily populated.

Thus, CPO in ^{6}Li atoms excited on the $2S_{1/2}(F_g = 3/2){\rightarrow}2P_{3/2}$ transition may lead to suppression or enhancement of fluorescence depending on the polarizations of the bichromatic light.

3.3. Modeling and Discussion

A model based on a semiclassical treatment of the atom-light interaction using optical Bloch equations and developed by A. Lezama in a series of papers [46, 47] was used for the numerical simulation of the probe absorption spectra of ^{6}Li atoms excited by a bi-chromatic light in the presence of a magnetic field [60].

Probe absorption spectra were calculated under the assumption that the reduced Rabi frequency of the pump component is $\Omega_D = 0.1\Gamma$, while the probe component is negligibly weak; the ground level relaxation is 0.001Γ and a Zeeman shift due to an applied transverse magnetic field is 0.02Γ.

Numerical simulation of EIT absorbtion spectra for the $2S_{1/2}(F_g = 3/2){\rightarrow}2P_{1/2}(F_e = 3/2)$ transition reproduces the occurrence of the magnetically insensitive resonance for orthogonal polarizations and positions of the magnetically dependent resonances. However the relative amplitudes of the calculated resonances differ substantially from the experimental fluorescence observations.

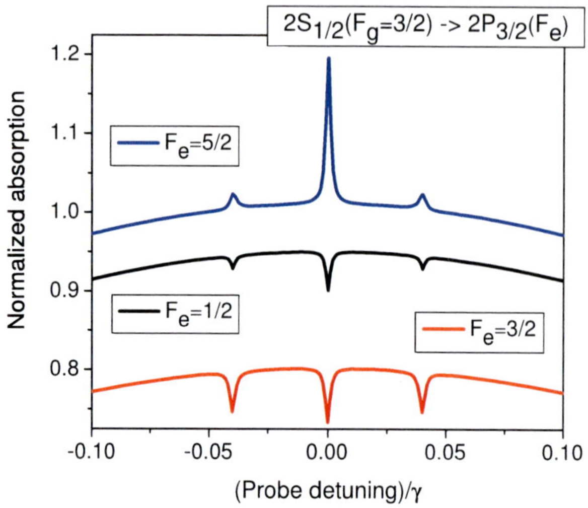

Figure 12. Numerical modeling of probe absorption spectra for different transitions within the ^{6}Li D_2 lines.

EIT and EIA resonances calculated separately for the three $2S_{1/2}(F_g = 3/2)\rightarrow 2P_{3/2}(F_e = 5/2, 3/2, 1/2)$ transitions for the case of orthogonal linear polarizations are shown in Fig. 12. The relative strengths of the transitions and optical pumping to the $F_e = 1/2$ hyperfine ground level were taken into account. The vertical axis represents absorption normalized to maximum linear absorption for a given transition.

Generally speaking, the experimental appearance of EIA for the ^{6}Li D_2 line is somewhat surprising, due to the competing processes taking place in the $2S_{1/2}(F_g = 3/2)\rightarrow 2P_{3/2}$ transition. Firstly, there is incoherent hyperfine optical pumping. This process can be considered as a leak to a non-resonant hyperfine level, which imposes an additional limit on the ground-level lifetime. Secondly, atoms may be trapped in a non-absorbing dark coherent state prepared in the $2S_{1/2}(F_g = 3/2)\rightarrow 2P_{3/2}(F_e = 1/2, 3/2)$ transitions. Additionally a more absorbing coherent superposition can be generated for the cycling transition $2S_{1/2}(F_g = 3/2)\rightarrow 2P_{3/2}(F_e = 5/2)$. The bichromatic light is almost equally resonant with all these transitions. This is a unique situation for alkali atoms. For the Cs or Rb D_2 lines, for example, monochromatic resonant radiation tuned to transitions from the upper ground hyperfine level interacts with atoms contained in a vapour cell in three different velocity groups. Atoms from two velocity groups may be pumped into non-absorbing dark states, while a more absorbing state may be prepared in the third group as was shown in [63]. The contributions from those groups to the total probe absorption are opposite, but absorbing properties of atoms from the same group are very similar. They can be separated in an atomic beam or magneto-optical trap. This EIT/EIA separation is not possible in the case of the ^{6}Li D_2 line, because the hyperfine splitting of the level $2P_{3/2}$ is smaller than the natural linewidth. This means that laser light can be resonant with the three optical transitions $2S_{1/2}(F_g = 3/2)\rightarrow 2P_{3/2}(F_e = 1/2, 3/2, 5/2)$ simultaneously and ^{6}Li atoms can be excited on different transitions with comparable probabilities. This makes the experimental observation of EIA more interesting and modelling of the atomic response of ^{6}Li atoms more difficult.

It is worth noting that similar doublet and triplet structures of sub-natural resonances in the probe absorption for the ^{85}Rb D_2 line were obtained both experimentally and numerically in [47]. However, no connection of the magnetically insensitive sub-natural resonances to CPO was mentioned.

These results suggest that in atomic media excited by bi-chromatic resonant light the contribution of optical coherence (population oscillations between ground and excited state) to the sub-natural resonance is significant in many experimental situations and its role may have been underestimated.

4. Steep Dispersion of Coherent Atomic Media

Strong dispersion of an atomic medium is usually associated with an optically dense absorbing sample. However, even a moderate variation of absorption within a sufficiently narrow spectral region, such as that associated with EIT/EIA, results in extremely steep dispersion. According to the classical theory of dispersion the real and imaginary parts of the complex refractive index of a low-pressure two-level gas may be written in the form:

$$(n - i\kappa) = 1 + (Ne^2/4\pi m\nu_0)(\nu_0 - \nu + i\gamma/2\pi)^{-1}, \tag{1}$$

where N is the atomic density, and e and m the charge the mass of an electron. ν, ν_0 and γ are the frequencies of the optical field, the resonant frequency of an optical transition and the width of the line, respectively. At resonance ($\nu = \nu_0$), the dispersion $dn/d\nu$ and the linear absorption coefficient $\alpha = 2k_0\kappa$ (where k_0 is the vacuum wave number and $I(z) = I_0 exp(-\alpha z)$) depend on the linewidth γ and the atomic density N, and the dispersion may be expressed as a function of the linear absorption:

$$dn/d\nu = \alpha(\nu_0)/k_0\gamma. \tag{2}$$

Considering that the dispersion is inversely proportional to the width γ, even a small but narrow absorption resonance results in a steep dispersion. At resonance, the dispersion of an atomic medium with an absorption peak is negative or anomalous ($dn/d\nu < 0$), while a medium with an absorption dip possesses normal dispersion ($dn/d\nu > 0$).

A systematic study of the dispersive properties of an atomic medium with long-lived Zeeman coherence was performed using an RF heterodyne technique [63, 73] in which sensitivity to acoustic noise is dramatically reduced compared to the common optical homodyne method based on a Mach-Zehnder interferometer. A detailed description of the method can be found in [74]. This RF technique allows direct refractive index measurements at the same time as absorption measurements. The refractive index spectral dependence $n(\nu)$ can be extracted from the variation of the phase difference $\Delta\Phi$ of two RF beats produced by mixing the probe component with an off-resonant auxiliary wave

$$\Delta\Phi = 2\pi(n-1)L/c, \tag{3}$$

where L is the length of the atomic sample.

Transmission profiles and phase variations of the probe component in Cs vapour with *bright* and *dark* Zeeman superpositions as a function of the frequency offset δ are shown

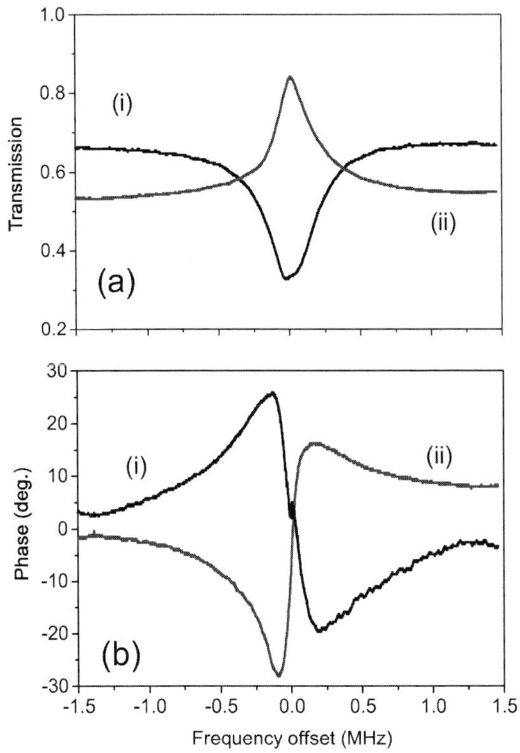

Figure 13. Transmission of probe component (a) and phase shift between two RF beat signals (b) as a function of the frequency offset δ on a two-photon Raman resonance in Cs vapour with long-lived *bright* (curves (i)) and *dark* (curves (ii)) Zeeman superpositions. The laser frequency was tuned to the $6S_{1/2}(F_g = 4) \rightarrow 6P_{3/2}(F_e = 5)$ transition for EIA, and the $6S_{1/2}(F_g = 3) \rightarrow 6P_{3/2}(F_e = 2)$ transition for EIT.

in Fig. 13. The dispersion of the medium $dn/d\nu$ can be estimated quite precisely from the spectral dependence $n(\nu)$. For example, in the EIA case the phase variation at the Raman resonance corresponds to $dn/d\nu = -25 \times 10^{-12}$ Hz^{-1}.

The dispersion of an atomic medium with ground level Zeeman coherence can be easily controlled by the drive intensity and the optical frequency. The intensity dependence of the anomalous dispersion was numerically modeled and experimentally studied in [73].

In summary, extraordinarily steep anomalous dispersion, unattainable by other means, has been obtained in Rb vapour with a *bright* superposition state [73]. It is also possible to realize anomalous dispersion on the slope of an EIT resonance or in a Λ-type atom with a double-gain structure [75, 76]. There are several ways to prepare anomalous dispersion in atomic media without generating long-lived coherent superpositions [77, 78, 79]. However, as far as we know, in all of these cases the magnitude of the anomalous dispersion is smaller.

5. Light Pulse Propagation

During the last decade the propagation of light through media with steep dispersion has received much attention, triggered by several 'slow light' experiments [32, 33] and driven by possible applications in optical telecommunications and quantum information processing, as well as by renewed interest in the paradox of superluminal light propagation.

5.1. Fast Light: Brief Historical Survey

The wavefronts of an electromagnetic wave move in a medium with phase velocity $v_{ph}=c/n$. However phase velocity has limited physical meaning because any optical signal has a finite duration and phase velocity does not describe real pulse propagation in a medium. If the index of refraction depends on the optical frequency $n(\nu)$, the various components of multi-chromatic light travel through the medium with different phase velocities, which may be greater or less than c. As a result of interference within a wave packet, the peak of the pulse, which could carry information, moves through a medium with a different velocity. By definition the group velocity of a wave packet

$$v_g = 2\pi(d\nu/dk) = c/[n + \nu(dn/d\nu)], \tag{4}$$

depends on the refractive index n and the dispersion, $dn/d\nu$. The peak moves at the group velocity, while the individual frequency components move at the phase velocity. When this definition was introduced in the 19th century, the group velocity of light was considered as the velocity of the propagation of radiant energy and information [80].

For anomalous dispersion ($dn/d\nu < 0$) the group velocity of a light pulse can exceed the speed of light in a vacuum c, or it can even be negative if the dispersion is sufficiently steep ($| \nu(dn/d\nu) | \gg 1$). The propagation time T through the medium of length L, defined as $T = L/v_g$, is less than through the same length of vacuum: $\Delta T = L/v_g - L/c < 0$. This is the phenomenon of superluminal light propagation or 'fast light'.

At the beginning of the 20th century this was seen to appear to be in serious conflict with the principle of causality and the fundamental axiom of special relativity that no signal can be transmitted faster than c. The problem was clarified by Sommerfield and Brillouin, who pointed out that Einstein's theory requires the speed of a signal to be limited by c, rather than the group velocity of a light pulse, and that the group velocity becomes meaningless due to strong absorption and distortion in a medium with steep anomalous dispersion [81]. However, no commonly accepted definition of the signal velocity has been suggested.

5.2. Fast Light Experiments in EIA Media

A pulse of light propagates in a medium with group velocity v_g determined by the refractive index $n(\nu)$ and the dispersion $dn/d\nu$. Pulse advance ΔT can be easily observed even for $L \simeq 1$ cm if $|dn/d\nu| > 1 \times 10^{-12} Hz^{-1}$.

Propagation of a smooth pulse of light with negative or 'superluminal' group velocity ($v_g = -c/23$) was first demonstrated in a GaP:N crystal by Chu *et al* [82]. However, the value of the anomalous dispersion of an atomic medium with long-lived 'bright' Zeeman superposition is much higher, as mentioned above. The first pulse propagation in a EIA

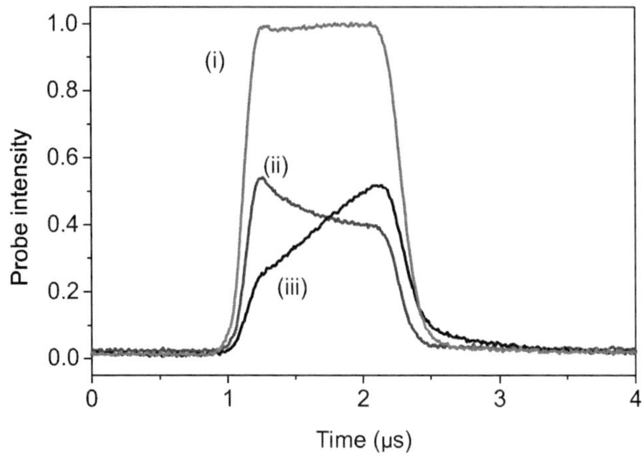

Figure 14. Distortion of a sharp-edge probe pulse in Cs vapour. Curve (i) represents the off-resonant pulse propagating with $v_g \simeq c$. Curves (ii) and (iii) correspond to the optical frequency tuned to the $6S_{1/2}(F_g = 4) \rightarrow 6P_{3/2}(F_e = 5)$ and $6S_{1/2}(F_g = 3) \rightarrow 6P_{3/2}(F_e = 2)$ transitions, respectively.

medium with a negative group velocity ($v_g \simeq -c/3600$) was reported in [83]. Many publications have appeared in the field of superluminal light propagation in recent years. These include the direct observation of optical precursors in a region of anomalous dispersion [84] and the almost simultaneous demonstration by three groups that the leading edge of the transmitted pulse never precedes that of the incident pulse, confirming that information cannot propagate with a velocity exceeding c [63, 76, 77].

Here we present experimental data related to the problem of superluminal pulse propagation in a highly dispersive atomic medium. Gaussian-shaped as well as sharp-edge square probe pulses were applied. Opposite circular polarizations of the probe and drive components were set with polarizers and quarter-wave plates. The drive and probe components produced coherence between magnetic states belonging to the same ground hyperfine level as a result of two-photon Raman transitions involving one photon from the drive and another from the probe field. Neither component is able to connect magnetic states in the ground level on its own.

The medium with long-lived Zeeman coherence affects the shape of the pulse of resonant light during propagation, as shown in Fig. 14. In the spectral region where EIA occurs and absorption is enhanced, the transmitted intensity decreases with time due to the establishment of the more-absorbing coherent superposition, curve (ii). It causes a shift of the centroid of the pulse toward the leading edge of the pulse. In the EIT media prepared on the $6S_{1/2}(F_g = 3) \rightarrow 6P_{3/2}$ transitions the situation is reversed, curve (iii). At the beginning of the pulse the transmitted intensity is lower. Thus, the centroid of the pulse is delayed with respect to the reference.

To avoid distortion of the pulse shape all monochromatic components of the pulse should be within the region of constant dispersion. This imposes a lower limit on the pulse duration and a restriction on the pulse shape. In the case of a 200 kHz-wide absorption resonance this means that the smooth probe pulse duration should be longer than

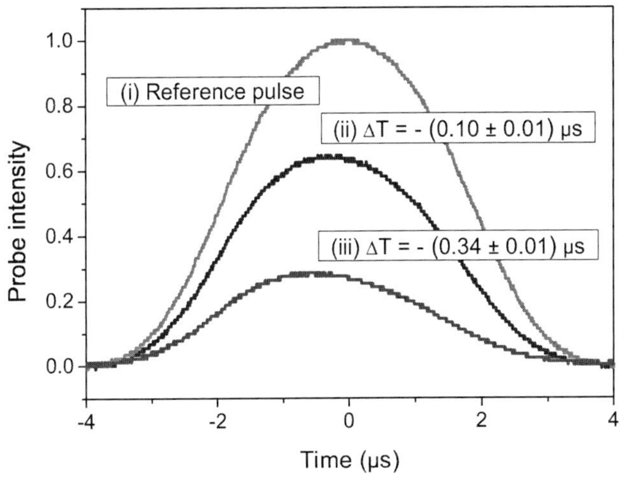

Figure 15. Probe pulse advancement in Cs vapor with steep anomalous dispersion. ΔT is the time advancement of the pulse peak relative to that of the non-resonant reference pulse (i) propagating with $v_g \simeq c$. Curves (ii) and (iii) correspond to the optical frequency tuned to the $6S_{1/2}(F_g{=}4)$-$6P_{3/2}(F_e{=}5)$ transition at atomic densities of $1.5 \times 10^{11} cm^{-3}$ and $7 \times 10^{11} cm^{-3}$, respectively.

$1/2\pi\Delta\nu \simeq 0.7\mu s$.

Propagation of a Gaussian-shaped probe pulse with negative v_g through a medium with 'bright' Zeeman superposition is shown in Fig. 15. The pulses of resonant light are advanced relative to the reference pulse. The maximum fractional advancement $\Delta T/\Delta t$, where $\Delta t \simeq 1.8\ \mu s$ is the pulse duration, is as large as 18%. There is little distortion of the pulse. The time advance is larger for higher Cs density because the dispersion is proportional to the atomic density: $dn/d\nu \sim N$. The maximum advance observed in our experiment with coherently prepared Cs vapor at 20% probe pulse transmission corresponds to a negative group velocity of about $-c/6000$.

There is a simple relationship between the dispersion $dn/d\nu$ and the advance ΔT:

$$dn/d\nu = c\Delta T/L\nu. \tag{5}$$

For the fast-light Cs vapour the estimated value of the anomalous dispersion based on the maximum observed advance is $dn/d\nu = -14 \times 10^{-12}$ Hz^{-1} which is close to the previously obtained value using direct refractive index measurement.

Figure 16 shows that for a sharp-edge probe pulse the rising slope of the resonant pulse is delayed despite the visible shift of the center of mass of the pulse toward the leading edge. This means that there are no photons travelling faster than in the off-resonant case.

We would like to emphasize a distinguishing feature of Zeeman coherence. The sign of the steep dispersion can be changed easily within milliseconds by tuning the laser frequency to another optical transition, a feature that could be important for some potential applications. The demonstration of such a dramatic change of the dispersive properties of Cs vapour with long-lived Zeeman coherence is presented in Fig. 17. The probe pulse propagates through the atomic sample in the *bright* coherent superposition 0.33 μs faster than

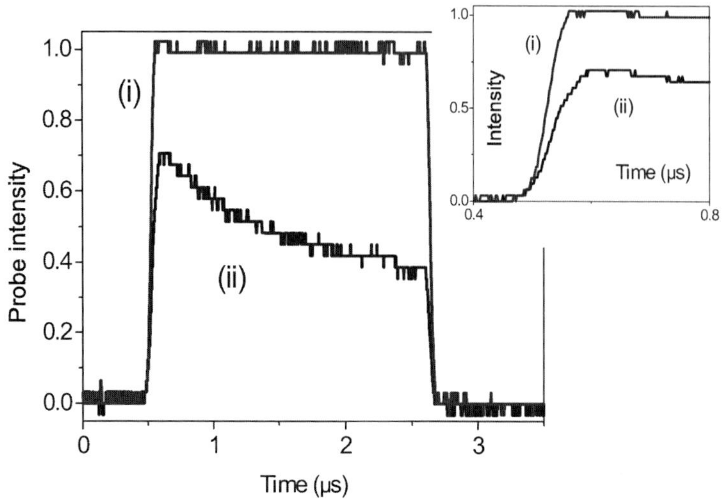

Figure 16. Sharp-edge probe pulse propagation through Cs vapour with steep anomalous dispersion. Curves (i) and (ii) represent a off-resonant reference and signal pulses, respectively. Insert shows the rising slope of probe pulses in detail.

an off-resonant reference pulse, but it is delayed by 0.42 μs in the case of the *dark* coherent state. The curves were taken under the same experimental conditions such as atomic density, polarization and light intensity. These conditions were not chosen to optimize either the pulse delay or advancement; indeed, maximum delay and advancement do not occur under the same experimental conditions.

Reducing the width of an EIA resonance by making the lifetime of the ground-level Zeeman coherence longer is a natural way to increase anomalous dispersion. As far as we know the largest advance has been obtained by Mikhailov *et al* [85, 86] in Rb vapour with a buffer gas corresponding to a negative group velocity as low as $v_g \simeq -c/10^6$.

All of these results are consistent with the common observation [75, 77, 85, 86] that it is much harder to achieve the same fractional advance of a light pulse as fractional delay, particularly with low distortion. While delays of several times the pulse width have been observed, Macke and Segard [87] have estimated the largest advance that could be attained in a realistic experiment, and claim that in any experimental system advances exceeding two times the full width at half maximum of the pulse intensity profile are unattainable.

After recent experimental progress in producing steep anomalous dispersion and small negative group velocities [75, 83, 88, 89, 85] the problem has been analyzed in a very general framework by Macke and Ségard [87] and by Bukhman [90].

To summarize this section, it is well established that the peak of a smooth pulse can propagate faster than c in a medium with anomalous dispersion, but information cannot be transmitted faster than c. In spite of the counterintuitive nature of light propagation with negative group velocity, there is no contradiction with the principle of causality. Light propagation with $v_g < 0$ is a result of interference within a wave packet of spectral components moving at different phase velocities $v_{ph} = c/n$ and this phenomenon can be entirely described within classical electrodynamics. Thus, our experiment has nothing in common

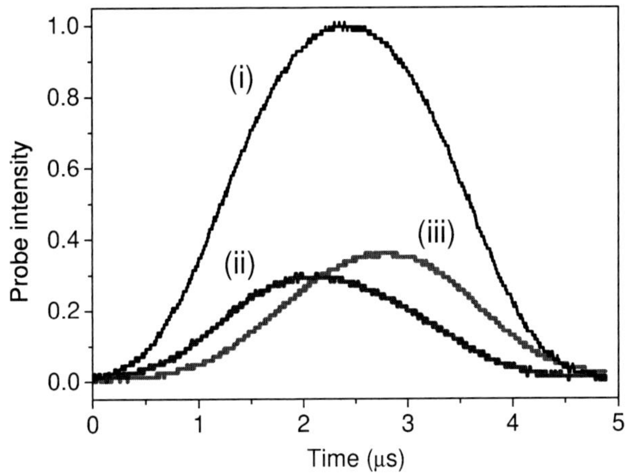

Figure 17. Gaussian-shaped probe pulse propagation through Cs vapour with steep anomalous (ii) and normal (iii) dispersion. Off-resonance pulse (i) is a reference.

with the ongoing search for optical quasi-tachion behavior [91].

5.3. Energy Exchange between Probe and Drive Waves

An interesting intensity exchange between the probe and drive components was discovered in coherently prepared Cs vapour [88]. This effect was investigated using the propagation of a sharp-edge probe pulse in the presence of a constant drive component.

First we consider Zeeman EIT prepared in Cs vapour on the $6S_{1/2}(F_g = 3) \to 6P_{3/2}$ transitions. Atoms in the velocity group resonant with the closed transition $6S_{1/2}(F_g = 3) \to 6P_{3/2}(F_e = 2)$ make the main contribution to the enhanced dispersion, while atoms in other velocity groups decay spontaneously into the non-resonant hyperfine level $6S_{1/2}(F_g = 4)$ and no longer interact with the laser light.

Bichromatic radiation that includes a dc component of the probe field partially pumps atoms from the resonant group into the coherent dark superposition. Incoherent optical pumping causes atoms to accumulate predominantly in the extreme magnetic states $m = +2, +3$ due to σ^+ transitions driven by the more intense drive component. The sharp-edge probe pulse modifies this steady-state distribution. The atoms from $m = +2$ and $m = +3$ are excited to the $6P_{3/2}$ level by the front edge of the pulse. They return to the ground level after emitting stimulated photons into the drive field; thus the transmitted intensity of the drive field rises. The absorption at the front edge of the pulse is rather strong, but decreases due to pumping into the modified dark state, corresponding to a new ratio between the drive and probe components, Fig. 18a. This transient process leads to a relatively slow increase of the transmitted intensity. Steady state can be reached within 0.1 to 1 μs. The relaxation time depends on the intensity of the bichromatic radiation.

A rather long tail appears in the transmitted probe intensity after switching off the pulse. A fraction of the drive intensity leaves the cell as probe light. This can be explained in terms of imprinting of information about the pulse into the Zeeman coherence. The ground level Zeeman coherence can be considered as a phase grating and this intensity exchange is

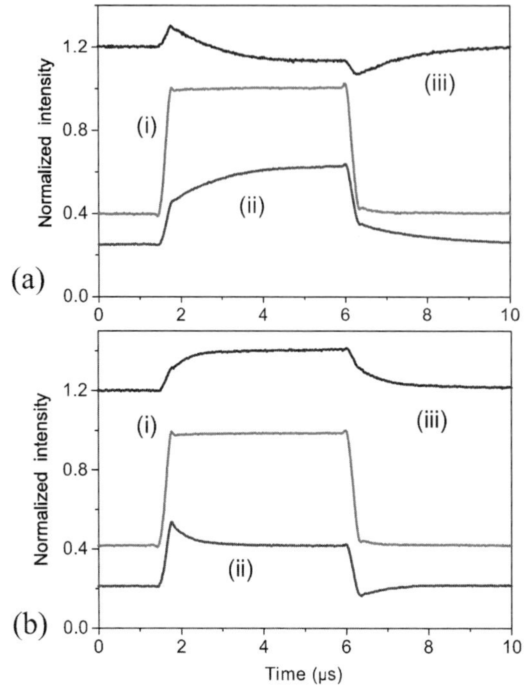

Figure 18. Transient effects in the transmitted probe and drive components. The intensities are normalized to the off-resonant probe intensity. (a) and (b) correspond to the optical frequency tuned to the $6S_{1/2}(F_g = 3) \rightarrow 6P_{3/2}(F_e = 2)$ and $6S_{1/2}(F_g = 4) \rightarrow 6P_{3/2}(F_e = 5)$ transitions, respectively.

evidence of transient behavior in the stimulated scattering from the modified phase grating. Thus, more efficient nonlinear mixing (or stimulated scattering) is responsible for some reduction of the drive intensity. Clearly, intensity transfer from the probe to the drive field and then back to the probe takes place.

It was also found that this process exists not only for the dark coherent superposition, but for the bright superposition as well. However, in the latter case the exchange is in the opposite direction. Probe pulse transmission through an EIA medium is shown in Fig. 18b. The front edge of the transmitted pulse is enhanced due to relatively low absorption during the pulse rise time. After switching off the pulse the drive intensity has a positive tail. The process of stimulated scattering for the bright superposition has opposite direction; that is, probe intensity is transferred into the drive intensity. The relaxation time (about 1 μs) also depends on the light intensity.

This process is rather similar to the optical information storage effect [92]. The possibility of storing information in an atomic medium with 'bright' coherent superposition is discussed in a later section.

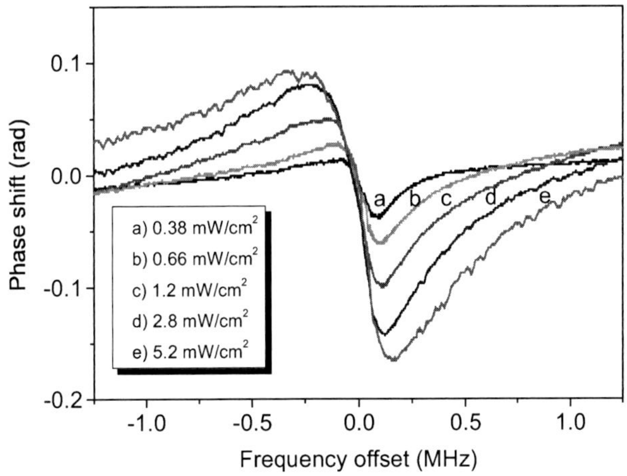

Figure 19. The probe phase shift in Cs vapour with bright Zeeman superposition as a function of the probe-drive frequency offset at different intensities. The laser frequency is at the $6S_{1/2}(F_g = 4) \rightarrow 6P_{3/2}(F_e = 5)$ transition.

6. Enhanced Kerr Nonlinearity

The concept of few-photon nonlinear optics [93] based on EIT/CPT related phenomena has became attractive for many applications, such as new optical field generation [94], quantum communication [95] and the generation of non-classical atomic ensembles [96]. A comprehensive review of closely related nonlinear magneto-optical effects in resonant atomic media has been published [62].

Taking into account that the amplitude of the EIA resonance and the associated refractive index of the coherent medium depend on the intensity of the drive component, we can conclude that an atomic vapour with a long-lived *bright* coherent superposition should also possess an enhanced nonlinear susceptibility.

Here we describe experiments in which first measurements of the intensity-dependent refractive index of an atomic medium under EIA conditions were performed [63, 97]. They were done with Cs vapour contained in a 2-cm long glass cell with no buffer gas. The cell was slightly heated to increase the atomic density. The maximum linear absorption did not exceed 60%. The ambient magnetic field in the cell was attenuated to less than 5 mG with a μ-metal shield.

The diameter of the bichromatic beam in the cell was about 1 cm, while the maximum intensities of the drive and probe components were 6 and 0.5 mW/cm^2. Bright Zeeman superpositions were prepared in the Cs vapour on the $6S_{1/2}(F_g = 4) \rightarrow 6P_{3/2}(F_e = 5)$ cycling transition. The saturation intensity $I_S = \pi ch/3\lambda^3\tau$ (λ is the resonant wavelength and τ is the lifetime of the excited level) for this cycling transition is about 1.1 mW/cm^2. The maximum saturation parameter $S = I/I_S$ was less than six.

The intensity-dependent refractive index was measured using the RF heterodyne technique described above. The phase shift induced in the probe beam is plotted in Fig. 19 as a function of frequency offset δ for a range of drive intensities I_D. The magnitudes of variations of refractive index n of the atomic medium is readily obtained from the phase shift

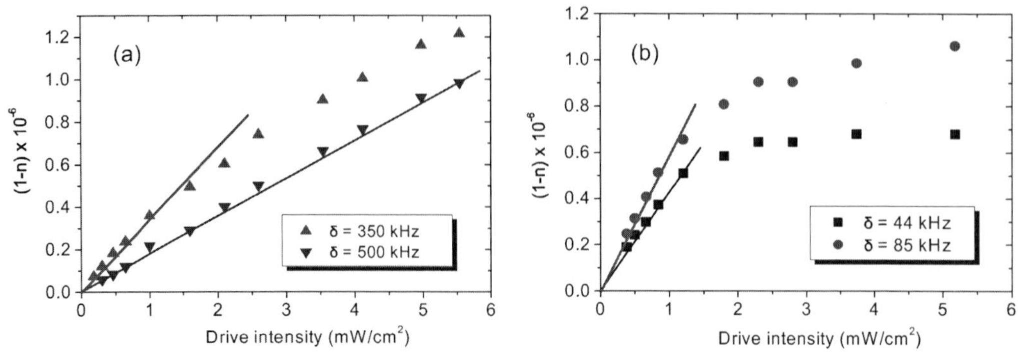

Figure 20. The refractive index as a function of the drive intensity in Cs vapour with bright long-lived coherence and linear fits at fixed probe-drive frequency offset. (a) Frequency offsets δ greater than the width of the EIA resonance. (b) Frequency offsets δ less than the width of the EIA resonance.

$\Delta\Phi$ using (3). Both the width and magnitude of the refractive index variations increase with the drive intensity I_D.

The refractive index variation with I_D depends on the frequency offset δ. Figure 20 shows the refractive index of Cs vapour with a bright long-lived Zeeman coherence as a function of I_D at a number of fixed frequency offsets δ. For both large positive and negative frequency offsets, where the dispersion is normal, the behavior of the refractive index is almost linear (Fig. 20a). Writing that the intensity dependent refractive index as a Taylor expansion

$$n(I_D) = n_0 + n_2 I_D + ..., \tag{6}$$

a polynomial fit to the experimental data allows the first-order nonlinear Kerr coefficient n_2 to be estimated. For example, for the positive offset δ=500 kHz (Fig. 20(a)), $n_2 \simeq (0.16 \pm 0.1) \times 10^{-6} cm^2/mW$. For comparison, the Kerr coefficient n_2 for quartz is $n_2 \simeq 4.5 \times 10^{-19} cm^2/mW$. For smaller frequency offsets, but still larger than the width of the coherence resonance, the Kerr coefficient n_2 is higher, although the dependence of $n(I_D)$ reveals some saturation at $I_D \geq 2\ mW/cm^2$. For frequency offsets less than the width of the bright resonance the dispersion is anomalous ($dn/d\nu < 0$). In this region of offsets $\delta < \Gamma$ the absolute value of the Kerr coefficient n_2 is larger ($n_2 \simeq (0.4 - 0.6) \times 10^{-6} cm^2/mW$) but only for the drive intensities $I_D \leq 1\ mW/cm^2$. At higher drive intensity ($I_D > 2\ mW/cm^2$) the refractive index shows little change (Fig. 20(b)).

It was found that the calculated value of the Kerr coefficient n_2 is an odd function of the frequency offset δ, with some asymmetry, as shown in Fig. 21. The absolute value of n_2 has extremes at offsets comparable with the half-width of the EIA resonance. For larger δ (in the spectral region of normal dispersion $dn/d\nu > 0$) the absolute value of the Kerr coefficient n_2 decreases steadily. Thus, n_2 (or the nonlinear susceptibility $\chi^{(3)}$) has a resonant character with sub-natural width.

A comparison of the enhanced Kerr nonlinearity in bright and dark Zeeman coherent states is of interest. Dark coherences in Cs atoms were prepared by the bichromatic light though the transitions $6S_{1/2}(F_g = 3) \rightarrow 6P_{3/2}(F_e = 2, 3)$. The phase variation of the

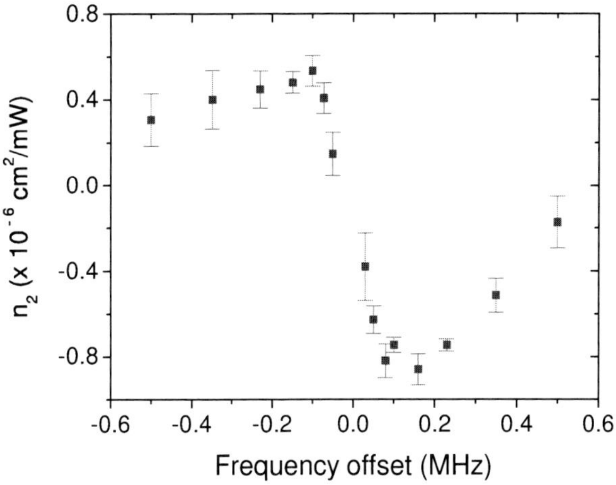

Figure 21. The first-order nonlinear Kerr coefficient n_2 of Cs vapour with bright Zeeman coherent superpositions as a function of frequency offset from the Raman resonance. Errors are fitting errors.

probe component for different drive intensities I_D as a function of the frequency offset δ is shown in Fig. 22. The spectral profiles for the dark superpositions look more shape-preserving at higher drive intensity compared to those for the bright superpositions (Fig. 19). In contrast to the case of the bright superpositions, however, the intensity dependences of the refractive index $n(I_D)$ for small and larger frequency detunings are not very different: both exhibit saturation at $I_D > 2\ mW/cm^2$. The spectral dependence of the calculated first-order nonlinear coefficient n_2 for dark Zeeman coherent superpositions is shown in Fig. 23.

Except for the sign, the spectral behaviour of the Kerr coefficients n_2 for the dark and bright coherent superpositions are similar. The larger value of the Kerr nonlinearity for the bright Zeeman coherence compared with the dark one appears to be reasonable because in the bright state the atom-light interaction is enhanced, while it is inhibited for the case of non-absorbing (thus, non-interacting) dark states. On the other hand, for a medium with dark coherence, the range of atomic densities over which the medium may produce Kerr nonlinearity is much wider, and obviously at higher density the nonlinear Kerr coefficient n_2 will be larger.

In spite of strong interest in the field of low intensity nonlinearity enhanced by long-lived atomic coherence, to our knowledge there are only a few direct measurements of Kerr coefficient n_2 enhanced by dark coherence [32, 98, 99] and we are not aware of any for atomic media with bright coherent superpositions.

The spectral dependences of the nonlinear Kerr coefficient for the long-lived dark coherence obtained in our experiments are consistent with the result reported by Wang et al [98]. The unprecedentedly high Kerr nonlinearity ($n_2 = 0.18\ cm^2/W$) reported by Hau et al [32] could be explained by a very high atomic density because they estimated n_2 for a cold Na ensemble at the temperatures required for a Bose-Einstein condensate.

Strong saturation of the refractive index dependences $n(I_D)$ in the region of small

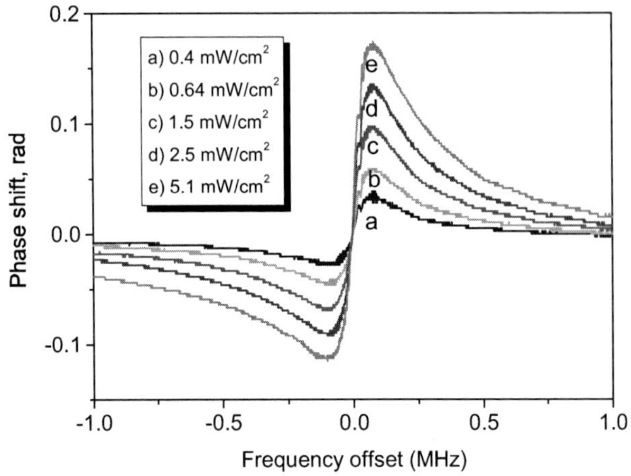

Figure 22. The probe phase shift in Cs vapour with dark Zeeman superposition as a function of frequency detuning at different intensities. The laser frequency is at the $6S_{1/2}(F_g = 3) \rightarrow 6P_{3/2}(F_e = 2)$ transition.

frequency detunings for the both dark and bright coherent superpositions may be due to higher-order nonlinearity, which is associated with the coherence between magnetic states of the ground level having $|\Delta m| > 2$ [97]. Contributions from higher-order multipoles to magneto-optical effects have been considered by Budker et al [62], and they have been observed in fluorescence in a number of experiments [100, 101, 102] and in forward scattering [103]. It would be of particular interest to compare the nonlinearities in atomic systems with different angular momenta and hence a different number of ground level Zeeman states, such as ^{87}Rb and ^{85}Rb. But it is more likely that the cause of this saturation is coherent population oscillations, which, as was shown above, can also produce absorption and, consequently, refractive index resonances with sub-natural width. Coherent population oscillations are readily generated by the linearly polarized fields in the present experimental configuration.

In conclusion, Zeeman coherence, both dark and bright, leads to enhanced nonlinear susceptibility of an atomic medium. The enhancement takes place in a spectral region of sub-natural width, suggesting that a long-lived ground-level coherence plays an essential role in the process. The intensity-dependent refractive index of Cs vapour with a long-lived Zeeman coherence has been measured at low light intensity ($0.3\ mW/cm^2 \leq I \leq 6\ mW/cm^2$). Direct measurement of the nonlinear components of the refractive index is important for optimizing and controlling the nonlinear optical processes.

It should also be mentioned that weak resonant light can form spatial solitons in an atomic medium if the nonlinearity is enhanced by a long-lived coherence. Recently, spatial solitons in an electromagnetically induced nonlinear waveguide have been considered [104]. Large bright state nonlinearity significantly extends the range of possibilities for controlling these solitons.

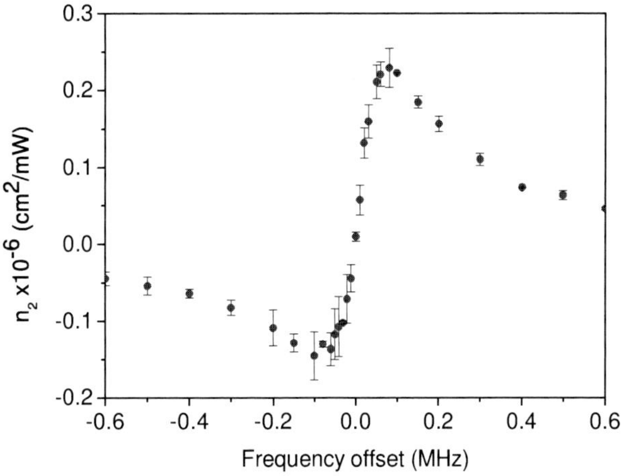

Figure 23. The first-order nonlinear coefficient n_2 of Cs vapour with dark Zeeman coherent superpositions as a function of frequency offset. Errors are fitting errors.

7. Four Wave Mixing

Enhanced nonlinear susceptibility of an atomic vapour provides a basis for efficient four-wave mixing (FWM). The demonstration of FWM at low light intensity ($I_1 + I_2 < 1\ mW/cm^2$) with efficiency approximately 1% [105] could be considered to be the first experimental evidence of enhanced nonlinear susceptibility of Rb vapour in a bright superposition state.

Bichromatic resonant light with a constant frequency offset δ from Raman resonance creates a coherent phase grating in the atomic sample, which diffracts optical waves, as shown in Fig. 24(a,b). The absorption of two photons from the drive field with frequency ν_1 and stimulated emission at ν_2 gives rise to a new wave of frequency $\nu_N = 2\nu_1 - \nu_2 = \nu_1 + \delta$ and the amplitude proportional to $\chi^{(3)} E_1^2 E_2$, where E_1 and E_2 are amplitudes of the components of resonant fields.

Figure 24(c) a shows the beat signal detected after the Rb cell for non-resonant and resonant light. Evidence of the presence of a new wave is clearly visible in the resonant case, Fig. 24(d). In addition to the beating at δ, a component at 2δ appears. This is the result of mixing of the new wave at $\nu_N = \nu_1 + \delta$ with the light component at $\nu_2 = \nu_1 - \delta$.

The wave mixing process in the case of ground-level Zeeman coherence can be both efficient at very low light intensity (about 1% at 0.8 mW/cm^2) and highly selective (FWHM less than 50 kHz) [105]. The efficiency is a maximum for orthogonal linear polarizations of the two components of the bichromatic beam. For intensities of the co-propagating components significantly above the saturation intensity (I_D, $I_P \gg I_S$), a comb-like structure of beat signals can be generated [63]. At the light intensities of 60 and 5 mW/cm^2 for the two components (produced by focussing the light to a 3-mm spot inside the cell) and a frequency offset δ = 350 kHz, up to five equally spaced components were observed.

The low-frequency refractive index oscillations due to time evolution of the coherent states are also responsible for periodic phase variations of both fields. The phase modulation in turn generates sidebands separated by δ, with amplitude distributed according to

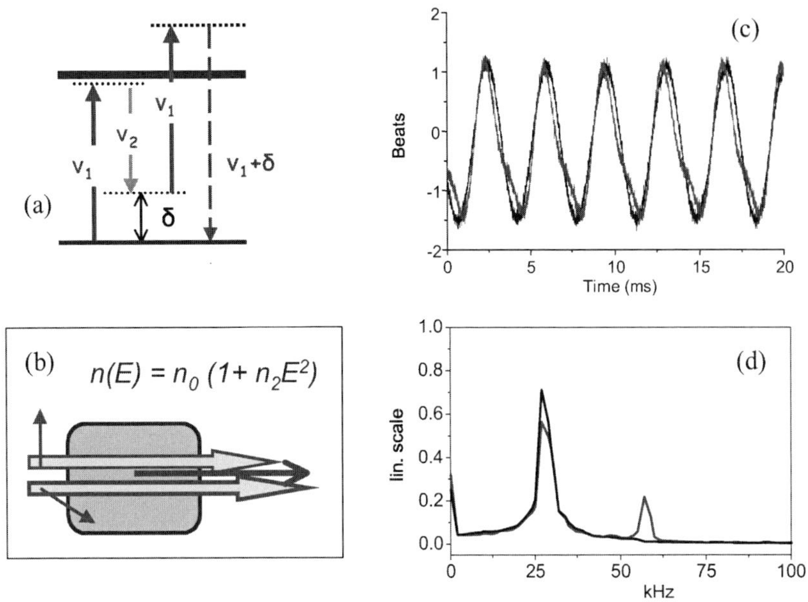

Figure 24. Four wave mixing in degenerate two-level atomic system with. (a) New wave generation at $\nu_N = \nu_1 + \delta$ by mixing two waves with frequency offset δ. (b) Scheme of the experimental observation of four wave mixing in a Rb vapour cell. c) The photodiode output detected after the Rb cell for off-resonance (i) and resonant (ii) light. (d) The Fourier transformation of the photodiode output for off-resonance (i) and resonant (ii) light, respectively.

a Bessel function. The spectral region covered by the equidistant peaks is less than the natural width of the optical transition, indicating high selectivity of the enhanced nonlinearity. Contributions to the comb-like spectrum from higher order nonlinearity should be distinguished from a careful study of the new waves.

8. 'Storage of Light' in a 'Fast-Light' Atomic Medium

'Storage of light' in atomic media, pioneered by two teams from Harvard University [92, 106], has generated much attention in the field of quantum optics because of possible applications in quantum communication [107, 108]. It has been suggested that the atomic medium can act as a temporary storage buffer for quantum states of light. Reversible mapping of a quantum state onto a collective atomic coherence was addressed in several experimental investigations [108, 109, 96] following proof-of-principle demonstrations. Even the storage and retrieval of classical information using atomic coherence could have important practical applications. For example, stored light can be considered as a candidate for optical buffering, which is needed to make all-optical internet processing possible [110].

The concept of a non-absorbing dark state due to CPT is usually the basis of storage of light. In fact storage of light is a process of information transfer from the signal light to an atomic medium and back to a regenerated light pulse by means of the so-called dark-state

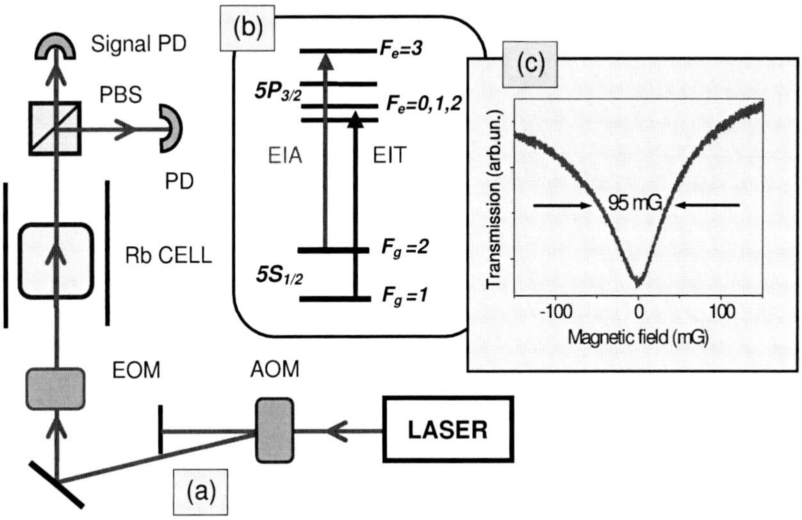

Figure 25. (a) Scheme of the experimental set-up and (b) laser frequency tunings for EIA and EIT media for ^{87}Rb. (c) EIA resonance for laser light tuned to the $5S_{1/2}(F_g = 2) \rightarrow 5P_{3/2}(F_e = 3)$ transition.

polariton [111]. This polariton is a quasi-particle, which can be understood as a superposition of atomic spin waves and a signal light pulse that can be modified by the control field [107, 108]. Turning off the control field causes the dark-state polariton to be converted into a purely atomic form (spin wave with zero group velocity), which is confined to the vapour cell. However, this process is reversible. Switching on the drive field again results in the conversion of the purely atomic form of the polariton ('frozen' spin wave) back into the combined form, and light can leave the atomic medium. Thus, information carried by the signal pulse is mapped onto long-lived atomic coherences and can be retrieved later.

In a recent review it has been stated that EIT has become firmly established as a powerful technique for coherent control of the resonant propagation of classical light pulses [108]. It is natural to consider the possibility of an extension of the physical basis for such control. Some experimental observations in coherently prepared atomic vapours suggest that the requirement of CPT (or EIT) is not a unique way to control the propagation process [88]. Recently an alternative explanation of the storage of light experiment in warm Rb vapour has been proposed [112]; numerical modelling based on light-induced anisotropy of the atomic medium under optical excitation with time-dependent polarization also agrees very well with the experimental data.

Here we review experimental results of light storage in an EIA atomic sample where the concept of the dark-state polariton does not apply and compare atomic responses from EIA and EIT media under similar experimental conditions [113, 114].

The experiments were carried out with rubidium vapour on the D_2 line. The experimental set-up shown in Fig. 25(a) was similar to the first storage of light experiment with monochromatic light [92]. The linearly polarized output from the AOM formed the control beam. An electro-optic modulator (EOM) generated a weaker probe component with orthogonal polarization. Thus the intensities of the control and the signal components were

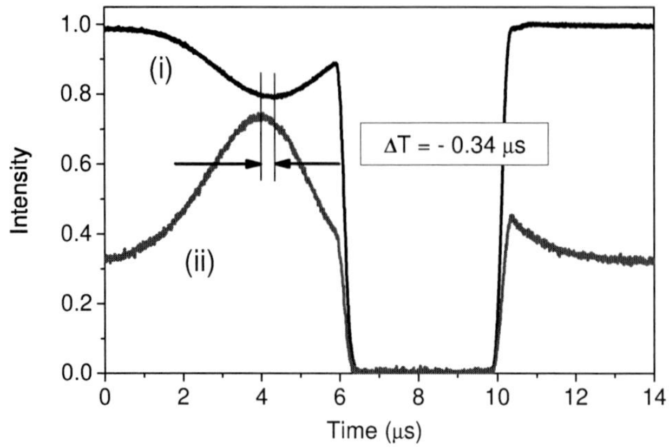

Figure 26. Propagation and retrieval of a Gaussian-shaped signal pulse in an EIA atomic medium. Curves (i) and (ii) represent intensity of the control and signal or probe components, respectively. The intensity of the signal component is scaled up approximately by a factor of two.

not independent: an intensity increase in the probe component (signal pulse) was accompanied by an intensity decrease in the control component. Gaussian-shaped and sharp-edge rectangular signal pulses on a constant background were used. The constant part of the probe component was roughly equal to the amplitude of the pulse.

A natural mixture of rubidium isotopes with no buffer gas was contained in a 5 cm-long glass cell. The ambient magnetic field in the cell was attenuated to less than 5 mG with a μ-metal shield. A controllable longitudinal magnetic field was produced by a solenoid mounted inside the magnetic shield. The cells were heated with a magnetic field-free method to give linear absorption of resonant light ranging up to 90 %.

For this experiment the interaction time between the atoms and light must be sufficiently long that the coherently prepared atoms are still within the laser beam when the light is turned back on after the dark interval. In a cell with no buffer gas the interaction time is determined by the transit time of atoms across the laser beam; so the beam was expanded by a telescope. The beam diameter inside the cell was about 1 cm, and the maximum intensities of the control and probe components were 2 and 0.3 mW/cm^2, respectively. The intensity ratio between the control and the probe components was typically 5 to 10. After the Rb cell the signal and control components were separated by a polarizing beam splitter and monitored simultaneously with fast photodiodes. The response time of the detection and modulation system was less than 0.5 μs.

EIA was prepared in a vapour of ^{87}Rb on the cycling transition $5S_{1/2}(F_g = 2) \rightarrow 5P_{3/2}(F_e = 3)$. The width of the EIA resonance, which is determined primarily by power broadening, was about 95 mG, as shown in Fig. 25(c).

At the EIA resonance the signal pulse in the probe component propagates with a negative group velocity because before the pulse enters the cell the medium has already been prepared by the constant part of the probe and the control components. The peak of the transmitted pulse experiences a negative delay (or advancement) relative both to non-resonant light and to the control field intensity minimum, as shown on the left side of Fig. 26. We

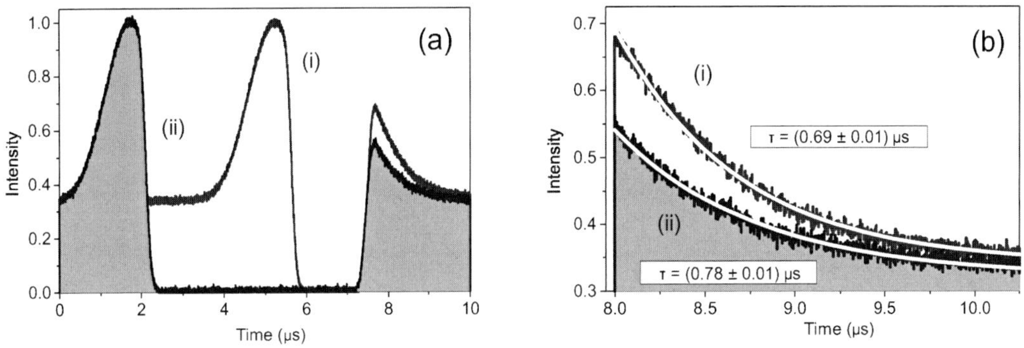

Figure 27. (a) Storage and retrieval of smooth signal pulses in the EIA atomic medium for two different dark intervals ΔT. The curves (i) and (ii) are normalized to the peak of the transmitted probe intensity. (b) Decaying slopes of the two retrieved pulses with exponential fit.

note that the group velocities for the control and probe components are different because the refractive index n and the dispersion at the EIA resonance are different for the intense control and weak probe components. The relative absorption variation for the weak component is higher and the dispersion $dn/d\nu$ of the medium seen by the probe components is steeper, so that the signal pulse has a smaller absolute value of negative v_g. The observed negative delay ΔT for the probe component of approximately 0.34 μs implies a limit for the absolute value of the negative group velocity of $v_g < L/\Delta T \simeq 1.5 \times 10^5$ m/s.

When the peak of the signal pulse has almost propagated through the cell both the control and probe components of the laser beam are shut off by the AOM for approximately 4 μs (Fig. 26). After the dark interval the laser light is turned on again and a sharp increase of light intensity is detected by the signal photodiode. This does not occur for non-resonant laser light or if the initial signal pulse is not present. Consequently, the feature indicates an atomic response to the switching on of the light that depends on the signal pulse having been present. This retrieved pulse results from the interaction of the control light and the coherently prepared atoms. A similar atomic response is also observed if the constant part of the probe component is absent in the preparation phase, but the relative amplitude of the retrieved pulse is smaller.

The shape of the retrieved pulse is asymmetric as shown in Fig. 27a. The rising edge is primarily determined by the switching and detection time constants while the decaying slope reflects the depletion of the ground level coherence and depends on the light intensity and the magnetic field. At zero magnetic field the decreasing slope can be well fitted by an exponential decay, as shown in Fig. 27b. The relaxation time τ = (0.74 $\pm$ 0.06) μs obtained from the exponential fitting is consistent with the width of the EIA resonance. The average relaxation time is shorter for higher light intensity, indicative of the stronger atom-light interaction destroying the ground level coherence more quickly. For nonzero magnetic fields the time evolution of the atomic coherence leads to a damped modulation on the slope.

The amplitude of the retrieved pulse decreases as the dark interval increases. An exponential decay fitted to the experimental data yields a decay time, $T = (6.0 \pm 1.5)$ μs,

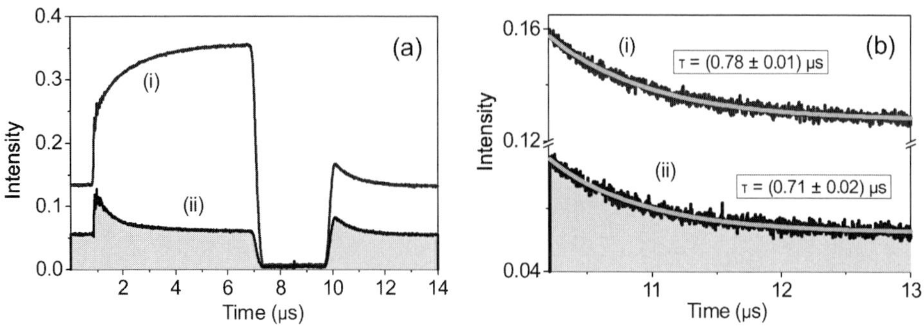

Figure 28. (a) Sharp-edge (square) signal pulse propagation, storage and retrieval in EIT and EIA atomic media. (b) Decaying slopes of the two retrieved pulses with exponential fit.

consistent with the transit time of atoms across the interaction zone. This time, which can be considered as the atomic memory time, is much longer than the relaxation time τ, mentioned above. The relaxation time of the atomic coherence is longer during the dark interval than under exposure to light. However, the obtained ground level coherence decay time T is much shorter than the decay times reported previously for the storage of light experiments (0.5 and 0.9 ms) [92, 106] because we use a Rb cell with no buffer gas. The interaction time, and consequently the ground level relaxation time, for a laser cooled atomic sample or an atomic vapour in a cell with buffer gas or with paraffin coated walls can be much longer.

This experiment shows that the main features of light storage demonstrated for an EIT media can be obtained in an atomic medium with *bright* coherent superposition. It is of interest to compare atomic responses from EIA and EIT media under experimental conditions that are as close as possible. To do this, an EIT medium was prepared in the cell without buffer gas on the Rb D_2 line by tuning the laser frequency to the transitions $5S_{1/2}(F_g = 1) \rightarrow 5P_{3/2}(F_e = 0, 1, 2)$.

The observed delay for the signal pulse in the probe component relative to the non-resonant reference pulse of $\Delta T \simeq 0.7\ \mu s$ corresponds to a group velocity of $v_g < L/\Delta T \simeq 0.7 \times 10^5\ m/s$. Although the spatial extent of the pulse is compressed in the atomic medium by the ratio $c/v_g \simeq 4 \times 10^3$, only a small fraction of the pulse can be in the cell at any time. Under such conditions we might not expect strong coupling between the spin and photon components according to the dark-state polariton model. Nevertheless, the retrieved pulse observed after the dark interval is rather large, having a maximum amplitude of about 46% of the transmitted signal pulse, which is even higher than the first observation [92].

The similarities and differences between atomic media with steep normal and anomalous dispersion are highlighted by using sharp-edge signal pulses. A significant part of the spectrum of a square pulse is outside the region of constant steep dispersion, so that such pulses experience strong distortion during propagation through the medium. The left side of Fig. 28(a) shows typical distortion of sharp-edge signal pulses for the EIT and EIA cases at the same atomic density and light intensity. For the EIT medium the initial transmission of the probe component is about 2.5 times higher than for the EIA medium, and after arrival of the signal pulse it grows to a new steady-state level as the coherent nonabsorbing state is established. For the EIA medium the transmission is high at the front edge of the pulse, but

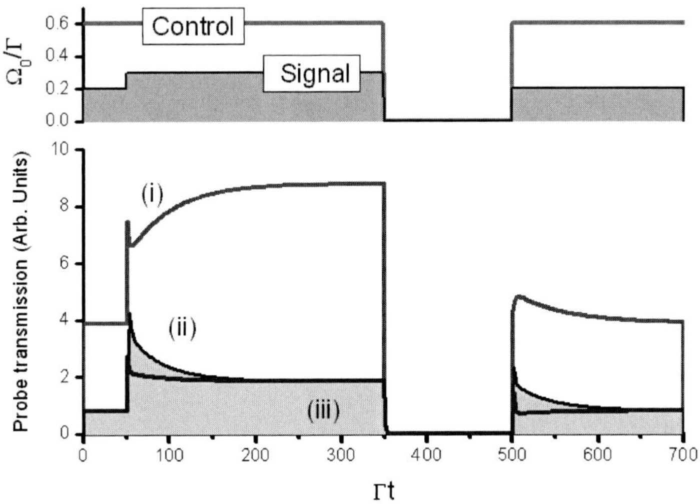

Figure 29. Top: Temporal sequence for the control and probe components. Bottom: Calculated transmission for the probe component $[\gamma/\Gamma = 0.001,\ B = 0]$ for (i) EIT medium, transition $5S_{1/2}(F_g{=}1)$-$5P_{3/2}(F_e{=}0)$ of ^{87}Rb and for (ii) EIA medium, transition $5S_{1/2}(F_g{=}2)$-$5P_{3/2}(F_e{=}3)$ of ^{87}Rb and π phase shift of the probe field $(50 < \Gamma t < 350)$. Curve (iii) corresponds to the same transition with zero phase shift of the probe field.

decreases with time as the new, more strongly absorbing state is established. Such distortion is consistent with the observed delay and advance for the smooth pulses. Two retrieved pulses after identical dark intervals appear in the right-hand part of Figure 28(a). In both cases the atomic responses after the dark interval are found to be remarkably similar. The slow decaying slopes of the retrieved pulses are well fitted by an exponential function, as shown in Fig. 28(b).

Both the shape and the obtained relaxation times of the response for both cases (0.78 μs for EIT and 0.71 μs for EIA) are very close to the results for the smooth pulses. From this we conclude that there is no correlation between the shapes of the retrieved and signal pulses under the present experimental conditions, which were far from the optimum for the shape of the signal pulse to be reproduced [115].

A numerical simulation of the observed transients based on the optical Bloch equations in the rotating wave approximation was done by Lezama [113, 114]. The probe transmission was calculated for the two closed transitions $5S_{1/2}(F_g = 1) \rightarrow 5P_{3/2}(F_e = 0)$ and $5S_{1/2}(F_g = 2) \rightarrow 5P_{3/2}(F_e = 3)$ of ^{87}Rb giving rise to EIT and EIA, respectively. The corresponding Rabi frequencies were calculated taking into account the relative strength of the two transitions. The same sample density was assumed for the two transitions. The calculated temporal sequence is indicated at the top of Fig. 29. The drive and probe fields have perpendicular linear polarizations. In the case of the EIA transitions two different phase differences between the probe pulse $(50 \leq \Gamma t \leq 350)$ and the constant background were considered.

The modelling presented in Fig. 29 reproduces the essential features of the experimental profiles of Fig. 28(a). In the case of EIA the shape of the retrieved pulse depends on the phase of the probe pulse relative to the constant probe field background. The shape of the calculated pulse agrees very well with experimental observations if the phase shift is π. A zero phase shift of the probe field with dc background results in a 'negative' retrieved pulse. A similar transient effect in the probe intensity has been observed in an EIA medium previously [88].

In conclusion, storage of light effects have been demonstrated in atomic media with both EIA and EIT. There is remarkable similarity in the responses of the EIT and EIA media, despite their dramatically different optical properties. These observations strongly indicate that there is common physics of light storage in the two cases and a more general description for media with different absorptive and dispersive properties is possible.

9. Conclusion

Light propagation in highly nonlinear atomic media with very steep sign-reversible dispersion continues to generate interest because of its fundamental aspects and promising applications, including quantum information processing.

Our study of light-induced ground-level coherence in degenerate two-level atomic systems demonstrates how easily such coherence can be prepared and how strongly it can affect the optical properties of resonant media, such as dispersion and Kerr nonlinearity.

Coherent atom-photon interactions depend on many parameters, including level structures, ground and excited level relaxation rates, light intensities and polarization, atomic density and applied magnetic fields. In describing the processes involved, all aspects of atom-light interaction should be taken into account. There remains a need for a complete theoretical model applicable to a wide range of experimental parameters, and for numerical simulations to assist in optimizing important parameters such as dispersion or nonlinear susceptibility. As well, comparisons with calculations will help to control the desired properties of the medium.

The presented experimental studies may stimulate a deeper understanding of light propagation, storage and retrieval using long-lived atomic coherences. This could also result in the development of new schemes and methods for more efficient control of resonant light in atomic media.

Acknowledgments

The authors are grateful to Arturo Lezama, Valery Yudin and Alexei Taichenachev for numerous valuable discussions.

References

[1] Agap'ev B.D.; Gornyi M.B.; Matisov B.G.; Rozhdestvenskii Y.V.; *Sov. Phys. Usp.* 1993; 36, 763–793.

[2] Arimondo E.; *Progress in Optics, (ed. Wolf, E.)* 1996; 35, 257–354.

[3] Harris S.E.; *Phys. Today* 1997; 50(7), 36–42.

[4] Marangos J.P.; *J. Mod. Opt.* 1998; 45, 471–503.

[5] Ficek Z.; Swain S.; *Quantum Interference and Coherence*; Springer Series in Optical Science, Vol. 100, New York 2005.

[6] Fleischhauer M.; Imamoglu A.; Marangos J.P.; *Rev. Modern Phys.* 2005; 77, 633–673.

[7] Alzetta G.; Gozzini A.; Moi L.; Orriols G.; *Nuovo Cimento B* 1976; 36, 5.

[8] Whitley R.M.; Stroud C.R.; *Phys. Rev. A* 1976; 14, 1498.

[9] Arimondo E.; Orriols G.; *Lett. Nuovo Cimento* 1978; 17, 333.

[10] Orriols G.; *Lett. Nuovo Cimento* 1979; 53, 1.

[11] Gray H.R.; Whitley R.M.; Stroud C.R.; *Opt. Lett* 1978; 3, 218–220.

[12] Akulshin A.M.; Celikov A.A.; Velichansky V.L.; *Opt. Commun.* 1991; 84, 139–143.

[13] Gawlik W.; Kowalski J.; Neumann R.; Trager F.; *Phys. Lett. A* 1974; 48, 283–284.

[14] Kanorsky S.I.; Weis A.; Wurster J.; Hänsch T.W.; *Phys. Rev. A* 1993; 47, 1220–1226.

[15] Budker D.; Kimball D.F.; Rochester S.M.; Yashchuk V.V.; *Phys. Rev. Lett.* 1999; 83, 1767–1770.

[16] Oreg J.; Hioe F.T.; Eberly J.H.; *Phys. Rev. A* 1984; 29, 690.

[17] Gaubatz U.; Rudecki P.; Becker M.; Schiemann S.; Kulz M.; Bergmann K.; *Chem. Phys. Lett.* 1988; 149, 463.

[18] Agrawal G.P.; *Phys. Rev A* 1983; 28, 2286–2295.

[19] Mlynek J.; Köster E.; Kolbe J.; Lange W.; *J. Opt. Soc. Am B* 1984; 1, 532.

[20] Köster E.; Mlynek J.; Lange W.; *Opt. Commun.* 1985; 53, 53.

[21] Bloembergen N.; Zou Y.H.; Rothberg L.J.; *Phys. Rev. Lett.* 1985; 54, 186.

[22] Aspect A.; Arimondo E.; Kaiser R.; Vansteenkiste N.; Cohen-Tannoudji C.; *Phys. Rev. Lett.* 1988; 61, 826–829.

[23] Scully M.O.; *Phys. Rev. Lett.* 1991; 67, 1855–1858.

[24] Kocharovskaya O.A.; Khanin Y.I.; *JETP Lett.* 1988; 48, 630.

[25] Scully M.O.; Zhu S.Y.; Gavrielides A.; *Phys. Rev. Lett.* 1989; 62, 2813.

[26] Harris S.E.; *Phys. Rev. Lett.* 1989; 62, 1033–1036.

[27] Harris S.E.; Field J.E.; Imamoglu A.; *Phys. Rev. Lett.* 1990; 64, 1107.

[28] Harris S.E.; Field J.E.; Kasapi A.; *Phys. Rev. A* 1992; 46, R29.

[29] Xiao M.; Li Y.; Jin S.; Gea-Banacloche J.; *Phys. Rev. Lett.* 1995; 74, 666–669.

[30] Schmidt O.; Wynands R.; Hussein Z.; Meschede D.; *Phys. Rev. A* 1996; 53, R27.

[31] Kasapi A.; Jain M.; Harris S.E.; *Phys. Rev. Lett.* 1995; 74, 2447–2450.

[32] Hau L.V.; Harris S.E.; Dutton Z.; Behroozi C.H.; *Nature* 1999; 397, 594–598.

[33] Kash M.M.; Sautenkov V.A.; Zibrov A.S.; Hollberg L.; Welch G.R.; Lukin M.D.; Rostovtsev Y.; Fry E.S.; Scully M.O.; *Phys. Rev. Lett.* 1999; 82, 5229–5232.

[34] Walls D.F.; Zoller P.; *Opt. Commun.* 1980; 34, 260.

[35] Mlynek J.; Mitschke F.; Deserno R.; Lange W.; *Appl. Phys. B* 1982; 28, 135–135.

[36] Thomas J.E.; Ezekiel S.; Leiby C.C.; Picard R.H.; Willis C.R.; *Opt. Lett.* 1982; 6, 298.

[37] Thomas J.E.; Hemmer P.R.; Ezekiel S.; Leiby C.C.; Picard R.H.; Willis C.R.; *Phys. Rev. Lett.* 1982; 48, 867–870.

[38] Wynands R.; Nagel A.; *Appl. Phys B* 1999; 68, 1–25.

[39] Budker D.; Romalis M.; *Nature Phys.* 2007; 3, 227–233.

[40] Duan L.M.; Lukin M.D.; Cirac J.I.; Zoller P.; *Nature* 2001; 414, 413.

[41] Eisaman M.D.; Andre A.; Massou F.; Fleischhauer M.; Zibrov A.S.; Lukin M.D.; *Nature* 2005; 438, 837.

[42] Jenkins S.D.; Matsukevich D.N.; Chanelire T.; Lan S.Y.; Kennedy T.A.B.; Kuzmich A.; *J. Opt. Soc. Am. B* 2007; 24, 316–323.

[43] Smirnov V.S.; Tumaikin A.M.; Yudin V.I.; *Sov. Phys. JETP* 1989; 89, 913.

[44] Renzoni F.; Maichen W.; Windholz L.; Arimondo E.; *Phys. Rev. A* 1997; 55, 3710–3718.

[45] Akulshin A.M.; Barreiro S.; Lezama A.; *Phys. Rev. A* 1998; 57, 2996–3002.

[46] Akulshin A.M.; Barreiro S.; Lezama A.; *Phys. Rev. A* 1999; 59, 4732–4735.

[47] Lezama A.; Barreiro S.; Lipsich A.; Akulshin A.M.; *Phys. Rev. A* 2000; 61, 013801.

[48] Dancheva Y.; Alzetta G.; Cartaleva S.; Taslakov M.; Andreeva C.; *Opt. Commun.* 2000; 178, 103–110.

[49] Letokhov V.S.; Chebotaev V.P.; *Nonlinear Laser Spectroscopy*; Springer Series in Optical Science, Vol. 4, Springer-Verlag, Berlin 1977.

[50] Renzoni F.; Zimmermann C.; Verkerk P.; Arimondo E.; *J. Opt. B: Quantum Semiclass. Opt.* 2001; 3, S7–S14.

[51] Taichenachev A.V.; Tumaikin A.M.; Yudin V.I.; *Phys. Rev. A* 2000; 61, 011802R.

[52] Goren C.; Wilson-Gordon A.D.; Rosenbluh M.; Friendmann H.; *Phys. Rev. A* 2003; 67, 033807.

[53] Baklanov E.V.; Chebotaev V.P.; *Zh. Eksp. Theor. Fiz.* 1971; 61(3), 922–929.

[54] Baklanov E.V.; Chebotaev V.P.; *Sov. Phys. JETP* 1972; 34, 490.

[55] Baklanov E.V.; Chebotaev V.P.; *Zh. Eksp. Theor. Fiz.* 1971; 60(2), 552–568.

[56] Haroche S.; Hartmann F.; *Phys. Rev. A* 1972; 6, 1280.

[57] Boyd R.W.; Raymer M.; Narum P.; Harter D.; *Phys. Rev. A* 1981; 24, 411–423.

[58] Boyd R.W.; Mukamel S.; *Phys. Rev. A* 1984; 29, 1973–1983.

[59] Bigelow M.S.; Lepeshkin N.N.; Boyd R.W.; *Science* 2003; 301, 200–202.

[60] Fuchs J.; Duffy G.J.; Rowlands W.J.; Lezama A.; P P.H.; Akulshin A.M.; *J. Phys. B: At. Mol. Opt. Phys.* 2007; 40, 1117–1129.

[61] Kastler A.; *Nuc. Inst. and Meth.* 1973; 110, 259–263.

[62] Budker D.; Gawlik W.; Kimble D.K.; Rochester S.M.; Yashchuk V.V.; Weis A.; *Rev. Mod. Phys.* 2002; 74, 1153.

[63] Akulshin A.M.; Cimmino A.; Sidorov A.I.; McLean R.; Hannaford P.; *J. Opt. B: Quantum Semiclass. Opt.* 2003; 5, S479–S485.

[64] Renzoni F.; Lindner A.; Arimondo E.; *Phys. Rev. A* 1999; 60, 450–455.

[65] Kim K.; Kwon M.; Park H.D.; Moon H.S.; Rawat H.S.; An K.; Kim J.B.; *J. Phys. B: At. Mol. Opt. Phys.* 2001; 34 (23), 4801–4808.

[66] Molella L.S.; Rinkleff R.H.; Danzmann K.; *Phys. Rev. A* 2005; 72, 041802(R).

[67] Dimitrijevic J.; Grujic Z.; Mijailovic M.; Arsenovic D.; Panic B.M.; Jelenkovic B.; *Optics Express* 2008; 16, 1343–1353.

[68] McAlexander W.I.; Abraham E.R.I.; Hulet R.G.; *Phys. Rev. A* 1996; 54, R5–R8.

[69] Mikhailov E.E.; Novikova I.; Rostovtsev Y.V.; Welch G.R.; *Phys. Rev. A* 2004; 70, 033806.

[70] Braje D.A.; V. Balic G.Y.Y.; Harris S.E.; *Phys. Rev. A* 2003; 68, 041801(R).

[71] Sargsyan A.; Sarkisyan D.; Staedter D.; Akulshin A.M.; *Opt. Spectroscopy* 2006; 101, 762–768.

[72] Javan A.; Kocharovskaya O.; Lee H.; Scully M.; *Phys. Rev. A* 2002; 66, 013805.

[73] Akulshin A.M.; Barreiro S.; Lezama A.; *Phys. Rev. Lett.* 1999; 83, 4277–4280.

[74] Müller G.; Wicht A.; Rinkleff R.H.; Danzmann K.; *Opt. Commun.* 1996; 127, 37–43.

[75] Wang L.J.; Kuzmich A.; Dogariu A.; *Nature* 2000; 406, 277–279.

[76] Stenner M.D.; Gauthier D.J.; Neifeld M.A.; *Nature* 2003; 425, 695–698.

[77] Tanaka H.; Niwa H.; Hayami K.; Furue S.; Nakayama K.; Kohmoto T.; Kunitomo M.; Fukuda Y.; *Phys. Rev. A* 2003; 68, 053801.

[78] Brown W.; McLean R.; Sidorov A.; Hannaford P.; Akulshin A.; *J. Opt. Soc. Am. B* 2008; 25, C82–C86.

[79] Akulshin A.; Singh M.; Sidorov A.; Hannaford P.; *Optics Express* 2008; 16, 15463–15468.

[80] Lombardi A.M.; *J. Opt. B: Quantum Semiclass. Opt.* 2002; 4, S473–S476.

[81] Brillouin L.; *Wave Propagation and Group Velocity*; Academic, New York 1960.

[82] Chu S.; Wong S.; *Phys. Rev. Lett.* 1982; 48, 738–741.

[83] Akulshin A.M.; Cimmino A.; Opat G.I.; *Quantum Electron.* 2002; 32, 567–569.

[84] Jeong H.; Dawes A.; Gauthier D.; *Phys. Rev. Lett.* 2006; 96, 143901.

[85] Mikhailov E.E.; Sautenkov V.A.; Rostovtsev Y.V.; Welch G.R.; *J. Opt. Soc. Am. B* 2004; 21, 425–428.

[86] Mikhailov E.E.; Sautenkov V.A.; Novikova I.; Welch G.R.; *Phys. Rev. A* 2004; 69, 063808.

[87] Macke B.; Ségard B.; Wielonsky F.; *Phys. Rev. E* 2005; 72, 035601(R).

[88] Akulshin A.M.; Cimmino A.; Sidorov A.I.; Hannaford P.; Opat G.I.; *Phys. Rev. A* 2003; 67, 011801(R).

[89] Kim K.; Moon H.S.; Lee C.; Kim S.K.; Kim J.B.; *Phys. Rev. A* 2003; 68, 013810.

[90] Bukhman N.S.; *Opt. Spectroscopy* 2004; 96, 626–631.

[91] Hollowood T.J.; Shore G.M.; *Nucl. Phys. B* 2008; 795, 138–171.

[92] Phillips D.F.; Fleischhauer A.; Mair A.; Walsworth R.L.; Lukin M.D.; *Phys. Rev. Lett.* 2001; 86, 783–786.

[93] Harris S.E.; Hau L.V.; *Phys. Rev. Lett.* 1999; 82, 4611.

[94] Zibrov A.S.; Matsko A.B.; Kocharovskaya O.; Rostovtsev Y.V.; Welch G.R.; Scully M.O.; *Phys. Rev. Lett.* 2003; 88, 103601.

[95] Duan L.M.; Lukin M.D.; Cirac J.I.; Zoller P.; *Nature* 2001; 414, 413–418.

[96] van der Wal C.H.; Eisaman M.D.; Andre A.; Walsworth R.L.; Phillips D.F.; Zibrov A.S.; Lukin M.D.; *Science* 2003; 301, 196–200.

[97] Akulshin A.M.; Sidorov A.I.; McLean R.J.; Hannaford P.; *J. Opt. B: Quantum Semiclass. Opt.* 2004; 6, 491–494.

[98] Wang H.; Goorskey D.; Xiao M.; *Phys. Rev. Lett.* 2001; 87, 073601.

[99] Kang H.; Zhu Y.; *Phys. Rev. Lett.* 2003; 91, 093601.

[100] Ducloy M.; *Phys. Rev. A* 1974; 9, 1319–1342.

[101] Fischer A.; Hertel I.V.; *Z. Phys. A* 1982; 304, 103–117.

[102] McLean R.J.; Ballagh R.J.; Warrington D.M.; *J. Phys. B: At. Mol. Phys.* 1986; 19, 3477–3491.

[103] Gawlik W.; Kowalski J.; Neumann R.; Trger F.; *Opt. Commun.* 1974; 12, 400–404.

[104] Hong T.; *Phys. Rev. Lett.* 2003; 90, 183901.

[105] Akulshin A.M.; Barreiro S.V.; Lezama A.; *Quantum Electron.* 2000; 30, 189–190.

[106] Liu C.; Dutton Z.; Behroozi C.H.; Hau L.V.; *Nature* 2001; 409, 490–493.

[107] Lukin M.D.; *Rev. Mod. Phys.* 2003; 75, 457–472.

[108] Andre A.; Eisaman M.D.; Walsworth R.L.; Zibrov A.S.; Lukin M.D.; *J. Phys. B: At. Mol. Opt. Phys.* 2005; 38, S589–S604.

[109] Kuzmich A.; Bowen W.P.; Boozer A.D.; Boca A.; Chou C.W.; Duan L.M.; Kimble H.J.; *Nature* 2003; 423, 731–734.

[110] Matsko A.B.; Strekalov D.V.; Maleki L.; *Optics Express* 2005; 13, 2210–2223.

[111] Fleischhauer M.; Lukin M.D.; *Phys. Rev. Lett.* 2000; 84, 5094–5097.

[112] Kozlov G.G.; Aleksandrov E.B.; Zapasskii V.S.; *Optics and Spectroscopy* 2004; 97, 909–914.

[113] Akulshin A.M.; Lezama A.; Sidorov A.I.; McLean R.J.; Hannaford P.; *J. Phys. B: At. Mol. Opt. Phys.* 2005; 38, L365–L374.

[114] Lezama A.; Akulshin A.M.; Sidorov A.I.; Hannaford P.; *Phys. Rev. A* 2006; 73, 033806.

[115] Grigoryan G.; Nikoghosyan G.; *Phys. Rev. A* 2005; 72, 043814.

In: New Trends in Quantum Coherence...
Editor: Rafael Drampyan, pp. 45-82

ISBN 978-1-60741-025-6

Chapter 2

NONLINEAR FARADAY EFFECT AND ITS APPLICATIONS

Wojciech Gawlik * ***and Szymon Pustelny*** †
Center for Magneto-Optical Research,
M. Smoluchowski Institute of Physics, Jagiellonian University
Reymonta 4, 30-059 Kraków, Poland

Abstract

This chapter provides introduction to the important method of contemporary magneto-optics, the nonlinear Faraday effect. It starts with a theoretical background linking the nonlinearity of the effect with quantum coherences of atomic states. The discussion of methods enabling analytical and numerical calculation of nonlinear magneto-optical rotation are given. Next, Essential aspects of a typical experimental apparatus used for investigation of the effect are described. Finally, the most important applications of the phenomenon are reviewed, such as in magnetometry, nuclear magnetic resonance, magnetic resonance imaging, magnetic particle detection and quantum-state engineering.

1. Introduction

Nonlinear Faraday Effect (NFE) is a nonlinear magneto-optical phenomenon in which the light-induced coherences, i.e. quantum superpositions of Zeeman sublevels, of atoms constituting the sample contribute significantly to the rotation of the polarization-plane of a light propagating across the sample placed in a longitudinal magnetic field [1]. Early works in which the nonlinear polarization rotation was observed were the experiments with gas-lasers equipped with Brewster windows. Interplay between strong polarization anisotropy and NFE resulted in strong variations of laser output power with magnetic field [2, 3]. First more systematic studies of NFE were performed after the advent of tunable lasers which allowed working with atomic vapors close to their resonance frequency [4].

* E-mail address: gawlik@uj.edu.pl
† E-mail address: pustelny@uj.edu.pl

Both the linear Faraday Effect (LFE) and NFE allow applications to magnetometry but the latter offers much higher sensitivity. This sensitivity enhancement is one of the reasons why NFE attracts so much interest. Application of the effect for optical magnetometry was first suggested by Barkov *et al.* [5] and the idea has been verified experimentally by Budker and co-workers [6]. The authors demonstrated very high sensitivity of the magnetometric method for detection of a weak magnetic field within a narrow dynamic range around zero. These constraints were lessened with the introduction of a light modulation technique [7,8], as described in Sec. 4.1. It enabled measuring much stronger magnetic fields with ultra-high sensitivity.

The possibility of measuring nonzero magnetic fields with high sensitivity has already found its applications in several different fields. Applications include detection of magnetic micro- and nano-particles [9], nuclear magnetic resonance [10], and magnetic resonance imaging [11]. These applications are reviewed in Secs. 4.2.–4.4. of this chapter.

Another attractive and growing application of NFE is the quantum-state engineering. Just like long-living superpositions (coherences) of atomic states are essential for applications in high-sensitivity optical magnetometry, such superpositions are also a key element for quantum information. Therefore, the same experimental techniques which are used for control and modification of the coherences for NFE magnetometry are also useful for quantum information. These application of NFE for quantum state engineering are described in Sec 4.5. of this Chapter.

In the following sections we discuss main physical mechanisms of NFE and present the fundamentals of its theoretical description, describe typical experimental setup and discuss application of NFE.

2. Theoretical Description of Nonlinear Faraday Effect

Theoretical description of NFE is performed using the density-matrix formalism. The knowledge of the matrix allows one to calculate the induced polarization $\mathbf{P}$ which characterizes the response of the medium to the electric field of light $\mathbf{E}$

$$\mathbf{P} = \mathrm{Tr}(\rho \mathbf{d}), \tag{1}$$

where $\mathbf{d}$ is the electric dipole moment operator. In our consideration, the matrix ρ is given in the mm' representation, where m and m' are the magnetic quantum numbers. In this representation, the populations and the coherences of the $|F, m\rangle$ atomic states, where F is the total angular momentum of an atom, are used to determine the overall quantum state of the system. Throughout this chapter it is assumed that the quantization axis is aligned with an external magnetic field. In such a case, the field shifts energies of the magnetic states $|F, m\rangle$ but does not mix them. Linearly polarized light propagates along the magnetic field and its polarization is perpendicular to the quantization axis. Therefore, the light is σ-polarized and can be decomposed into two circular components, the right- (σ^+) and left-handed (σ^-) components. We apply the convention in which the σ^+-polarized light generates transitions between magnetic sublevels of $\Delta m = 1$ and the σ^--polarized light of $\Delta m = -1$.

The evolution of the density matrix is described with the equation of motion [12]

$$\dot{\rho} = -\frac{i}{\hbar}[H, \rho] + \tilde{\Gamma}(\rho), \tag{2}$$

where H is the full Hamiltonian of the system (sum of the unperturbed Hamiltonian H_0 and the Hamiltonian of interaction V), $\tilde{\Gamma}$ is the operator describing relaxation and repopulation of the states, and $[\ ,\ ]$ denotes the commutator operator. Two approaches to solve Eq. (2) are presented in this chapter. In the first, perturbative approach, discussed in details in Sec. 2.1., the density matrix is expanded in a power series in the amplitude of the light field E and successive terms of the expansion are calculated iteratively. The approach is very useful because it enables analytical solutions and detailed analysis of the light-atom interaction, but it also has some limitations. One significant drawback of the method is its applicability to only weak and/or off-resonance interactions. In order to solve Eq. (2) for any light intensity and/or frequency, a nonperturbative approach is required. In the nonperturbative approach, a set of complex self-coupled equations, one for each density matrix element, needs to be evaluated precisely. Usually that cannot be done analytically and the equations need to be calculated numerically which is feasible for modern computers. However, the simplicity of the numerical approach is usually at the price of full understanding the physical background of the phenomena. The nonperturbative approach is discussed in Sec. 2.2.

2.1. Perturbative Approach

As discussed, for example, in Ref. [12], the density matrix ρ can be expanded in powers of the amplitude of electric field of light E

$$\rho = \sum_{n=0}^{\infty} \rho^{(n)} E^n. \tag{3}$$

Substituting Eq. (3) into Eq. (2) one can write a general relation for the density-matrix element $\rho_{\alpha\beta}$ in successive orders of perturbation

$$\dot{\rho}^{(n)}_{\alpha\beta} = -\frac{i}{\hbar}\left[H, \rho^{(n-1)}\right]_{\alpha\beta} + \tilde{\Gamma}\left(\rho^{(n)}\right)_{\alpha\beta}, \tag{4}$$

where superscript (n) indicates the n-th order of expansion. In the zeroth order of perturbation, i.e., in the absence of light, all atomic populations take their thermal equilibrium values and the density matrix consists only of diagonal elements representing the ground-state populations $\rho^{(0)}_{gg}$.[1] With no magnetic field, Zeeman sublevels of a given state are degenerate and equally populated. In the lowest-order of interaction, the off-diagonal elements $\rho^{(1)}_{ge}$ appear in the density matrix. These elements, called *optical coherences*, are associated with the linear susceptibility of a medium $\chi^{(1)}$, thus they determine linear absorption and dispersion. Once generated, the optical coherences evolve with time oscillating at a frequency which equals the unperturbed frequency of the transition ω_0 when atoms are not illuminated, or at the light frequency ω if the atoms interact with light (we assume $|\Delta\omega/\omega| \ll 1$, where

[1] It was assumed that the energy difference between the ground and excited state is big, e.g., of an optical range, thus the excited state is not populated in thermal equilibrium.

$\Delta\omega = \omega_0 - \omega$ is the detuning of the light from the transition). Within the second order of perturbation the excited-state populations $\rho_{ee}^{(2)}$ are generated. Additionally, the ground- and exited-state *Zeeman coherences*, described respectively with $\rho_{gg'}^{(2)}$ and $\rho_{ee'}^{(2)}$, are created. The Zeeman coherences oscillate at the frequency difference between the levels among which they are generated, thus the oscillations are much slower (from Hz to MHz) than those of optical coherences (on the order of 10^{14} Hz). At the second order of perturbation such phenomena as fluorescence and level-crossing can be described. In the next order of perturbation, the third-order optical coherences ($\rho_{ge'}^{(3)}$) can be calculated. As is shown below, the $\rho_{ge'}^{(3)}$ elements depend, among other quantities, on the ground-state Zeeman coherences, if such coherences can be generated in atoms. Within the third order, phenomena such as nonlinear absorption and dispersion, and hence NFE[2], as well as other optical phenomena like coherent population trapping [13], electromagnetically induced transparency [14] and absorption [15] can be explained. The fourth order of perturbation enables explanation of such phenomena as nonlinear fluorescence and nonlinear Hanle effect. Generally, processes which involve the first and second order of the density matrix expansion, namely $\rho^{(1)}$ and $\rho^{(2)}$, are considered as linear processes, while those related to higher orders are the nonlinear processes.

Successive terms of the perturbative expansion of the density matrix can be visualized graphically as shown in Fig. 1.

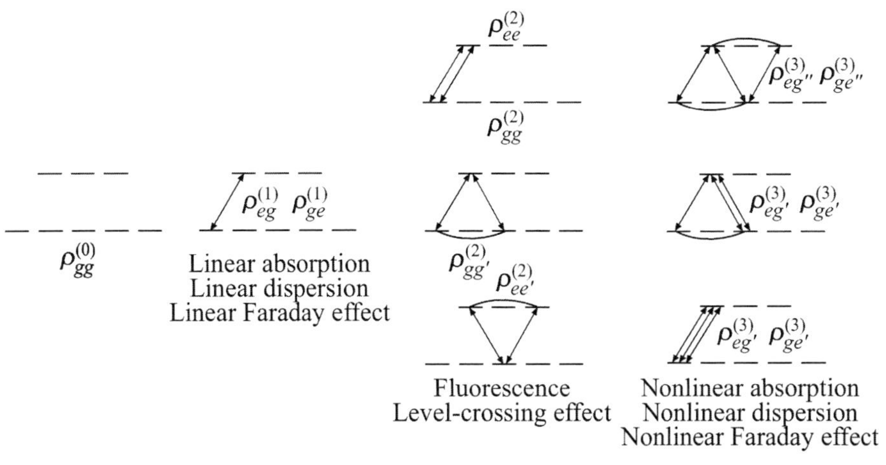

Figure 1. Visualization of the perturbative approach to the interaction of a two-level atom with an optical field. It is assumed that atomic levels have $F_g = 2$ and $F_e = 1$ and their degenerate magnetic sublevels are represented by horizontal bars. For a given order of perturbation the same subscript letters denote atomic populations or, if one of the letters is primed, the Zeeman coherences. Subscript letters e and g appearing together denote the optical coherences.

[2]Note that even if the coherences associated with the higher order of perturbation could contribute to NFE, their effect is usually much smaller than the third order contributions unless the light intensity is too high to justify the perturbative approach.

The basic system in which NFE can be analyzed is a three-level system, e.g., Λ, V. In this chapter we concentrate on the Λ system. That system can be realized by optical pumping on the $F_g = 1 \to F_e = 1$ transition with σ-polarized light. Optical pumping depletes the $m_g = 0$ sublevel and leaves whole atomic population in the $m_g = -1$ and $m_g = 1$ sublevels which, when interacting with the light, couple exclusively to the $m_e = 0$ sublevel. Since the $m_g = 0 \to m_e = 0$ transition is forbidden for $F_g = F_e$, a perfect Λ system is realized with $m_g = \pm 1$ and $m_e = 0$ sublevels.. The familiar $F_g = 1 \to F_e = 0$ system also approximates the Λ-system, if one neglects the $m_g = 0$ sublevel.

In order to theoretically describe NFE for a specific atomic system, for instance, the Λ system interacting with light and magnetic field, the general form of the total Hamiltonian H [Eq. (4)] is taken

$$H = H_0 + V = H_0 - \mathbf{E} \cdot \mathbf{d} - \boldsymbol{\mu} \cdot \mathbf{B} \tag{5}$$

where $\boldsymbol{\mu}$ is the magnetic dipole-moment operator. Unfortunately, Eq. (4) cannot be solved exactly for a real light field represented as

$$\mathbf{E} = \mathbf{e}_+ E \cos \omega t + \mathbf{e}_- E \cos \omega t, \tag{6}$$

where $\mathbf{e}_+$, $\mathbf{e}_-$ are the right- and left-handed unit vectors. This can be done, however, in the so-called rotating wave approximation wherein light is written in a complex form

$$\mathbf{E} = \mathbf{e}_+ E e^{-i\omega t} + \mathbf{e}_- E e^{-i\omega t}. \tag{7}$$

Combining Eqs. (4), (5), and (7), one obtains a set of equations for all density matrix elements in a given order of perturbation

$$\begin{aligned}
\dot{\rho}_{--}^{(n)} &= -i\Omega\left(\rho_{-0}^{(n-1)} - \rho_{0-}^{(n-1)}\right) - \gamma\left(\rho_{--}^{(n)} - \rho_{--}^{(0)}\right),\\
\dot{\rho}_{++}^{(n)} &= -i\Omega\left(\rho_{+0}^{(n-1)} - \rho_{0+}^{(n-1)}\right) - \gamma\left(\rho_{++}^{(n)} - \rho_{++}^{(0)}\right),\\
\dot{\rho}_{00}^{(n)} &= i\Omega\left(\rho_{+0}^{(n-1)} - \rho_{0+}^{(n-1)} + \rho_{-0}^{(n-1)} - \rho_{0-}^{(n-1)}\right) - \Gamma\rho_{00}^{(n)},\\
\dot{\rho}_{0-}^{(n)} &= -i\Omega\left(\rho_{00}^{(n-1)} - \rho_{--}^{(n-1)} - \rho_{+-}^{(n-1)}\right) - i\left(\omega_0 + \omega_L - i\Gamma_{coh}\right)\rho_{0-}^{(n)},\\
\dot{\rho}_{0+}^{(n)} &= -i\Omega\left(\rho_{00}^{(n-1)} - \rho_{++}^{(n-1)} - \rho_{-+}^{(n-1)}\right) - i\left(\omega_0 - \omega_L - i\Gamma_{coh}\right)\rho_{0+}^{(n)},\\
\dot{\rho}_{-+}^{(n)} &= -i\Omega\left(\rho_{-0}^{(n-1)} - \rho_{0+}^{(n-1)}\right) + i\left(2\omega_L + i\gamma_{coh}\right)\rho_{-+}^{(n)},
\end{aligned} \tag{8}$$

where the $\pm$ and 0 subscripts denote the states with $m_g = \pm 1$ and $m_e = 0$, respectively, $\Omega = Ed/\hbar$ is the Rabi frequency, $\omega_L = g\mu_B B/\hbar$ the Larmor frequency with μ_B being the Bohr magneton, g the Landé factor, and B the strength of the magnetic field, γ and Γ are the relaxation rates of the ground- and excited-state populations, respectively, Γ_{coh} the relaxation rate of the optical coherence and γ_{coh} the relaxation rate of the ground-state Zeeman coherences[3].

[3]For the sake of simplicity, it was assumed that the excited state does not relax to the ground state but deexcites to the other, uncoupled state. In such a case of an open system, the equations for $\dot{\rho}_{\pm\pm}^{(n)}$ do not depend on $\dot{\rho}_{00}^{(n)}$, i.e., there is no ground-state repopulation.

As discussed above, for nearly resonant light ($\Delta\omega \ll \omega$) optical coherences oscillate at the frequency ω. Therefore, in order to separate the fast oscillations of the coherences from much slower variation of their amplitudes, the slowly varying envelope $\sigma_{\alpha\beta}$ of the optical coherence $\rho_{\alpha\beta}$ is introduced

$$\rho_{\alpha\beta}(t) = \sigma_{\alpha\beta}(t)e^{-i\omega t}. \tag{9}$$

Introduction of $\sigma_{\alpha\beta}$ allows one to calculate the steady-state solution of Eqs. (8), i.e., the solution that is valid long after all transients associated with the light turn-on have died out. This solution is obtained by setting the left-hand side of Eqs. (8) equal to zero and the equations become

$$\begin{aligned}
\rho_{--}^{(n)} &= \rho_{--}^{(0)} - \frac{i\Omega}{\gamma}\left(\sigma_{-0}^{(n-1)} - \sigma_{0-}^{(n-1)}\right), \\
\rho_{++}^{(n)} &= \rho_{++}^{(0)} - i\frac{\Omega}{\gamma}\left(\sigma_{+0}^{(n-1)} - \sigma_{0+}^{(n-1)}\right), \\
\rho_{00}^{(n)} &= \frac{i\Omega}{\Gamma}\left(\sigma_{+0}^{(n-1)} - \sigma_{0+}^{(n-1)} + \sigma_{-0}^{(n-1)} - \sigma_{0-}^{(n-1)}\right), \\
\sigma_{0-}^{(n)} &= -\frac{\Omega}{A_-}\left(\rho_{00}^{(n-1)} - \rho_{--}^{(n-1)} - \rho_{+-}^{(n-1)}\right), \\
\sigma_{0+}^{(n)} &= -\frac{\Omega}{A_+}\left(\rho_{00}^{(n-1)} - \rho_{++}^{(n-1)} - \rho_{-+}^{(n-1)}\right), \\
\rho_{-+}^{(n)} &= \frac{\Omega}{2\omega_L + i\gamma}\left(\sigma_{-0}^{(n-1)} - \sigma_{0+}^{(n-1)}\right),
\end{aligned} \tag{10}$$

where $A_\pm = \Delta\omega \mp \omega_L - i\Gamma/2$, $\gamma_{coh} = \gamma$ and $\Gamma_{coh} = (\Gamma + \gamma)/2 \approx \Gamma/2$ for $\Gamma \gg \gamma$.

Explicit forms of the density-matrix elements calculated up to the third order of perturbation are shown in Table 1. Comparison of $\sigma_{0\pm}^{(1)}$ and $\sigma_{0\pm}^{(3)}$ reveals differences between linear and nonlinear optical processes. The first-order optical coherences $\sigma_{0\pm}^{(1)}$ are characterized with complex Lorentz functions. The widths of the associated dispersive and absorptive curves corresponding to the real and imaginary parts of $\sigma_{0\pm}^{(1)}$, respectively, are given by the relaxation rate of the optical coherences $\Gamma/2$, while the amplitudes of the contributions depend linearly on the amplitude of the light electric field ($\sigma_{0\pm}^{(1)} \propto \Omega$). Third-order coherences $\sigma_{0\pm}^{(3)}$ take more complicated forms than $\sigma_{0\pm}^{(1)}$. Not only are they characterized with a linear combination of several Lorentz functions of the width $\Gamma/2$ but also they acquire additional contribution associated with the ground-state Zeeman coherences. This contribution has much narrower width, which is determined by the relaxation rate of ground-state coherences γ ($\gamma \ll \Gamma$) and amplitude depending in a nonlinear way on the electric light field ($\sigma_{0\pm}^{(3)} \propto \Omega^3$). It is noteworthy that for very small γ the feature associated with ground-state coherences has very strong dependence on magnetic field which makes it attractive for magnetometric applications (Sec. 4.1.).

In Fig. 2, the real and imaginary parts of the first- $\sigma_{0+}^{(1)}$ and third-order $\sigma_{0+}^{(3)}$ optical coherences are shown as functions of the magnetic field. The plots reveal differences between the linear and nonlinear dispersion and absorption, in particular, the narrow contribution associated with the ground-state Zeeman coherences.

The perturbative approach yields relatively simple analytical formulae for different density matrix elements. These forms allow one to study various processes associated with

Table 1. Density matrix elements calculated for a given order of perturbation. It was assumed that there are three equally populated levels in the ground state $N/3$, where N is the number of atoms.

Perturbation order	Density matrix element
0	$\rho_{gg}^{(0)} = \frac{N}{3}$
1	$\sigma_{0\pm}^{(1)} = \frac{\Omega N}{2A_\pm}$
2	$\rho_{\pm\pm}^{(2)} = \frac{N}{3}\left[1 - \frac{i\Omega^2}{\gamma}\left(\frac{1}{A_\pm^*} - \frac{1}{A_\pm}\right)\right]$ $\rho_{00}^{(2)} = \frac{i\Omega^2 N}{2\Gamma}\left(\frac{1}{A_-^*} - \frac{1}{A_-} + \frac{1}{A_+^*} - \frac{1}{A_+}\right)$ $\rho_{-+}^{(2)} = \frac{\Omega^2 N}{2(2\omega_L + i\gamma)}\left(\frac{1}{A_-^*} - \frac{1}{A_+}\right)$
3	$\sigma_{0\pm}^{(3)} = \frac{\Omega N}{3A_\pm}\left[1 - \frac{i\Omega^2}{\gamma}\left(\frac{1}{A_\pm^*} - \frac{1}{A_\pm}\right) - \frac{i\Omega^2}{\Gamma}\left(\frac{1}{A_\mp^*} - \frac{1}{A_\mp}\right) + \right.$ $\left. + \frac{\Omega^2}{2\omega_L - i\gamma}\left(\frac{1}{A_\pm} - \frac{1}{A_\mp^*}\right)\right]$

light-atom interaction, and in such a way as to point out the differences between linear and nonlinear optical processes. One significant limitation of the approach is its applicability to low light-intensity and/or off-resonance interactions only. An additional drawback lies in the usage of the steady-state approximation which does not allow the dynamics of the process to be analyzed. These problems, however, can be addressed using the approach discussed in the following section.

2.2. Nonlinear Faraday Effect as a Three-Stage Process

Consideration of nonlinear optical phenomena as three-stage processes is a useful way of describing these effects. In such an approach, light first modifies medium properties by redistributing populations and generating coherences between atomic or molecular levels (the pumping stage) which then evolve under interaction with external fields leading to further change of medium properties (the probing stage). Finally, the medium affects the light (the probing stage) which depends also on the properties of the light. Despite the fact that usually all these stages occur simultaneously, the three-stage approach facilitates understanding of the nonlinear optical effects.

As discussed above, in the most general case, Eq. (2) describing the density matrix evolution cannot be solved analytically. Hence, in order to obtain exact solution of the equation one needs to apply numerical methods. In that case it is very useful to visualize the matrix ρ which facilitates study of its time evolution. The method described, for example, in Ref. [16] enables representation of the matrix in a form of a 3D surface[4] whose distance

[4]It should be noted that numerical and graphical representations of the density matrix are fully equivalent; for a given density matrix a specific surface can be drawn and a given surface enables extraction of the numerical form of the density matrix.

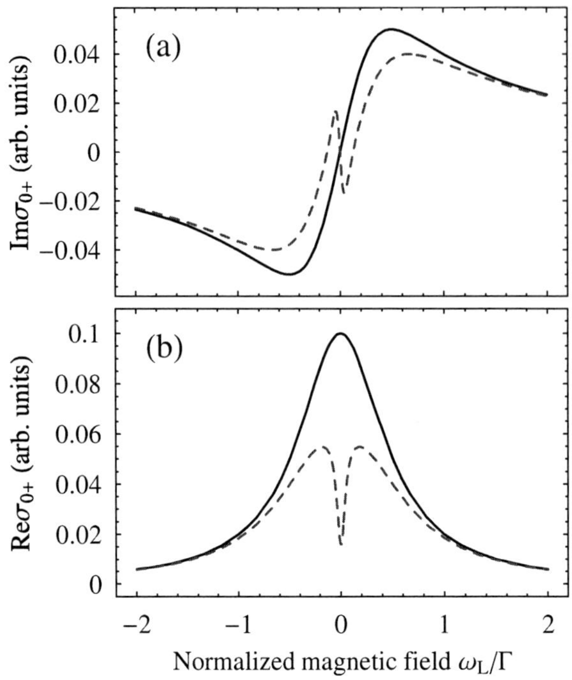

Figure 2. Comparison of real (a) and imaginary (b) parts of the optical coherences calculated in the first ($\sigma_{0+}^{(1)}$) (solid lines) and third ($\sigma_{0+}^{(3)}$) order of perturbation (broken lines).

from the origin in a given direction $r(\theta, \alpha)$ corresponds to

$$r(\theta, \alpha) = R^{-1}(\alpha, \theta, 0)\rho R(\alpha, \theta, 0)|_{FF}, \tag{11}$$

where $|_{FF}$ denotes the projection with the highest m and $R(\alpha, \theta, 0)$ is the quantum-mechanical rotation operator [17] in spherical coordinates, while θ and α denote the azimuthal and polar angles, respectively. Below we applied the method to analyze evolution of the density matrix during pumping, evolution, and probing stages in the nonlinear Faraday effect.

A sample of unpolarized Λ-type atoms, i.e., atoms equally populated in all ground-state sublevels, interacts with a linearly polarized light in the absence of a magnetic field. Under such conditions the density matrix was evaluated numerically using Eq. (2) and visualized with the method discussed above. The successive phases of the pumping stage are shown in Fig. 3. The light repopulates atomic levels and generates coherences among them which manifest as a change of the 3D surface used for the matrix visualization. The surface shape evolves from a symmetric sphere [Fig. 3(a)] to an anisotropic, peanut-like shape [Fig. 3(d)]. The axis of the induced density-matrix anisotropy, and hence the axis of the associated optical anisotropy of atoms, is aligned with the light polarization.

Application of the magnetic field causes evolution of the pumped-atom density matrix [Fig. 4(a)] which manifests as the rotation of the peanut-like shape around the field (the Lar-

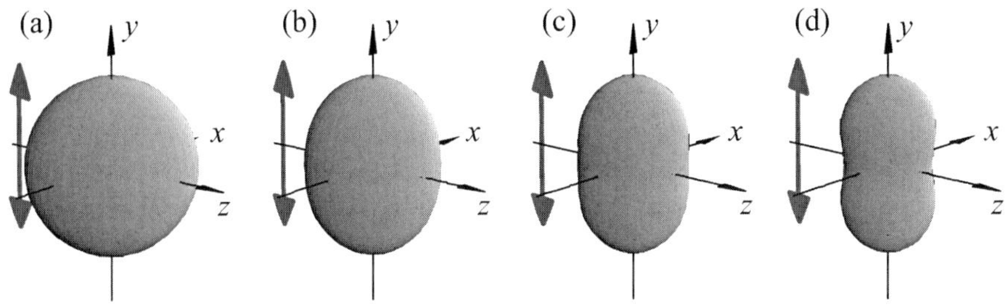

Figure 3. Pumping stage: subsequent phases of density-matrix evolution (a)-(d) under interaction with linearly polarized light (light polarization along y). The initially isotropic distribution of atomic angular momenta (a) evolves toward an anisotropic, peanut-like shape whose axis is aligned along light polarization (d).

mor precession) [Fig. 4(a)-(e)]. During the rotation a number of pumped atoms decreases due to relaxation which is seen as deterioration of the peanut-like shape amplitude (only density-matrix contribution associated with pumped atoms is shown in Fig. 4). Simultaneously with the precession and relaxation, light is continuously pumping new atoms of optical anisotropy axis aligned along its polarization. Anisotropy axes of all these atoms precess in the field, all with different phases. As a result, atoms described with density matrices corresponding to differently spatially oriented "peanuts" of different amplitudes coexist in the medium [Fig. 4(e)].

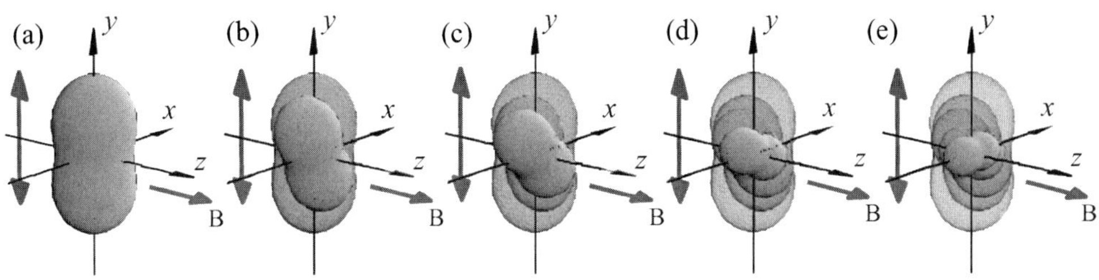

Figure 4. Evolution stage: subsequent phases (a)-(e) of the density-matrix evolution under interaction with y-polarized CW light and nonzero magnetic field $\vec{B}$. The initially pumped atoms precess in the field and relax in time which is seen as the rotation of the "peanut" and decrease of its amplitude over time (contribution from pumped atoms are only shown). Simultaneously with these processes, CW light pumps new atoms. The precessing atoms accumulate different phases than the atoms pumped earlier or later. As a result, atoms with all spatial orientations of the anisotropy axes exist in the medium (e).

Pumping, precession, and relaxation lead to creation of a net macroscopic anisotropy of the medium. In order to calculate this anisotropy, one needs to integrate over density matrices of all atoms existing in the medium. That may be reduced to simple averaging over all spatial orientations of corresponding 3D surfaces. The averaging results in appearance of a net 3D surface (Fig. 5) and hence macroscopic optical anisotropy of the medium with the axis orientation depending on the magnetic field, relaxation rate, and electric field of light.

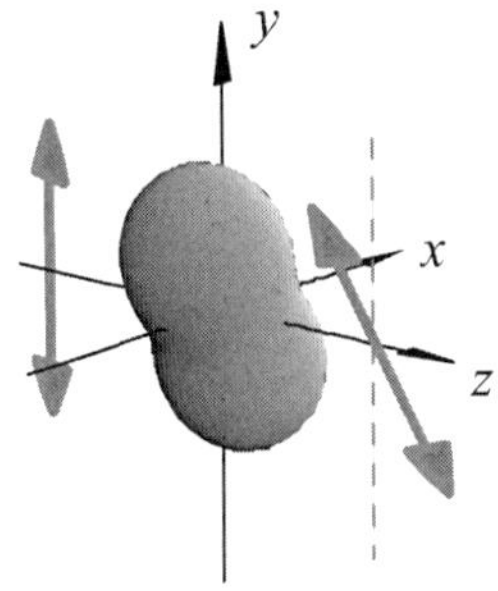

Figure 5. Net anisotropy of the medium generated under interaction with the linearly y-polarized light polarized light and the magnetic field directed along z. The orientation of the anisotropy is determined by the magnetic field, the rate of relaxation of atomic polarization and the light intensity.

Finally, the optical anisotropy of the medium is analyzed with the probe light[5]. In general, the axis of that anisotropy is not aligned along the the probe-light polarization so the medium is birefringent for the light. It results in the rotation of its polarization upon propagation through the medium. Since, as it was shown earlier, net polarization of the medium depends, among other quantities, on the pumping light intensity, the rotation of the probe-light polarization is also its function. This light-intensity dependence of the rotation is thus the main signature of nonlinear nature of this effect.

As shown, the visual approach is a very powerful and useful tool to describe of non-linear optical phenomena. It allows NFE to be presented in a simple and intuitive way. Its important advantage over perturbative approach is the ability to describe a number of interesting, especially dynamic, effects.

2.3. Calculation of the Polarization-Plane Rotation

In the two preceding subsections, the methods of evaluation of the density matrix were discussed. As shown in Eq. (1), the knowledge of the matrix allows one to calculate the induced polarization $\mathbf{P}$. Complex amplitude P of the induced polarization $\mathbf{P}$

$$\mathbf{P} = \hat{e}_E P e^{-i\omega t}, \tag{12}$$

where $\hat{e}_E$ is the unit vector along light polarization, enables formulation of the relation between the polarization amplitude P and the amplitude of the electric field of light E

[5] In many experimental arrangements pumping and probing is performed with the same light beam – the so-called single-beam arrangement.

$$P = \epsilon_0 \chi_{eff} E, \tag{13}$$

with ϵ_0 being the electric permittivity of free space and χ_{eff} the effective electric susceptibility of the medium[6]. Using Eqs. (1), (12), and (13), one can show that

$$\chi_{eff} = \frac{\sum_{eg} d_{eg}\rho_{ge}}{E}, \tag{14}$$

where d_{eg} is the dipole moment of a transition between states e and g. The susceptibility of a medium is related to the relative permittivity ϵ_r by

$$\epsilon_r = 1 + \chi_{eff}. \tag{15}$$

The complex refractive index η can be calculated through the relation

$$\eta = n + i\kappa = \sqrt{\epsilon_r} = \sqrt{1 + \chi_{eff}} \approx 1 + \frac{1}{2}\chi_{eff}, \tag{16}$$

where n is the refractive index, κ denotes the absorption coefficient, and it was assumed that $\chi_{eff} \ll 1$ which is true for low density atomic vapors. Combining Eqs. (14) and (16), one can write relations for the refractive index n and the absorption coefficient κ

$$\begin{aligned} n &= 1 + \mathrm{Re}\frac{\sum_{eg} d_{eg}\rho_{ge}}{2E}, \\ \kappa &= \mathrm{Im}\frac{\sum_{eg} d_{eg}\rho_{ge}}{2E}. \end{aligned} \tag{17}$$

For the considered Λ-type atoms, the nonzero magnetic field removes degeneration of the ground-state sublevels which differentiates the complex refractive indices, $\eta_+ \neq \eta_-$, of the two circular polarizations σ^+ and σ^-, respectively. A result of this light-induced birefringence is that each of the circular components of linearly polarized light propagate in the medium with different velocities, so after a sample of length l the accumulated phase difference $\Delta\varphi$ is

$$\Delta\varphi = \frac{\omega l}{c}\left(n_+ - n_-\right). \tag{18}$$

[6]Typically in perturbative nonlinear optics, the amplitude of induced polarization P is defined as $P = \epsilon_0 \Sigma_{n=1}^{\infty} \chi^{(n)} E^n$, where $\chi^{(n)}$ is the n-th order electric susceptibility. For that definition, all $\chi^{(n)}$ are independent of the electric field E. By comparing this definition with Eq. (13), one finds that the effective electric susceptibility $\chi_{eff} = \Sigma_{n=1}^{\infty} \chi^{(n)} E^{(n-1)}$ is light dependent.

The phase difference $\Delta\varphi$ causes rotation of the polarization plane of linearly polarized light by the angle ϕ

$$\phi = \frac{\Delta\varphi}{2} = \frac{\omega l}{2c}(n_+ - n_-) = \frac{\omega l}{4Ec}\sum_{eg}\left[d_{eg}\rho_{ge}\delta_{m_e - m_g, 1} - d_{eg}\rho_{ge}\delta_{m_e - m_g, 1}\right], \tag{19}$$

where $\delta_{x,y}$ is the Kronecker delta function. Equation (19) allows the rotation of the polarization plane of linearly polarized light to be calculated using the density matrix formalism.

2.4. Nonlinear Faraday Effect with Modulated Light

There are several techniques of light modulation that can be applied for synchronous pumping of atoms. The one described in this chapter employs amplitude modulation of the pumping light. More information on other techniques, in particular frequency modulation of light, can be found in Refs. [18–20].

In order to account for amplitude modulation of light, Eq. (7) describing the electric field of light in rotating-wave approximation needs to be rewritten into the form

$$\begin{aligned}\mathbf{E} &= \left(\mathbf{e}_+ E e^{-i\omega t} + \mathbf{e}_- E e^{-i\omega t}\right)\left[1 + a_m\cos(\Omega_m t)\right] = \\ &= \mathbf{e}_+ E\left(e^{-i\omega t} + \frac{a_m}{2}e^{-i(\omega - \Omega_m)t} + \frac{a_m}{2}e^{-i(\omega + \Omega_m)t}\right) - \\ &- \mathbf{e}_- E\left(e^{-i\omega t} + \frac{a_m}{2}e^{-i(\omega - \Omega_m)t} + \frac{a_m}{2}e^{-i(\omega + \Omega_m)t}\right),\end{aligned} \tag{20}$$

where Ω_m is the modulation frequency and a_m the numerical factor responsible for the modulation depth ($a_m \in (0, 1]$). As seen, amplitude modulation of light leads to a change of light spectrum which, in addition to the carrier frequency, consists of two sidebands at $\omega - \Omega_m$ and $\omega + \Omega_m$ of the amplitudes $a_m/2$. Substituting Eq. (20) into Eq. (2) enables the calculation of the density matrix of atoms interacting with amplitude-modulated light. In such a case, however, the rotating-wave approximation does not allow one to fully extract time dependence from the density-matrix equation of motion. Due to this reason, matrix ρ is expanded into a Fourier series over the modulation frequency Ω_m

$$\rho = \sum_{k=-\infty}^{\infty} \rho^{[k]} e^{ik\Omega_m t}, \tag{21}$$

where $\rho^{[k]}$ is the k-th Fourier coefficient. Using the Fourier expansion of the density matrix and collecting terms oscillating at the same frequencies ($r\Omega_m$, where r is the integer) one obtains the infinite set of self-coupled equations

$$
\begin{aligned}
\rho_{--}^{[k]} &= \rho_{--}^{(0)}\delta_{0,k} - \frac{i\Omega}{\gamma}\left[\sigma_{-0}^{[k]} - \sigma_{0-}^{[k]} + \frac{a_m}{2}\left(\sigma_{-0}^{[k-1]} - \sigma_{0-}^{[k-1]} + \sigma_{-0}^{[k+1]} - \sigma_{0-}^{[k+1]}\right)\right],\\
\rho_{++}^{[k]} &= \rho_{++}^{(0)}\delta_{0,k} - \frac{i\Omega}{\gamma}\left[\sigma_{+0}^{[k]} - \sigma_{0+}^{[k]} + \frac{a_m}{2}\left(\sigma_{+0}^{[k-1]} - \sigma_{0+}^{[k-1]} + \sigma_{+0}^{[k+1]} - \sigma_{0+}^{[k+1]}\right)\right],\\
\rho_{00}^{[k]} &= \frac{i\Omega}{\Gamma}\left[\sigma_{+0}^{[k]} - \sigma_{0+}^{[k]} + \sigma_{-0}^{[k]} - \sigma_{0-}^{[k]} + \frac{a_m}{2}\left(\sigma_{+0}^{[k-1]} - \sigma_{0+}^{[k-1]} + \sigma_{-0}^{[k-1]} - \sigma\rho_{0-}^{[k-1]} +\right.\right.\\
&\quad \left.\left. +\sigma_{+0}^{[k+1]} - \sigma_{0+}^{[k+1]} + \sigma_{-0}^{[k+1]} - \sigma_{0-}^{[k+1]}\right)\right],\\
\sigma_{0-}^{[k]} &= -\frac{\Omega}{A_-}\left[\rho_{00}^{[k]} - \rho_{--}^{[k]} - \rho_{+-}^{[k]} + \frac{a_m}{2}\left(\rho_{00}^{[k-1]} - \rho_{--}^{[k-1]} - \rho_{+-}^{[k-1]} +\right.\right.\\
&\quad \left.\left. +\rho_{00}^{[k+1]} - \rho_{--}^{[k+1]} - \rho_{+-}^{[k+1]}\right)\right],\\
\sigma_{0+}^{[k]} &= -\frac{\Omega}{A_+}\left[\rho_{00}^{[k]} - \rho_{++}^{[k]} - \rho_{-+}^{[k]} + \frac{a_m}{2}\left(\rho_{00}^{[k-1]} - \rho_{++}^{[k-1]} - \rho_{-+}^{[k-1]} +\right.\right.\\
&\quad \left.\left. +\rho_{00}^{[k+1]} - \rho_{++}^{[k+1]} - \rho_{-+}^{[k+1]}\right)\right],\\
\rho_{-+}^{[k]} &= \frac{\Omega}{2\omega_L + i\gamma}\left[\sigma_{-0}^{[k]} - \sigma_{0+}^{[k]} + \frac{a_m}{2}\left(\sigma_{-0}^{[k-1]} - \sigma_{0+}^{[k-1]} + \sigma_{-0}^{[k+1]} - \sigma_{0+}^{[k+1]}\right)\right].
\end{aligned}
\tag{22}
$$

New terms arising in relations for $\rho^{[k]}$ appear due to coupling between the $[k-1]$ and $[k+1]$ density-matrix elements with the sidebands $\omega - \Omega_m$ and $\omega + \Omega_m$ and the strength of this coupling is determined by parameter a_m.

In general, evolution of the density-matrix elements given by Eqs. (22), requires solution of an infinite number of the self-coupled equations. In practice, however, the series given by Eq. (21) is replaced by a finite Fourier series with the cut-off at k_c, i.e., for $|k| > k_c$ $\rho^{[k]} \equiv 0$. Such an approximation is valid because k–1-th and k+1-th order elements of the density matrix contribute to $\rho^{[k]}$ by, at most, half of the contribution from different k-th order elements of the matrix; the higher-order coefficients reflect response of a medium to higher harmonics of the modulation frequency which are weaker than that to the lower ones. Due to this fact, it is usually enough to truncate the Fourier expansion of the density matrix at $k = 7$.

The idea of synchronous pumping of atoms, which is fundamental for NFE with modulated light, can be easily understood using the density-matrix visualization method. In Fig. 6 subsequent phases of pumping with amplitude-modulated light are shown. At time $\tau = 0$ [Fig. 6(a)], the atoms are pumped by the light and the peanut-like density matrix is generated. Similarly to the CW case, the "peanut" rotates in the magnetic field (a)-(e) but in this case the rotation occurs with strongly attenuated or completely turned off pumping light. The light is on again when the peanut-like density matrix is rotated by 180° [Fig. 6(a)], that is, at time $\tau = \pi/\omega_L$. In such a case, the anisotropy axis of newly pumped atoms is aligned with the axis of atoms pumped earlier. If the precession frequency ω_L is much faster than the relaxation rate γ, the decrease of the "peanut" amplitude in a single pumping-precession cycle is negligible and eventually most of the atoms precess in the field with the same phase. This leads to generation of macroscopic dynamic anisotropy of the

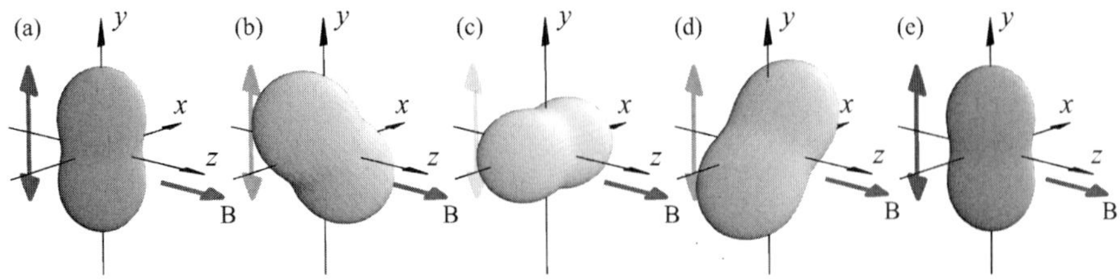

Figure 6. Visualization of interaction of amplitude-modulated light (polarization along y) with atoms subjected to the magnetic field $\vec{B}$ directed along z. If the pumping frequency ω_m is synchronized with the Larmor frequency ω_L (see text), the anisotropy axes of the atoms pumped during the different pumping cycles are aligned and the macroscopic dynamic polarization of the medium is generated. Time dependence of the amplitude-modulated light intensity is illustrated by different shading of the double-head arrows which mark oscillations of the light electric vector.

medium whose anisotropy axis rotates in time. This results in the time-dependent rotation of the polarization plane of unmodulated probe light propagating through the medium at twice the Larmor frequency[7]. Amplitude of this polarization rotation depends on the efficiency of generation of the mediums anisotropy which is maximal when the modulation frequency Ω_m coincides with twice the Larmor frequency Ω_L, $\Omega_m = 2\Omega_L$.

A detailed theoretical treatment of nonlinear optical processes, especially NFE, presented in this section provides a good background for further discussions performed in this chapter.

3. Experimental Arrangements

The basic experimental arrangement for investigation of NFE is essentially the same as for all magneto-optical studies. It involves a light source, an appropriate sample, e.g. a vapor cell, and a detection system. In the studies of NFE the light comes from a tunable laser. For development and laboratory tests of the NFE magnetometers, their sensing samples (vapor cells) are subjected to well controlled magnetic fields produced by a coil system and protected from uncontrolled stray fields. The schematic of a typical setup in which modulated light is used is shown in Fig. 7.

In the following sections, we outline the most essential features of the principal elements of the NFE experimental setup.

3.1. Light Sources

Unlike in LFE, the NFE requires light that is sufficiently coherent and intense to create atomic Zeeman coherences. This means that laser sources are necessary for such experiments. The lasers need to be tunable and their frequency should be stabilized to the desired

[7]The "peanut" oriented at 0° is indistinguishable from the one oriented at 180° therefore the light-polarization is modulated at $2\omega_L$, rather than ω_L.

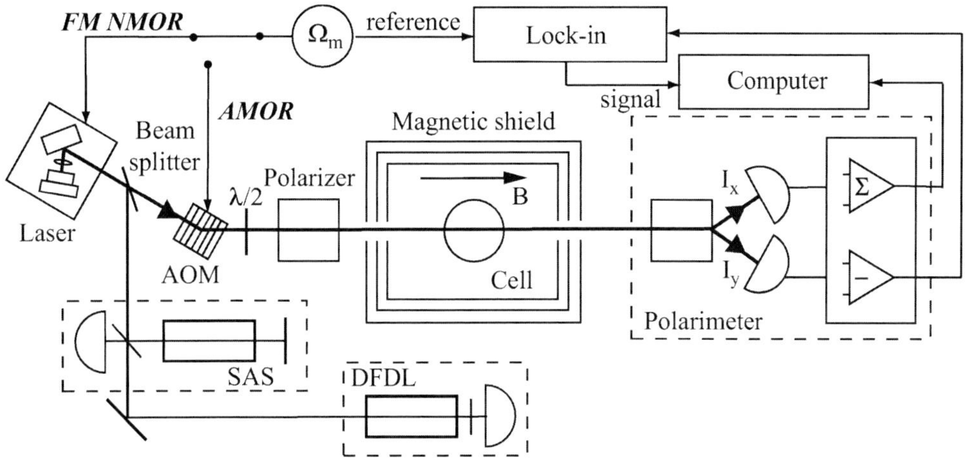

Figure 7. Typical setup for NFE experiments with modulated light. A laser light can be either frequency- or amplitude-modulated with a switch at the FM NMOR or AMOR position, respectively. The light of adjustable intensity traverses atomic vapors placed in the longitudinal magnetic field B. A multi-layer magnetic shield protects the atoms from external, uncontrollable magnetic fields. The polarization rotation is measured with a polarimeter and demodulated at a given harmonic of modulation frequency Ω_m. The signal is stored on a computer. SAS denotes saturated absorption frequency reference and DFDL the Doppler-free dichroic lock used for stabilizing laser frequency, $\lambda/2$ is a half-wave plate, and AOM the acousto-optical modulator. I_x and I_y are the intensities of the two orthogonally polarized outputs of the analyzing beam splitter. For balancing of the polarimeter the axis of polarizer prior the cell needs to be oriented at 45° relative to the x and y axes (see Sec. 3.4.).

atomic transitions. For example, in the setup shown in Fig. 7 the laser frequency is stabilized by a Doppler-free dichroic lock and additional frequency reference is provided by a saturated-absorption spectroscopy in an additional vapor cell. In the experiments where light modulation is used (FM NMOR or AMOR), the light frequency or intensity is varied[8]. Besides frequency control, laser sources for NFE may have rather modest parameters from the point of view of power and linewidth. Usually some mW output light power is sufficient, so inexpensive diode lasers are quite adequate. In more demanding applications external-cavity diode lasers may be applied with their frequency stabilized to atomic transition. For applications where power consumption and/or small size become is an issue, the vertical cavity surface emitting lasers (VCSELs) can be used.

3.2. Samples

In most of the existing applications, the NFE sensors are atomic gases contained in glass cells. Recently, novel materials were investigated for their possible use in quantum and nonlinear optics, like atoms inside hollow fibers [21,22] and NV color centers in diamond crystals [23]. As outlined in Sec. 4.1., one crucial parameter for magnetometric applications

[8]In case of FM NMOR, the frequency of light is scanned around fixed light frequency.

that determines the device sensitivity is the linewidth of the resonance which is associated with the relaxation time of the coherences. Long relaxation times are also crucial for applications in quantum state engineering. Thus, it is essential that the sample possesses appropriate long-living states which can be affected by magnetic fields and connected by an easily accessible optical transition.

For ground-state coherence of atoms contained in a gas cell, the relaxation times is limited by several factors. One is caused by thermal movement of atoms and finite dimensions of the light beam probing the coherence. This is known as the transit relaxation. Usually, for typical cell temperatures and beam sizes, this effect limits the relaxation times to some μs which results in resonance widths of tens of kHz. Such linewidths do not allow very high magnetometric sensitivities, so various techniques are used to overcome the transit relaxation. One known technique is application of buffer, usually noble, gases. The atoms in a gas cell undergo elastic collisions with the buffer gas atoms which limit atomic free path and lengthen their interaction time with laser beams. At the same time, the elastic collisions do not substantially destroy the atomic coherences and, consequently, very narrow NFE resonances are obtained resulting in high magnetometric sensitivities. Another way of slowing down the relaxation is coating the internal cell walls with special anti-relaxation, most often made of paraffin, layer which protect atomic coherences from destruction during atom-wall collision. Thanks to such coating, an atom can survive about 1000 wall collisions before it decoheres/depolarizes [24,25] and the coherence lifetime can be extended up to hundred of ms.

3.3. Magnetic Fields and Magnetic Shielding

Studying Faraday Effect requires good control over all magnetic fields acting upon the sample. The problem becomes particularly important with NFE, due to its high sensitivity to even very weak magnetic fields. Since the rotation of the polarization plane is determined by the magnitude of the field, its inhomogeneity within the cell leads to broadening and/or deformation of the NFE signals [26]. Moreover, the field orientation is also very important. Improper orientation reduces the rotation which causes systematic errors (heading errors) when measuring magnetic fields. Additionally, transverse components of the magnetic field mix the Zeeman sublevels which deteriorates Zeeman coherences and is very undesirable for applications in quantum-state engineering.

Familiar sources of homogenous magnetic fields are Helmholtz coils and solenoids. For studies of NFE in very low fields it is essential to reduce the effect of external fields, both the AC and the DC ones. While DC stray fields can be compensated by proper coils, the AC fields are most efficiently eliminated by shields made of appropriate ferromagnetic material of high permeability, e.g. mu-metal and ferrites. Well designed shields not only reduce the stray fields but, additionally, improve the homogeneity of the effective field acting on a sample. For example, by using the magnetostatic image method, it can be shown that the field generated by a current loop situated at $x = 0$ oriented parallel to the surface of the high permeability material placed at $x = a$, is the same as one produced by a pair of parallel loops separated by the distance $2a$. Thanks to that, the effective field produced by a coil in a magnetic shield is the same as from an infinite chain of coils. This fact may substantially improve homogeneity of the effective field acting upon the sample.

The choice of the shield material is of importance, too. The shields need to have possibly high magnetic permeability but they also must sustain the fields produced by the internal coils and not saturate. Besides that, the material should be free of fluctuations caused by random orientations of magnetic domains and thermal currents which could generate extra magnetic noise. For most demanding applications multilayered shields are designed with the most inner layer being made of ferrite and the others of other material such as mu-metal [27].

3.4. Detection of the Polarization Rotation

Measuring optical rotation requires sensitive polarimetric detection. While very different methods have been developed for various magneto-optical experiments, the detection technique most often employed for studies of NFE is based on balanced polarimetry.

The basic design of a balanced polarimeter consists of a polarizing beam splitter (e.g. Glan, Glan-Thompson, or Wollastone prism) which separates the analyzed beam transmitted by the sample into two orthogonally polarized beams, say along the x and y directions, and two identical photodetectors which measure the beam intensities (see Fig. 7). The whole system is oriented in such a way that for zero rotation the transmitted beam has its $\mathbf{E}$ vector at $45°$ with respect to the x and y axes, so the detectors record equal light intensities $I_x = I_0 \sin^2(\pi/4) = I_y = I_0 \cos^2(\pi/4)$. If, however, the light polarization rotates by the angle ϕ, the detectors record different light intensities $I_x = I_0 \sin^2(\pi/4 + \phi)$ and $I_y = I_0 \cos^2(\pi/4+\phi)$ which for small ϕ become $I_x \approx I_0(1+\phi)^2/2$ and $I_y \approx I_0(1-\phi)^2/2$. The difference of these signals, $I_x - I_y$, is then proportional to the rotation angle. When divided by doubled sum, $2(I_x + I_y)$, the difference signal becomes independent of the incident light intensity I_0 and yields directly the rotation angle. Moreover, fluctuations of I_0 in each detection channel cancel out and do not contribute directly to the noise of a rotation measurement. Such an arrangement is very useful for measuring small rotations with not-too-strong light intensities. For higher intensities, problems may occur since each detector is illuminated by a strong average intensity and resolving small intensity variations may be limited by the detector dynamic range and saturation.

The limitations caused by dynamic range of the detectors are less important for another detection technique which employs one detector only and where the sample is placed between two crossed polarizers. This method is known as the forward scattering [28]. The main advantage of the forward-scattering detection is its simplicity and elimination of a background which could deteriorate the signal-to-noise ratio and sensitivity: only the photons which contribute to the signal are detected for an ideal system. In a system with ideal polarizers, low optical thickness and no dichroism of the sample, optical rotation by angle ϕ yields the detector signal $I_0 \sin^2 \phi$ which for small angles is proportional to ϕ^2. In this way, a square of the rotation angle can be measured with high sensitivity but the signal does not allow one to determine the sign of the rotation angle. Such a homodyne detection is widely used in a Doppler-free polarization spectroscopy [29].

An interesting technique of measuring small rotations with very high sensitivity is based on homodyning the forward scattering signal with a coherent background. Such background is most easily obtained from non-scattered primary light beam by small uncrossing of the polarizers from their perfectly crossed position. If the uncrossing angle ε is small, but still much bigger than the measured signal ϕ, $\phi \ll \varepsilon$, the detected intensity of a forward scattered light becomes $I = I_0 \sin^2(\varepsilon + \phi) \approx I_0(\varepsilon + \phi)^2 \approx I_0\varepsilon^2 + 2I_0\varepsilon\phi$. This signal consists of a constant background $I_0\varepsilon^2$ and a contribution proportional to ϕ. In this way, the sign of ϕ can be easily determined and the amplitude of the magneto-optical rotation is substantially increased by its homodyning with a coherent background which is responsible for multiplicative factor 2ε.

The heterodyne method can be modified to eliminate the constant background by modulating the coherent background and application of the phase-sensitive detection. In addition to the measured rotation ϕ this introduces a time dependent background $\varepsilon_0 \cos \Omega t$, where ε_0 is the amplitude and Ω the frequency of the introduced background. After demodulation of the resulting time-dependent light intensity, $I(t) = I_0 \sin^2(\varepsilon_0 \cos \Omega t + \phi) \approx I_0(\varepsilon_0 \cos \Omega t + \phi)^2$, only the $2\varepsilon_0\phi$ term is recorded with a high signal-to-noise ratio.

This simple discussion assumes optically thin samples with no dichroism and ideal quality polarizers. For more realistic cases with optically thick samples and imperfect polarizers corrections are necessary to the above expressions. The effect of dichroism, which introduces some elipticity of the transmitted light, may be taken into account thanks to different spectral responses of the dichroic and birefringent contributions. In particular, for a well resolved optical transition, the dichroic spectrum is characterized by asymmetric lineshape whereas the birefringence spectrum is symmetric with respect to optical transition frequency.

3.5. Electronic System

A standard signature of the Faraday effect, LFE or NFE, is the dependence of the angle of polarization rotation ϕ on the magnetic field B. It is represented by an asymmetric curve, centered at $B = 0$, such as the one shown in Fig. 8(a). Application of modulated light leads to the appearance of additional resonances at non-zero magnetic fields [Fig. 8(b)]. The positions of these extra (high-field) resonances are strictly determined by the light-modulation frequency through the relation

$$B = \frac{n_i \omega_m \hbar}{2 g \mu_B}, \tag{23}$$

where n_i is an integer. Application of modulated light for studying NFE was pioneered by Budker and co-workers who developed the technique of Nonlinear Magneto-Optical Rotation with Frequency-Modulated light (FM NMOR) [7]. In FM NMOR, frequency modulation of light resulted in modulation of the pumping rate of atoms. When modulation of the pumping light was synchronized with the precession of atomic polarization in a magnetic field (see Sec. 2.4.), the macroscopic, dynamic polarization of the medium is generated.

Frequency modulation of laser light can be realized particularly easily with diode lasers through modulation of the laser current. Current modulation, however, is not perfect as it is always associated with some amplitude modulation of the light emitted from the laser diode. Moreover, even pure frequency modulation may introduce light-shifts which have undesirable consequences, for instance, from a point of view of applications of NFE for magnetic-field measurements (systematic errors). On the other hand, the modulation of

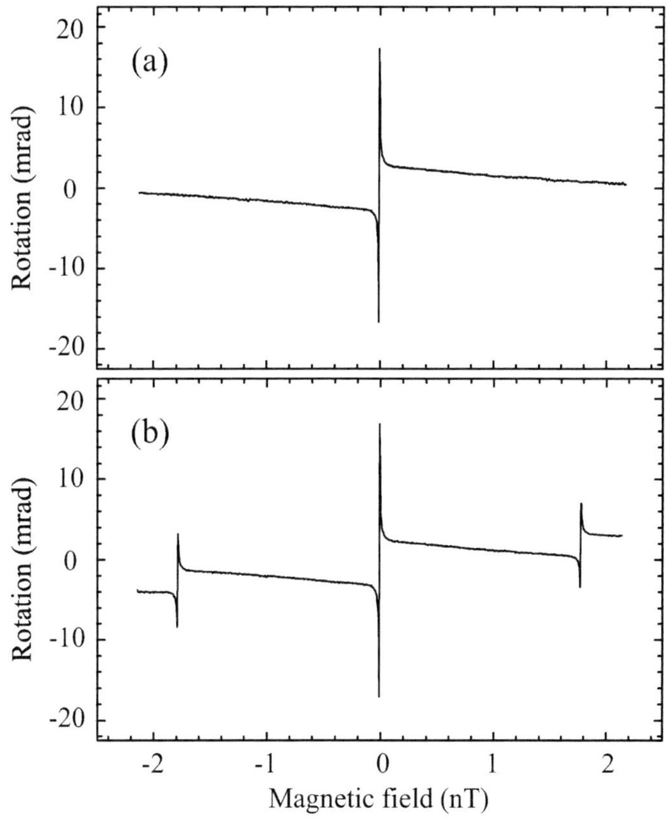

Figure 8. (a) Rotation of polarization plane vs. magnetic field observed with unmodulated light. (b) In-phase magneto-optical rotation vs. magnetic field recorded with amplitude modulated light ($\omega_m = 2\pi \times 25000$ 1/s).

the pumping rate by pure amplitude modulation of the laser intensity reduces the risk of light-shifts because the laser frequency may be made exactly resonant with the atomic transition. The NFE technique which relies on the modulation of laser intensity, known as Amplitude-Modulated Magneto-Optical Rotation (AMOR) has been introduced in Ref. [8]. An intrinsic advantage of AMOR technique is the ability to choose arbitrary modulation waveforms. In particular, in Ref. [8] it was shown that a square-pulse modulation of light intensity creates a comb of extra resonances occurring at multiplicities of $\omega_m\hbar/2g\mu_B$. Pure amplitude modulation is most easily applied with the help of acousto-optical modulator (AOM).

FM NMOR and AMOR have much in common with the method of synchronous pumping already used in early optical-pumping experiments [30, 31].

NFE signals observed with the modulation techniques (FM NMOR and/or AMOR) are time-dependent and must be demodulated at a particular frequency. This is usually performed with the phase-sensitive (lock-in) detection which allows one to register both signals related to the modulation, the in-phase and quadrature signals, shifted respectively by 0° and 90° relative to the modulation signal.

An interesting alternative of a single-beam NFE arrangement is the two-beam setup, where one beam acts as a pump and the second, unmodulated beam as a probe [32]. This gives one the ability of optimization of NFE signals by varying pump- and probe-light intensities, detunings, and polarizations.

4. Applications of Nonlinear Faraday Effect

In addition to the fact that NFE is a very interesting optical phenomenon on its own, it finds a number of interesting scientific and commercial applications. One group of applications is associated with ultra-high sensitive magnetometry. Magnetic-field measurements with such high sensitivity may be employed, for instance, in prospecting for natural resources or detecting distant magnetic objects, but also in testing of fundamental symmetries of nature. The other group of applications is the quantum-state engineering. In that case, NFE is exploited for controllable generation and modification of quantum states of the atomic system. Below the most important NFE applications are reviewed.

4.1. Magnetometry

Applications of nonlinear magneto-optical phenomena for magnetic field measurements [33] have a long history. Since first experiments in the 1960s [34, 35], different configurations of optical magnetometers have been used. Recent interest in the field is driven by technological developments in opto-electronics, especially, the invention of small, tuneable, and inexpensive diode lasers. It enabled construction of various kinds of optical magnetometers with the sensitivity rivaling [6, 36, 37], or even exceeding [38] the sensitivity of magnetometers based on properties of superconductors (SQUIDs), long recognized as the most sensitive magnetometers. Currently, the demonstrated sensitivity of the most sensitive magnetometer, the, so-called SERF (spin-exchange relaxation-free magnetometer) [38], reached a level of a fraction of 10^{-15} T/$\sqrt{\text{Hz}}$ with its projected fundamental limit below 10^{-17} T/$\sqrt{\text{Hz}}$.

One fundamental challenge of contemporary magnetometry is miniaturization. Luckily, there are no principal constraints for making such devices small; accomplishments of opto-electronics and microfabrication of mechanical systems allow impressive reduction of the dimensions of optical magnetometers. Recently, optical magnetometers with volumes of a few mm^3 were demonstrated [39, 40]. These devices exploit microfabrication techniques for production of atomic vapor cells and optical elements for shaping and generation of light. Moreover, their low power consumption enables charging of the devices with batteries. The best micromagnetometers reach a magnetic-field measurement sensitivity below 7×10^{-14} T/$\sqrt{\text{Hz}}$ [41]. Exciting miniaturization possibilities are also opened by novel photonic materials, the diamonds with NV-color centers [42–44].

4.1.1. Characteristics of Optical Magnetometers

One of the most important characteristics of any magnetometer is its sensitivity. For all optical magnetometers, and hence for NFE magnetometers, quantum mechanics sets a fundamental limit on the magnetic-field-measurement sensitivity. This limit is associated with fluctuations in the numbers of atoms[9] and photons that contribute to the measured optical signal. Since the number of contributing atoms N is described with the Poisson statistics, fluctuations in atomic number is given by $\sqrt{N}$. Thus the fundamental sensitivity limit due to the atomic shot-noise is given by

$$\delta B_{at} = \frac{\hbar}{g\mu_B}\sqrt{\frac{1}{NT\tau}}, \tag{24}$$

where $T = 1/\gamma$ is the coherence lifetime and τ is the time of the performed measurement[10].

Another contribution to the sensitivity limit is associated with fluctuations in the number of detected photons N_{ph}. If the measured quantity is the rotation of polarization, as it is for NFE magnetometers, the shot-noise limit on that measurement is given by

$$\delta\phi_{ph} = \frac{1}{2\sqrt{N_{ph}\tau}}, \tag{25}$$

which corresponds to the limit on the magnetic-field-measurement sensitivity

$$\delta B_{ph} = \frac{\hbar}{g\mu_B}\frac{\gamma}{\phi_N}\frac{1}{2\sqrt{N_{ph}\tau}}, \tag{26}$$

where ϕ_N is the NFE signal amplitude.

Since fluctuations in the number of atoms and photons are uncorrelated, these noises add in quadrature determining the fundamental quantum limit on the sensitivity

$$\delta B_{ql} = \sqrt{\delta B_{at}^2 + \delta B_{ph}^2}. \tag{27}$$

Depending on particular experimental realization, either atomic or photon shot-noise may dominate. However, in an optimized magnetometer both noises are comparable [45] which gives $\delta B_{ql} = \sqrt{2}\delta B_{at}$.

Very often magnetometers suffer from different kinds of magnetic noise that limit ultra-precise measurements of the field. A remedy for such strong noise is the, so-called, gradiometer mode. In this mode, the magnetic field is measured by two spatially separated magnetometers whose independent readouts are subtracted. In that way, the inhomogeneous magnetic field generated by a local source can be precisely measured whereas the magnetic field background common for both magnetometers may be eliminated. The gradiometer arrangements find many applications (see, for example, Secs. 4.3. and 4.4.).

An important characteristic of a magnetometer is its dynamic range. In the NFE magnetometers with CW excitation, the dynamic range of the magnetic-field measurements ΔB is limited by the width of the recorded resonance, $\Delta B \approx \hbar\gamma/g\mu_B$. While for $B < |\Delta B|$ there is an unambiguous relation between the rotation angle and the magnetic field, for

[9]In fact, the limit on the sensitivity is set by the uncertainty in determination of projections of total angular momenta of atoms. Since projections of the total angular momentum are noncommuting operators, $[F_x, Fy] = iF_z$, where F_i is the total angular momentum projection on i direction, thus the spin projection-noise arises due to $\delta F_x \delta F_y \geq |F_z|/2$.

[10]Please note that some parameters on the order of unity, such as the total value of angular momentum, were omitted in Eq. (24).

the stronger fields, $|B| > \Delta B$, the same magneto-optical rotation corresponds to two different values of the field, and for even stronger fields $|B| \gg \Delta B$ there is no rotation [see Fig. 8(a)]. The relaxation rate γ determines thus the sensitivity, as well as the dynamic range of the magnetometer; the narrower the resonance, the higher the sensitivity but smaller the dynamic range of the magnetometer. In order to achieve the magnetometer of the desired sensitivity and dynamic range, the width of the resonance needs to be adjusted. It can be done, for example, by varying the resonance width by adjusting light intensity and power broadening of the resonance.

In order to alleviate the limitations on the small dynamic range of the magnetic-field measurements, modulation techniques were employed (see Sec. 2.). It enables expanding of the NFE magnetometers dynamic range to the Earth magnetic field, which is very important in numerous applications because it eliminates necessity of shielding or compensating external, uncontrollable magnetic fields.

Another relevant parameter of optical magnetometers is its bandwidth which reflects magnetometer response time to a change of a measured magnetic field. Although the Larmor precession has no inertia and atoms instantaneously react to the field change, it takes a finite time to measure the difference in the recorded signal. It can be shown that the response of a CW (low-field) magnetometer can be modeled as that of a low-pass filter and hence the natural magnetometer bandwidth is equal to $1/2\pi T$ [46]. The bandwidth of a magnetometer can be increased by amplifying its output signal but this also affects the output noise of the magnetometer [47]. Thus, the increase of the magnetometer bandwidth is usually at the price of reduction of the signal-to-noise ratio.

The same relation for the bandwidths that holds for the CW magnetometers also works for the NFE magnetometers with modulated light; the bandwidth of these magnetometers is also determined by the NFE resonance width. It is noteworthy, however, that for the magnetic-field measurements with phase-sensitive detection, more stringent limitation on the bandwidth may appear. In such a case, the bandwidth of the magnetometer may be determined by the time constant τ_c of the lock-in in which case the bandwidth is given by $1/2\pi\tau_c$,

Other important characteristic of a magnetometer is whether it enables measurements of static and quasi-static fields or it allows fast-oscillating magnetic fields to be detected. Although most optical magnetometers are used for measuring DC fields, there is a class of devices that can be employed for measuring AC magnetic fields [48, 49]. The significant advantage of the AC optical magnetometers is their tuneability for measurements of the fields oscillating at specific frequencies. Application of a bias field allows one to shift the magnetometer resonance frequency and in such a way detect selectively the field at the desired frequency. The measurements of the oscillating fields up to 100 kHz were already demonstrated and there are no physical limitations preventing significant expansion of the magnetometer frequency range.

An important feature of a magnetometer is whether it enables a measurement of field magnitude only (scalar magnetometer) or if it also provides information about the field direction (vector magnetometer). Most optical magnetometers, including the NFE-based devices, are scalar. An advantage of such magnetometers is their insensitivity to the field orientation, in particular, heading errors, which may be important for devices operating at moving platforms. Nevertheless, there are standard techniques that enable transformation of scalar magnetometers into the vector ones by application of small oscillating fields in three orthogonal directions [50, 51]. Since small fields along with, and transverse to, the measured magnetic field contribute differently to the magnetometer readout, modulation of the total field in all three orthogonal directions with various frequencies and successive demodulation of the detected signal at these frequencies enables extraction of the information about the direction of the measured field. An alternative method, exploited in the optical magnetometers, is observation of amplitudes of two resonances detected at given multiplicities of the Larmor frequency which vary with a direction of light propagation with respect to the magnetic field [52, 53]. The ratio of these amplitudes can be evaluated from simple analytical formula. It is insensitive to the measured field magnitude but does depend on the field orientation. This provides a convenient way of measuring the field direction. The method enabled reaching angular sensitivity of 4×10^{-3} degree/$\sqrt{\text{Hz}}$ [52].

4.1.2. Low-magnetic Field Measurements

Applicability of NFE with CW light for measuring magnetic fields was first suggested in Ref. [5]. Yet, it was only in Ref. [45] when the first detailed studies of this magnetometric technique was experimentally analyzed. The authors of Ref. [45] observed NFE with rubidium vapor contained in an anti-relaxation-coated buffer-gas-free cell. Analysis of the polarization rotation versus different physical parameters yielded optimal conditions for the observation of NFE. The recorded signals were 10 mrad in amplitudes and had widths of 100 pT. By taking advantage of such narrow resonances, magnetic-field measurements with a sensitivity of 4×10^{-16} T/$\sqrt{\text{Hz}}$ were demonstrated. This sensitivity is comparable with the sensitivity of SQUIDs. Therefore, in many applications NFE magnetometers can replace superconductive devices, eliminating the problem with a necessity of cryogenic cooling. The intrinsic drawback of such NFE magnetometers is their limitation for detection of low fields only. In particular, the dynamic range of the device described in Ref. [45] was ± 100 pT. Measurements of such fields demand passive and/or active shielding of external, uncontrollable magnetic fields which usually are orders of magnitude stronger than the detected field.

4.1.3. High-magnetic Field Measurements

As already noted above, the limitation of NFE magnetometers for detection of weak fields was only successfully relaxed by application of a frequency- [7] and amplitude-modulated [8] light. As shown in Sec. 2.4., the modulation of light leads to the appearance of strong magneto-optical resonances, whose positions vs. magnetic field are strictly determined by the modulation frequency. The high-field resonances have amplitudes and widths not unlike zero-field resonances (see Fig. 8), thus the sensitivity of stronger-field measurements is comparable with the ones achieved for the zero-field measurements (see Sec. 4.1.2.). Reaching higher fields eliminates necessity of shielding or compensating external, uncontrollable magnetic fields.

High-field NFE magnetometers operate in either passive or self-oscillating mode. In the passive mode, an external oscillator modulates the light while the rotation of the light polarization is measured at a given harmonic of the modulation frequency. Figure 8(b)

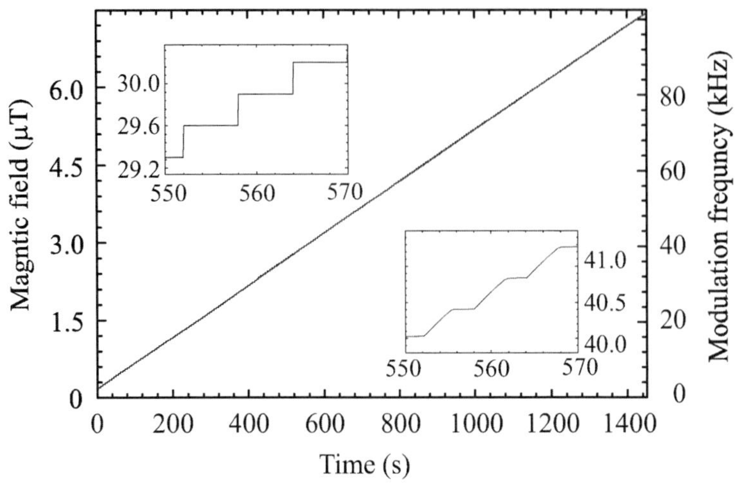

Figure 9. Magnetic-field tracking with the NFE magnetometer operating in the passive mode. Every 9 s, the total magnetic field was increased by 15 nT and the modulation frequency was subsequently adjusted with the feedback system such that the condition for the effective of the higher-field resonance was fulfilled.

demonstrates that a given modulation frequency and magnetic field correspond to a particular high-field NFE resonance. Thus, each change of the field magnitude causes a modification of the measured NFE signal. In the passive-mode magnetometers, one may track the magnetic field by keeping a fixed value of the signal, $\phi = \text{const}$. It can be done, for example, by continuous fulfilling the condition for effective generation of the high-field resonance, for instance, $\omega_m = 2\Omega_L = 2g\mu_B/\hbar B$, for which the amplitude of the polarization rotation is maximal; the field change is followed by such an adjustment of the modulation frequency Ω_m in a feedback loop that NFE signal is maximal.

A significant advantage of the passive-mode magnetometers is their insensitivity to various types of noise. Application of the phase-sensitive detection strongly suppresses the noise at frequencies other than the reference frequency. For instance, one can reduce the 50-Hz or 60-Hz noise associated with AC lines. The drawback of the mode, however, is the magnetometer low bandwidth related to a finite time constant of a lock-in amplifiers. Therefore, the best application of passive-mode magnetometers is for detection of static or slowly varying magnetic fields.

In Fig. 9 the magnetic-field tracking signal obtained with the passive-mode NFE magnetometer employing amplitude-modulated light is shown [54]. Every 9 s, the magnetic field was increased by 15 nT and the modulation frequency was adjusted according to the scheme described above. Determining a new value ot the magnetic-field magnitude usually took 2-3 s, which depended on the field-change magnitude, the lock in time constant, etc. While maintaining a sensitivity of 4.3×10^{-13} T/$\sqrt{\text{Hz}}$, the device was able to track magnetic fields in a range of a few μT. Moreover, in Ref. [54], extension of the dynamic range toward Earth magnetic field and further increase of the device's sensitivity were predicted.

In general, the self-oscillation mode of optical magnetometers, is based on using the detected optical signal for modulation of optical [34, 55, 56] or RF excitation [34, 57] of atoms. Since the signal oscillates at the Larmor frequency or its harmonic, in such an arrangement the condition for the efficient generation of the higher-field magneto-optical resonances is automatically fulfilled. When the detected signal is amplified sufficiently, the oscillation builds up spontaneously (seeded by noise) at the resonant frequency. Due to no inertia in precession of spins, a change of the magnetic field is immediately reflected in the frequency of oscillations of the optical signal and hence the light modulation frequency.

Operation of NFE magnetometers in the self-oscillating mode is achieved by feeding amplified and phase shifted non-demodulated, time-dependent NFE signal directly to the light modulation controller, e.g., AOM driver [55,56]. The signal is phase shifted such that a 90° phase shift between the polarization rotation and the modulation frequency is achieved. While broad bandwidth is a big advantage of the self-oscillation mode, its drawback is lower sensitivity than sensitivity of the passive-mode magnetometers. Since no filters are applied to the signal, whole-spectrum noise contributes to the measured signal reducing signal-to-noise ratio. Despite the fact that the atoms themselves work as a narrow-band filter selectively transmitting only frequency $2\Omega_L$, the additional noise may introduce frequency jitter and distort measurements of magnetic fields.

A typical self-oscillation signal reported in Ref. [55] is shown in Fig. 10. For the purpose of a wide bandwidth demonstration, the NFE signal was heterodyned with a fixed reference with a time constant of 50 μs. The tracking of the field step is quasi-instantaneous,

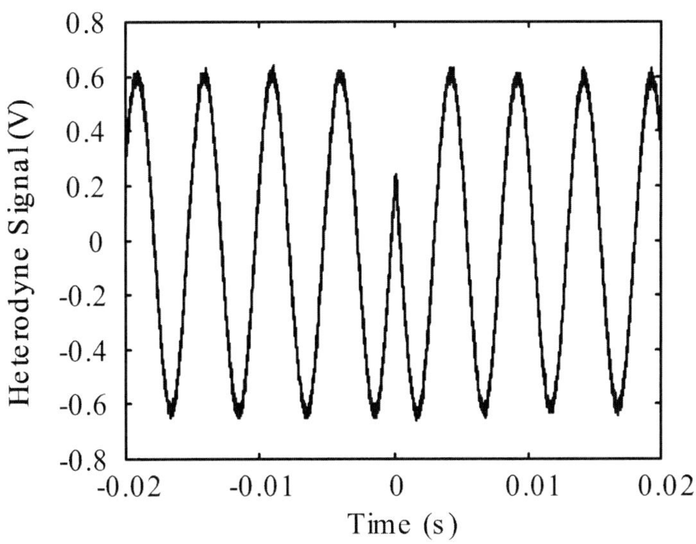

Figure 10. Time-dependent rotation of the light polarization observed with NFE magnetometer in the self-oscillating mode. In order to demonstrate high speed of the magnetometer the signal was heterodyned with the reference signal. At $t = 0$ a sudden change of 25 nT of the total magnetic field, resulting in a jump of the oscillation phase, was applied. The bandwidth of this magnetometer was above 1 kHz. From Ref. [55].

without apparent overshoot or ringing; the sudden change of about 25 nT applied to the bias field at $t = 0$ results in a phase-jump in the heterodyne signal. The demonstrated bandwidth of the magnetometer exceeded 1 kHz with a sensitivity of 300 fT for a measurement time of 1 s.

In Sec. 4.1., several schemes of NFE magnetometers were discussed. They differ in sensitivity, dynamic range, bandwidth, etc., hence each of them has own its advantages and disadvantages; while passive-mode magnetometers seem to be better for measuring static or quasi-static magnetic fields, self-oscillating magnetometers can be used for measuring varying fields.

4.2. Nuclear Magnetic Resonance

The first practical application of NFE magnetometers to detect the nuclear magnetic resonance (NMR) was described in Ref. [10]. For that purpose the used an NFE magnetometer with frequency-modulated light.

NMR is a powerful technique exploited in studies of structure and dynamics on micro- and macroscopic scale and its applications range from medicine to real-time process control [58]. A good example of the element extensively employed in NMR is ^{129}Xe. Its advantages include the polarization via spin-exchange collisions, strong chemical shifts, which makes it a good probe to study physicochemical environment, and long relaxation times.

For the NMR measurements, a typical NFE experimental apparatus such as one described in Sec. 3., was modified. In addition to the standard setup elements like laser, magnetic shield, vapor cell, etc., a solenoid piercing the shield was mounted in the vicinity of the vapor cell. The solenoid was longer than the shields thus magnetic field lines associated with the solenoid were closed outside the shield. It enabled generation of a homogenous magnetic field of about 50 μT inside the solenoid and almost no field in other parts of the shield, particularly, within the vapor cell (suppression factor on the order of 10^5). The solenoid field was necessary to orient the polarization of the Xe sample introduced into the shield via the solenoid. The 1.7 cm^3 sample containing natural abundance xenon at 3.7 torr pressure, spin-polarized (1% polarization) by spin-exchange collisions with optically pumped atoms outside the shield. A magnetic field of 14 nT perpendicular to the solenoid axis was generated inside the shield with a set of magnetic-field coils. In such a way, the detector measured a net magnetic field, that is, the field from the coils and polarized Xe but not from the solenoid. The application of that field enabled shifting of the NFE resonance position towards higher fields leading to significant reduction of the $1/f$ noise and a possibility of using the light-modulation technique for tracking of the samples demagnetization. The magnetometer worked in the passive mode (Sec 4.1.3.).

In Fig. 11 a typical data of Xe magnetization decay vs. time is shown. Each point was measured for about 3 s which enabled achieving S/N ratio at a level of 10. According to the authors, the S/N ratio could be improved by at least 2 orders of magnitude by better geometry of the experiment. In the performed experiment, the magnetic field generated within the vapor cell by polarized Xe was roughly 5000 times smaller then the one inside the sample.

In order to distinguish magnetic-field changes resulting from the decay of the Xe polarization from slow drifts of the other fields, for instance, related to the instability of current sources used for the magnetic-field generation, thermal current flowing in the shield, etc.,

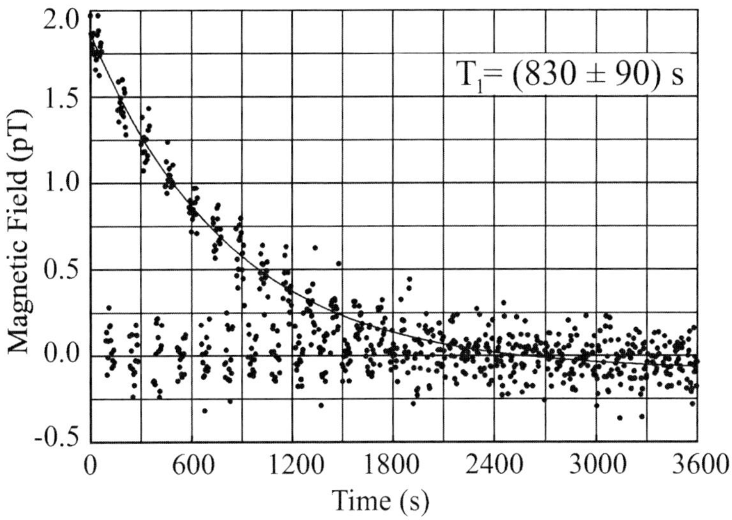

Figure 11. Xe-magnetization signal recorded with the NFE magnetometer [10]. The upper trace was generated by polarized Xe, while the lower trace is the reference enabling to eliminate the effect of slow drifts of the magnetic field. The traces were obtained by moving the xenon sample into and out of the shield in 1-minute intervals.

differential measurements were performed. In that case, the Xe sample was mechanically moved in and out of the shield which modulates magnetometer readout.

As seen in Fig. 11, an exponential decay of the Xe polarization from ≈2 pT to 0 with a relaxation rate of $(7.6 \pm 0.87) \times 10^{-3}$ 1/s was measured.

4.3. Magnetic Particle Detection

Another interesting application of NFE magnetometers is detection of magnetic particles. Such particles are micro- or nano-objects made of ferro- or paramagnetics. When bonded to other physical objects or chemical compounds, these particles can selectively react with given types of materials and/or chemicals which enables detection of such material/substances at very low concentrations. This technique is commonly used in medicine and biosciences, e.g., for DNA-targeting. Moreover, application of external fields is a convenient way of manipulation of the magnetic particles, as well as bonded objects. The particles are also broadly used as contrast agents in NMR.

In order to characterize magnetic particles and hence monitor behavior of tagged objects, ultra-precise measurements of magnetic fields are needed. Such devices as SQUIDs [59] and giant magnetoresistive (GMR) sensors [60,61] were used for this sake. Each of the techniques has its advantages and drawbacks; SQUIDs have very high sensitivity but they require cryogenic cooling, while GMR sensors can be extremely small (on the order of μm) but the measured sample needs to be bring extremely close to the sensor. The NFE magnetometers reach sensitivity comparable to SQUIDs and do not require cryogenic cooling, hence they seem to be ideally suited for the magnetic-particle detection.

The first demonstration of the magnetic-particle detection with the NFE magnetometers was described in Ref. [9]. The layout of an experimental arrangement is shown in Fig. 12. An important modification of an experimental apparatus with respect to setups tradition-

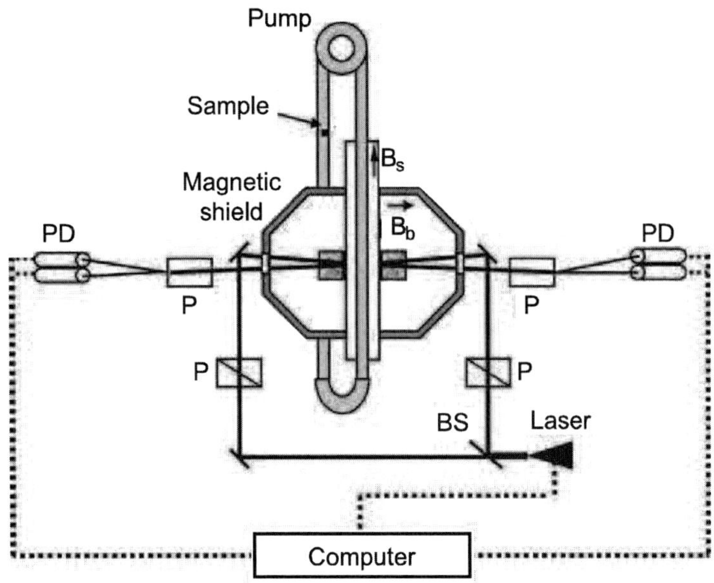

Figure 12. Experimental setup used for the magnetic particle detection. The magnetometer was operated in the gradiometer mode, i.e., the field was measured by two spatially separated magnetic-field sensors, which enabled detection of the magnetization of the magnetic particles flowing through the shield. The orientation of the particles' magnetization was assured with the piercing solenoid generating a field of $B_s = 50\ \mu$T. The smaller bias field $B_b = 70$ pT applied perpendicularly to the solenoid enabled detection of the magnetization using the passive-mode NFE magnetometers. P stands for the polarizers, PD for the photodiodes, and BS for the beam splitter. From Ref. [9].

ally used in NFE measurements (see Fig. 7) was, in addition to the piercing solenoid (see Sec. 4.2.), an application of two vapor cells enabling gradiometric measurements (Sec. 4.1.). The 1-cm^3 cells were placed symmetrically with respect to the solenoid in a distance of 1.5 cm. The monitored particles within the solenoid were spin-oriented by $B_s = 50\ \mu$T field. When there were no particles within the solenoid, the gradiometer cells were subjected only to a bias field of $B_b = 70$ pT, perpendicular to the solenoid axis, yielding zero differential signal. Introduction of the particles into the solenoid resulted in a field change in each cell; in one cell the sample magnetization B_e added up to and in the other subtracted from the bias field which results in non-zero gradiometer signal. That arrangement allowed a sensitivity of 100 fT/$\sqrt{\text{Hz}}$ to be achieved [11].

In Ref. [9], the study of three types of magnetic particles: multi- and single-domain cobalt particles of 150 μm diameter, 1 μm superparamagnetic magnetite particles, and ferro-

[11] For more details on magnetometer apparatus see Ref. [62].

magnetic cobalt nanoparticles of 120 kA/m magnetization were described. All the particles were suspended in water circulating through the solenoid within the shields. An example of the results obtained with the cobalt multi-domain particle is shown in Fig. 13. Each time

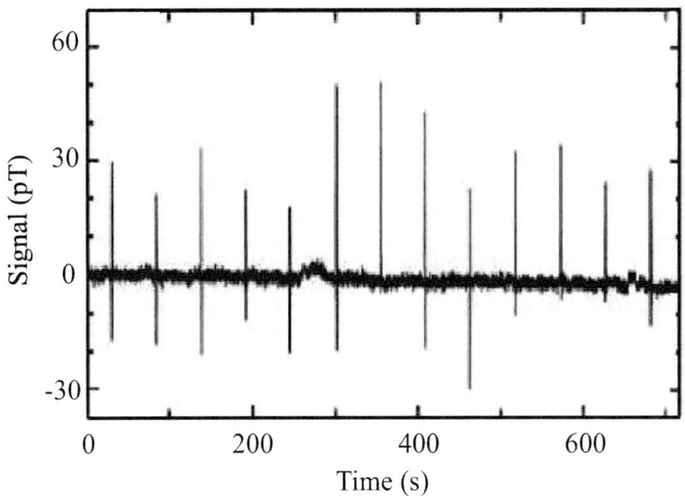

Figure 13. Magnetic detection of the cobalt multi-domain particle suspended in water flowing through the shield with a flowing rate 30 ml/min. Each time the particle traverses the magnetometers the strong gradiometer signal was recorded. The scattering in the amplitude of the signal was due to different positions of the particle inside the water tube each time it flew in vicinity of vapor cells. From Ref. [9].

the particle passed the gradiometer, the spike-like signal was observed. The amplitude of the signal depended on the magnetic field generated by the particle, as well as the water-flow rate, i.e., the time that the particle spent in a vicinity of the cells. Fluctuations in the amplitudes of the signal originated from the spread of particle's positions while traversing between the cells. Estimates performed in Ref. [62] showed that the detection limit on a size of the cobalt particle is about 20 μm for a multi-domain particle and about 5 μm for a single-domain particle.

Very interesting results were obtained for detection of amin-coated magnetite superparamagnetic particles of about 1 μm. Such particles are used for tagging, capturing, and extracting DNA from living organisms [63]. The authors of Ref. [9] proved the possibility of real-time detection of a sample containing 4.5×10^5 particles and of 5×10^3 particles with averaging.

Exciting alternative for the discussed applications may be a use of millimeter-scale cells [64]. Application of these cells would enable bringing a sample closer to the magnetic-field sensor which would enhance the detection sensitivity. This possibility, however, has not been yet experimentally verified.

4.4. Magnetic Resonance Imaging

Very appealing application of NFE magnetometers and a natural extension of the previously described applications is magnetic resonance imaging (MRI). MRI is a method enabling extraction of unique information on spatial distribution of specific chemicals within a sample. It is widely used in medicine, material science, geology, etc. Traditionally, MRI is based on detection of a change in the magnetic flux generated by nuclear spins precessing in an external magnetic field. It is routinely realized with induction pick-up coils. While very simple, this technique suffers strongly from the $1/f$ noise which implies use of higher precession frequencies corresponding to very strong magnetic field (above 1 T). This is a serious obstacle, especially taking into account a required field homogeneity at the ppm level. Such fields are generated by specially prepared solid or superconductive magnets which limits MRI applications to relatively small, nonmagnetic samples only, reduces mobility and makes standard MRI devices very expensive. Therefore, development of the low-field MRI technique is very much desired.

In order to realize MRI at low magnetic fields, a sensitive technique of magnetic-field detection needs to be used. One solution is application of SQUIDs [65] which enable reaching much higher sensitivities than standard pick-up-coil technique. Since they require cryogenic cooling and complex electronics for multi-channel detection their application also in this case is limited. Alternative technique exploits NFE magnetometers [11]. Unlike other MRI techniques, it is based on detection of DC rather than AC magnetic fields. It is done in the, so-called, remote-detection arrangement, in which information about an analyte is transferred onto a mobile spin-polarized substance, and is successively read out at a different location [66, 67]. The signal detected by the NFE magnetometer is the net, longitudinal magnetization of the spin-polarized sample. An important advantage of the remote detection scheme is an ability of independent optimization of conditions for encoding and detection regions and continuous measurements of MRI signals.

The experimental setup used for MRI in Ref. [11] was very similar to the one shown in Fig. 12 and discussed in Sec. 4.3.. The most important difference in comparison to two previous measurements was introduction of the prepolarization and encoding regions in the water-flow system. The water was prepolarized by a 0.3 T magnet situated a couple of meters away from the shield. The polarized water flew into the encoding region where it interacted with an encoding field of 3.1 mT and a sequence of magnetic-fields pulses. The first pulse of the sequence, the $\pi/2$ pulse, rotated the spins to a plane perpendicular to the encoding magnetic field, which resulted in their precession. Successively, a pair of orthogonal magnetic-field-gradient pulses was applied. Different precession frequencies corresponding to different geometrical positions allowed to store the spatial information in the phases of spins' rotation. After the application of a second $\pi/2$, the phase-encoded spatial information was stored in longitudinal magnetization of the sample. Phase-encoded water was then transferred into the magnetic shield via the piercing solenoid. Introduction of the water changed magnetic fields in both cells and hence readouts of the whole magnetometer.

An example of the MRI results obtained with the NFE magnetometer and a phantom consisting of two parallel cylindrical channels is shown in Fig. 14. The images were obtained by cycling phase of the second $\pi/2$ pulse, modifying gradient amplitudes along the encoding directions, and performing Fourier analysis of the signal. That procedure yielded

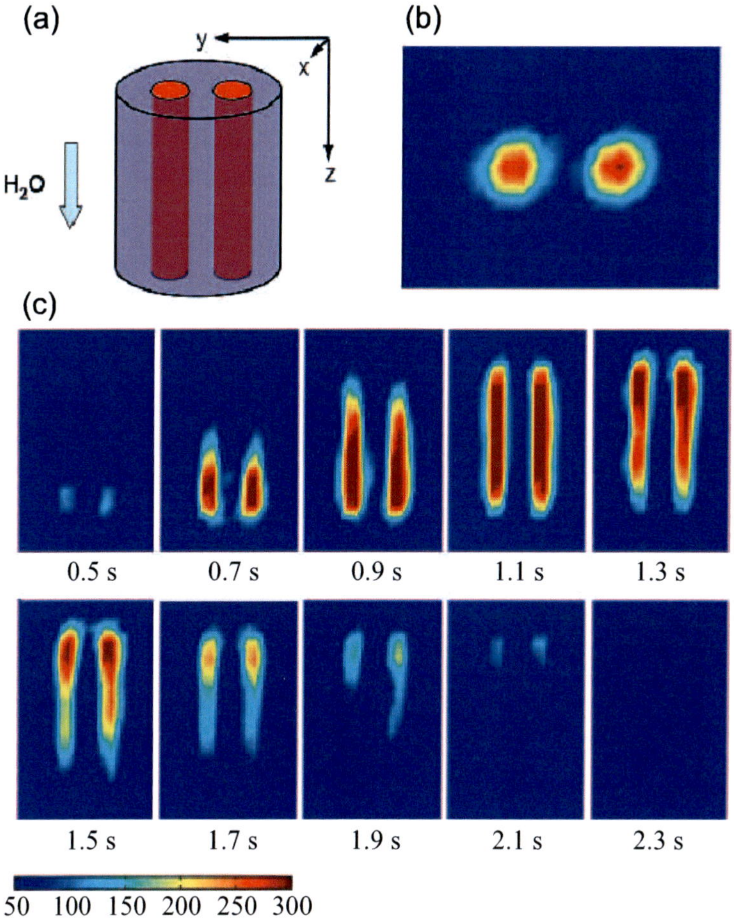

Figure 14. Magnetic resonance images obtained with the NFE magnetometer. (a) Phantom consisting of two channels of 3.2 mm in diameter and 25 mm length separated by 5.1 mm. (b) Image of the cross-section of the encoding volume in the xy-plane, perpendicular to the flow direction at $t = 1.1$ s. (c) Time-resolved images in the yz-plane, parallel to the flow (yz). From Ref. [11].

a spatial resolution of 1.6 mm perpendicular and 4.5 mm parallel to the encoding field and a temporal resolution of 0.1 s. In the experiment, the resolution was determined by the relaxation rate of the water polarization. The authors of Ref. [11] expected improvement of the resolution by shortening of water travel time from the prepolarization to the detection regions and/or by application of a stronger prepolarizing magnet.

Reference [11] demonstrated large potential of NFE magnetometers for MRI applications. Advantages of the technique are its high sensitivity, good spatial and temporal resolutions, and, most importantly, application of weak magnetic fields. Low fields eliminate problems associated with the field generation, appearance of magnetic-field artifacts, limitation of the technique for studying the nonmagnetic samples only, lack of the MRI device mobility, etc. These features are the reason for which the application of the NFE magnetometry for MRI gain increasing attention.

4.5. Selective Addressing of Atomic Coherences

In addition to the applications described above there is a rapid development of the NFE methodology for quantum-state engineering. This is caused by the fact that the same quantum superpositions which are responsible for the nonlinearity in the Faraday effect are also fundamental for storage of quantum information in atoms and quantum-state engineering. In particular, the techniques of slowing down coherence relaxation by paraffin-coating of cell walls or introduction of a buffer gas is the most effective way of extending the qubit lifetime. Moreover, the studies of NFE revealed that appropriate choice of modulation frequency with respect to the Larmor frequency enables selective generation of the atomic coherences between specific Zeeman sublevels of a given atomic state and successive modification of the state [32, 68, 69]. Since these coherences determine material properties, control over their creation and evolution allows one to generate media possessing unique physical, in particular, optical properties.

In Fig. 15 visualized density matrices containing different-order coherences, i.e., coherences between states with different values of the magnetic number m, are shown. As seen

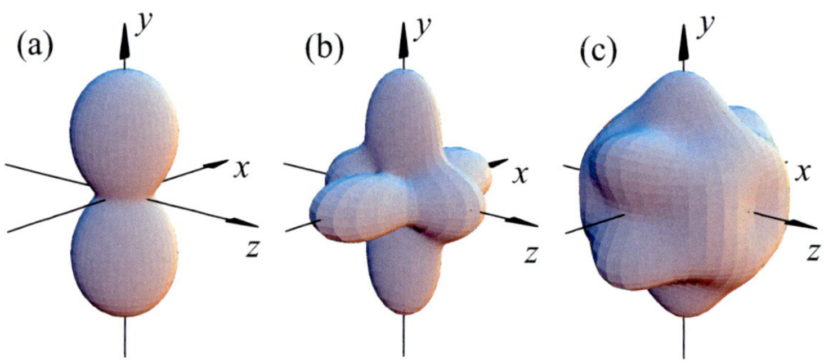

Figure 15. Visualized density matrices describing atomic systems with coherences of different orders: $\Delta m = 2$ (a), $\Delta m = 4$ (b), and $\Delta m = 6$ (c). The plots are drawn for the $F = 3$ state. From Ref. [32].

in Fig. 15(a), the $\Delta m = 2$ coherences correspond to the two-fold symmetric structure discussed above. Similarly, the $\Delta m = 4$ and $\Delta m = 6$ coherences are represented by the four- [Fig. 15(a)] and six-fold [Fig. 15(c)] symmetric shapes, respectively. In general, the symmetry of the visualized density matrix reflects an order of the coherences that exist in the atomic medium. Exploiting these symmetries enables generation of specific coherences. It was discussed in Sec. 2. that deviation from the sphericity of the visualized density matrix is a signature of an optical anisotropy of the medium. Thus, precessions of any of the shapes shown in Fig. 15, for example, in the external magnetic field, leads to the modulation of the probe-light polarization at different frequencies: $2\Omega_L$, $4\Omega_L$, or $6\Omega_L$. Demodulation of the NFE signal at these frequencies allows one to selectively detect the coherences of a given order.

The NFE signal with the resonances associated with the coherences of different orders are shown in Fig. 16 [32]. The resonance observed at twice the Larmor frequency

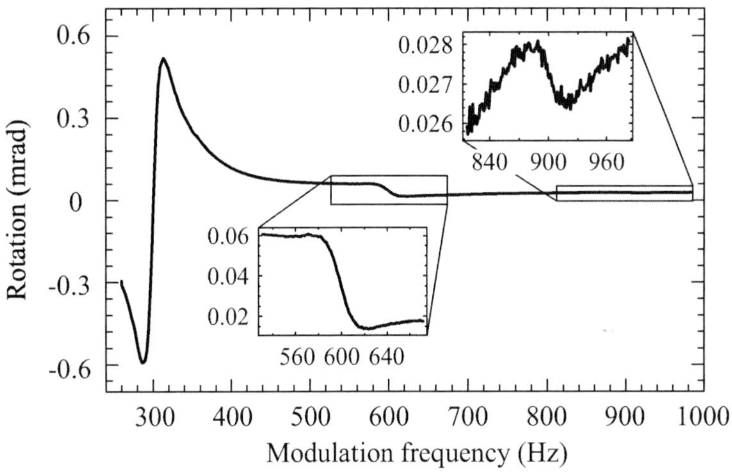

Figure 16. NFE signal recorded with frequency modulated light. The resonances at 300 Hz, 600 Hz, and 900 Hz are associated with $\Delta m = 2$, $\Delta m = 4$, and $\Delta m = 6$ coherences existing in atoms. From Ref. [32].

($\Omega_m = 2\Omega_L \approx 2\pi \times 300$ 1/s) is related to the $\Delta m = 2$ coherences ($\Omega_L \approx 2\pi \times 150$ 1/s). Similarly, the resonances at four ($\Omega_m \approx 2\pi \times 600$ 1/s) and six times ($\Omega_m \approx 2\pi \times 900$ 1/s) the Larmor frequency correspond to the $\Delta m = 4$ and $\Delta m = 6$ coherences, respectively. It was verified that all these resonances have different light-intensity dependences which indicates that different numbers of photons are needed for their generation and detection. In particular, the resonances associated with the higher-order coherences have significantly smaller amplitudes than those related to the lower-order coherences. This is expected since a photon is a spin-one particle and it can only generate coherences of $\Delta m \leq 2$. Generation of the higher-order coherences requires involvement of more photons which usually is less efficient than a single-photon interaction, thus the amplitudes of the NFE resonances associated with these coherences are smaller.

From a point of view of quantum-state engineering a better control over efficiency of coherence, especially the higher-order coherence, generation is needed. Therefore, a new, NFE based technique was elaborated [69]. With this technique, the higher-order coherences are generated in a multi-step process. In such a process, $\Delta m = 2$ coherence is first generated, and then the coherence is "updated" to $\Delta m = 4$ coherence by interaction with another photon. In order to selectively generate the $\Delta m = 4$ coherences, the second photon needs to be appropriately polarized with respect to the first photon; the two photons needs to have orthogonally polarized. Application of two orthogonally polarized photons allows the $\Delta m = 4$ coherences to be generated simultaneously causing destruction of the $\Delta m = 2$ coherences. As demonstrated in Ref. [69], the efficiency of higher-order coherence generation was significantly increased, which gave better control over creation of the multi-level quantum state was achieved.

Important feature of the described techniques is a possibility of coherence transfer between different atomic states. This was demonstrated in Ref. [32] by tuning the pump and probe light to two different hyperfine transitions of rubidium. In such a case, the pump light generated the coherences in the ground state which were successively excited to the upper atomic state. Since the excite state lifetime is 30 ns, the atoms instantaneously relax to one of the hyperfine ground states. The ones that relaxed to the same ground state from which they were excited can further interact with the light. However, those that decay to the other state do not interact with the pump light. Existence of the coherences in the latter state is reflected in modulation of the probe light. The modulation of the probe light tuned to the different than pump ground state, reported in Ref. [32], proved a possibility of transferring Zeeman coherence between different atomic states.

NFE provides an effective way of manipulation of atomic quantum states. The states can be engineered, i.e., the coherences may be selectively generated and the generated quantum state may be controllably modified. It should thus be possible to employ this technique in quantum computations for physical realization of multi-dimensional quantum "bits". Such multi-dimensional "qubits" would allow one to store more information than it is possible with "traditional" qubits. Simultaneous realization of the multidimensional "qubits" within the same atomic structure reduces one of the most severe limitations of quantum computations related to the exponential increase of the decoherence rate of a system of interacting qubits with an increase of its size. With the described technique the lifetime of the multi-dimensional "qubits" reaches tens of ms.

5. Conclusions

This chapter reviewed techniques and mechanisms related to nonlinear Faraday effect and presented some of its most important applications. We had discussed theoretical foundations of the phenomenon focusing on the relation between nonlinearity of the effect and the light-induced atomic coherences. Much attention was drawn to the modulation techniques (FM NMOR and AMOR) which substantially increase the range of possible applications. After describing the most essential experimental aspects, several specific applications were discussed: magnetometry, NMR, MRI and magnetic particle detection, as well as quantum state engineering.

Acknowledgements

The authors are grateful to Dmitry Budker for a long-standing cooperation and many fruitful discussions. They would like to express our gratitude to Simon Rochester for providing a powerful tool for the density-matrix visualization and Afrooz Family for his help in the preparation of the manuscript. Some of the research described in this work has been supported by the Polish Ministry of Sciences grants N N505 092033 and N N202 074135. SP acknowledges the Foundation for Polish Science.

References

[1] Budker, D.; Gawlik, W.; Kimball, D. F.; Rochester, S. M.; Yashchuk, V. V.; Weis, A. *Rev. Mod. Phys.* 2002, 74, 1153-1201.

[2] Hermann, G.; Scharmann, A. *Z. Phys.* 1968, 208, 367-380.

[3] Le Floch, A.; Le Naour, R. *Phys. Rev. A* 1971, 4, 290-295.

[4] Gawlik, W.; Kowalski, J.; Neumann, R.; Träger, F. *Opt. Commun.* 1974 12, 400-404.

[5] Barkov, L. M.; Melik-Pashayev, D. A.; Zolotorev, M. S. *Opt. Commun.* 1989, 70, 467-472.

[6] Budker, D.; Kimball, D. F.; Rochester, S. M.; Yashchuk, V. V.; Zolotorev, M. *Phys. Rev. A* 2000, 62, 043403.

[7] Budker, D.; Kimball, D. F.; Yashchuk, V. V.; Zolotorev, M. *Phys. Rev. A.* 2002, 65, 055403.

[8] Gawlik, W.; Krzemien, L.; Pustelny, S.; Sangla, D.; Zachorowski, J.; Graf, M.; Sushkov, A. O.; Budker, D. *Appl. Phys. Lett.* 2006, 88, 131108.

[9] Xu, S.; Donaldson, M. H.; Pines, A.; Rochester, S. M.; Budker, D.; Yashchuk, V. V. *Appl. Phys. Lett.* 2006, 89, 224105.

[10] Yashchuk, V. V.; Granwehr, J.; Kimball, D. F.; Rochester, S. M.; Trabesinger, A. H.; Urban, J. T.; Budker, D.; Pines, A. *Phys. Rev. Lett.* 2004, 93 160801.

[11] Xu, S.; Yashchuk, V. V.; Donaldson, M. H.; Rochester, S. M.; Budker, D.; Pines, A. *Proc. Nat. Acad. Sci. (USA)* 2006, 103, 12668-12671.

[12] Boyd, R. W. *Nonlinear Optics*, Academic Press, Amsterdam, 2003.

[13] Arimondo, E. In *Progess in Optics XXXV*, Elsevier Science, Edited by Wolf, E. 1996, 259-354.

[14] Harris, S. E. *Phys. Today* 1997, 50, 36-42.

[15] Akulshin, A. M.; Barreiro, S.; Lezama, A. *Phys. Rev. A* 1998, 57, 2996-3002.

[16] Rochester, S. M.; Budker, D. *Am. Journ. Phys.* 2001, 69, 450-454.

[17] Edmonds, A. R. *Angular Momentum in Quantum Mechanics*, Princeton University Press, Princeton, 1985.

[18] Nayak, N.; Agarwal, G. S. *Phys. Rev. A* 1985, 31, 3175-3182.

[19] Malakyan, Y. P.; Rochester, S. M.; Budker, D.; Kimball, D. F.; Yashchuk, V. V. *Phys. Rev. A.* 2004 69, 013817.

[20] Alexandrov, E. B.; Auzinsh, M.; Budker, D.; Kimball, D. F.; Rochester, S. M.; Yashchuk, V. V. *JOSA B* 2005, 22, 7-20.

[21] Benabid, F.; Knight, J. C.; Antonopoulos, G.; Russell, P. St. J. *Science* 2002, 298, 399-402.

[22] Light, P. S.; Benabid, F.; Couny, F.; Maric, M.; Luiten, A. N. *Opt. Lett* 2007, 32, 1323-1325.

[23] Hemmer, P. R.; Turukhin, A. V.; Shahriar, M. S.; Musser, J. A. *Opt. Express* 2001, 6, 361-363.

[24] Kanorsky, S. I.; Weis, A.; Skalla., J. *Appl. Phys. B* 1995, 60, S165-S168.

[25] Budker, D.; Hollberg, L.; Kimball, D. F.; Kitching, J.; Pustelny, S.; Yashchuk, V. V. *Phys. Rev. A* 2005, 71, 012903.

[26] Pustelny, S.; Jackson Kimball, D. F.; Rochester, S. M.; Yashchuk, V. V.; Budker, D. *Phys. Rev. A* 2006, 74, 063406.

[27] Kornack, T. W.; Smullin, S. J.; Lee, S.-K.; Romalis, M. V. *Appl. Phys. Lett* 2007, 90, 223501.

[28] Corney, A.; Kibble, B. P.; Series, G. W. *Proc. R. Soc. London, Ser. A* 1966, 70, 293-298.

[29] Wieman, C; Hänsch, T *Phys. Rev. Lett.* 1976, 36, 1170-1173.

[30] Bell, W.; Bloom, A. *Phys. Rev. Lett.* 1961, 6, 280-281.

[31] Corney, A.; Series, G. W. *Proc. Phys. Soc.* 1964, 83, 207-212.

[32] Pustelny, S.; Jackson Kimball, D. F.; Rochester, S. M.; Yashchuk, V. V.; Gawlik, W.; Budker, D. *Phys. Rev. A* 2006, 73, 023817.

[33] Budker, D.; Romalis, M. V. *Nat. Phys.* 2007, 3, 227-234.

[34] Bloom, A. L. *Appl. Opt.* 1962 1, 61-68.

[35] Dupont-Roc, J.; Haroche, S.; Cohen-Tannoudji, C. *Phys. Lett.* 1969, 28A, 638-639.

[36] Alexandrov, E. B.; Balabas. M. V.; Vershovski, A. K.; Pazgalev, A. S. *Tech. Phys.* 2004, 49, 779-783.

[37] Groeger, S.; Schenker, L. J.; Wynands, R.; Weis, A. *Euro. Phys. J. B* 2004, 33, 221-230.

[38] Kominis, I. K.; Kornack, T. W.; Allred, J. C.; Romalis, M. V. *Nature* 2003, 422, 596-599.

[39] Liew, L.-A.; Knappe, S.; Moreland, J.; Robinson, H.; Hollberg, L.; Kitching, J. *Appl. Phys. Lett.* 2004, 84, 2694-2696.

[40] Schwindt, P. D. D.; Knappe, S.; Shah, V.; Hollberg, H.; Kitching, J.; Liew, L.-A.; Moreland, J. *Appl. Phys. Lett.* 2004, 85, 6409-6411.

[41] Shah, V.; Knappe, S.; Schwindt, P. D. D.; Kitching, J. *Nat. Phot.* 2007, 1, 649-652.

[42] Taylor, J. M.; Cappellaro, P; Childress, L; Jiang, L; Budker, D; Hemmer, P. R.; Yacoby, A.; Walsworth, R.; Lukin, M. D. *Nature Phys.* 2008, 4, 810-816.

[43] Maze, J. R. ;Stanwix, P. L.; Hodges, J. S.; Hong, S.; Taylor, J. M.; Cappellaro, P.; Jiang, L.; Dutt, M. V. G.; Togan, E.; Zibrov, A. S.; Yacoby, A.; Walsworth, R. L.; Lukin, M. D. Nature 2008, 455, 644-647.

[44] Balasubramanian, G; Chan, I. Y.; Kolesov, R; Al-Hmoud, M.; Tisler, J.; Shin, C.; Kim, C.; Wojcik, A.; Hemmer, P. R.; Krueger, A.; Hanke, T.; Leitenstorfer, A.; Bratschitsch, R.; Jelezko, F.; Wrachtrup, J. *Nature* 2008, 455, 648-651.

[45] Budker, D.; Kimball, D. F.; Rochester, S. M.; Yashchuk, V. V.; Zolotorev, M. *Phys. Rev. A* 2000, 62, 043403.

[46] Allred, J. C.; Lyman, R. N.; Kornack, T. W.; Romalis, M. V. *Phys. Rev. Lett.* 2002, 89, 130801.

[47] Bechhoefer, J. *Rev. Mod. Phys.* 2005, 77, 783-836.

[48] Ledbetter, M. P., Acosta, V. M., Rochester, S. M., Budker, D.; Pustelny, S.; Yashchuk, V. V. *Phys. Rev. A* 2007, 75, 023405.

[49] Savukov, I. M.; Seltzer, S. J.; Romalis, M. V.; Sauer, K. L. *Phys. Rev. Lett.* 2005, 95, 063004.

[50] Alexandrov, E. B.; Balabas, M. V.; Kulyasov, V. N.; Ivanov, A. E.; Pazgalev, A. S.; Rasson, J. L.; Vershovski, A. K.; Yakobson, N. N. *Meas. Sci. Technol.* 2004, 15, 918-922.

[51] Seltzer, S.; Romalis, M. V. *Appl. Phys. Lett.* 2004, 85, 4804-4806.

[52] Pustelny, S.; Gawlik, W.; Rochester, S. M.; Jackson Kimball, D. F.; Yashchuk, V. V.; Budker, D. *Phys. Rev. A* 2006, 74, 063420.

[53] Bison, G.; Weis, A. *Phys. Rev. A* 2006, 74, 033401.

[54] Pustelny, S.; Wojciechowski, A.; Gring, M.; Zachorowski, A.; Gawlik, W. *J. Appl. Phys.* 2008, 103, 063108.

[55] Higbie, J. M.; Corsini, E.; Budker, D. *Rev. Sci. Instr.* 2006, 77, 113106.

[56] Pustelny, S.; Wojciechowski, A.; Kotyrba, M.; Sycz, K; Zachorowski, J.; Gawlik, W.; Cingoz, A.; Leefer, N.; Higbie, J. M.; Corsini, E.; Ledbetter, M. P.; Rochester, S. M.; Sushkov, A. O.; Budker, D. *Proc. of SPIE* 2007, 6604, 660404.

[57] Alexandrov, E. B.; Balabas, M. V.; Pasgalev, A. S.; Vershovskii, A. K.; Yakobson, N. N. *Laser Phys.* 1996, 6, 244-251.

[58] Blümich, B. *NMR Imaging of Materials*, Oxford University Press, Oxford, 2000.

[59] Tsukamoto, A.; Saitoh, K.; Suzuki, D.; Sugita, N.; Seki, Y.; Kandori, A.; Tsukada, K.; Sugiura, Y.; Hamaoka, S.; Kuma, H.; Hamasaki, N.; Enpuku, K. *IEEE Trans. Appl. Superconduc.* 2005, 15, 656-659.

[60] Wood, D. K.; Ni, K. K.; Schmidt, D. R.; Cleland, A. N. *Sens. Actuators, A* 2005, 120, 1-6.

[61] Pekas, N.; Porter, M. D.; Tondra, M.; Popple, A.; Jander, A. *Appl. Phys. Lett.* 2004, 85, 4783-4785.

[62] Xu, S.; Rochester, S. M.; Yashchuk, V. V.; Donaldson, M. H.; Budker, D. *Rev. Sci. Instrum.* 2006, 77, 083106.

[63] Nakagawa, T.; Tanaka, T.; Niwa, D.; Osaka, T.; Takeyama, H.; Matsunaga, T. *J. Biotechnol.* 2004, 118, 105-111.

[64] Balabas, M. V.; Budker, D.; Kitching, J.; Schwindt, P. D. D.; Stalnaker, J. E. *JOSA B* 2006, 23, 1001-1006.

[65] Wong-Foy, A.; Saxena, S.; Moule, A. J.; Bitter, H.-M. L.; Seeley, J. A.; McDermott, R.; Clarke, J.; Pines, A. *J. Magn. Reson.* 2002, 157, 235-241.

[66] Moule, A. J.; Spence, M. M.; Han, S.; Seeley, J. A.; Pierce, K. L.; Saxena, S.; Pines, A. *Proc. Natl. Acad. Sci. (USA)* 2003, 100, 9122-9127.

[67] Seeley, J. A.; Han, S.; Pines, A. *J. Magn. Reson.* 2004, 167, 282-290.

[68] Yashchuk, V. V.; Budker, D.; Gawlik, W.; Kimball, D. F.; Malakyan, Y. P.; Rochester, S. M. *Phys. Rev. Lett.* 2003, 90, 253001.

[69] Acosta, V.; Auzinsh, M.; Gawlik, W.; Grisins, P.; Higbie, J. M.; Jackson Kimball, D. F.; Krzemień, L.; Ledbetter, M. P.; Pustelny, S.; Rochester, S. M.; Yashchuk, V. V.; Budker, D. *Opt. Express* 2008, 16, 11423-11430.

In: New Trends in Quantum Coherence... ISBN: 978-1-60741-025-6
Editor: Rafael Drampyan, pp. 83-122

Chapter 3

OPTICAL PROCESSES IN MICRO- AND NANOMETRIC THIN CELLS CONTAINING ATOMIC VAPOR

David Sarkisyan[1] and Aram Papoyan[2]
Institute for Physical Research, National Academy of Sciences of Armenia, Ashtarak-2, 0203 Armenia

Abstract

Design, construction and operation regimes of novel micro- and nanometric thin cells containing atomic vapor of alkali metals, allowing smooth variation of atomic vapor thickness L in the range from 30 nm to several mm, are presented. Unusual behavior of resonant absorption, fluorescence, reflection, coherent and magneto-optical processes in the case when L is comparable with resonant radiation wavelength λ, and its strong dependence on L/λ ratio is described.

The advantages of an atomic frequency reference method based on nanometic cell compared with widely used saturated absorption technique are demonstrated.

So-called "λ-Zeeman technique" (LZT) for studies of individual transitions between Zeeman sublevels of atomic hyperfine structure in an external magnetic field of $B = 0.1$ mT ÷ 0.25 T is described. Particularly, implementation of LZT allows direct determination of frequency shift and transition probability of Zeeman components in a B-field, as well as revealing the appearance and dynamics of "forbidden" transitions in spectrum under magnetic field, with new selection rules for the quantum number F.

The peculiarities of electromagnetically induced transparency (EIT) phenomenon in micrometric and nanometric thin cells filled with pure atomic vapor of alkali metal are studied. Several remarkable distinctions in formation of EIT and velocity-selective optical pumping (VSOP) resonances compared with the case of ordinary cm- size cells are revealed and discussed. It is demonstrated that with the help of a nanocell with $L \sim 1$ μm, exploiting large frequency detuning of the coupling laser, it is possible to determine the atom-dielectric wall collision rate.

---

[1] E-mail address: david@ipr.sci.am
[2] E-mail address: papoyan@ipr.sci.am

I. Introduction

In the pioneering experimental works [1-3] it was experimentally proved that the resonant interaction of laser radiation with atoms in a thin film of alkaline vapor with thickness $L = 10 - 1000$ μm may underlie a novel method for sub-Doppler spectroscopy, as was earlier predicted theoretically in [4]. Distinguishing peculiarity of thin vapor films is caused by the fact that the duration of the atom-laser radiation interaction, governed by wall-to-wall trajectories, is anisotropic. The contribution of atoms flying nearly parallel to the cell window is strongly enhanced in the recorded response signal thanks to their longer interaction time with the laser field: $t_D = D/u$ (D being the diameter of the laser beam, u being the atomic velocity), whereas for atoms flying along the beam, the interaction time $t_L = L/u$ (L being the distance between windows) is much shorter since $D >> L$.

Particularly, under special experimental conditions in thin atomic vapor column, the so-called VSOP (velocity selective optical pumping / saturation) sub-Doppler resonance peaks of reduced absorption appear in transmission spectra at the exact positions of atomic resonances. Obviously, the VSOP resonances are contributed by atoms flying nearly parallel to the cell window, hence being insensitive to the Doppler shift. However, in order to have a good contrast and signal-to-noise ratio of the VSOP resonance when $L \sim 10 - 1000$ μm, some special experimental methods have to be implemented [1-3]. A number of experimental and theoretical papers concerning peculiarities of the optical processes in micrometric thin atomic vapor columns are presented in [5-7].

Further progress in this field was stimulated by recently developed nanocells, or so called extremely thin cells (ETC) of submicron thickness L of atomic vapor of alkali metals. These cells allow one to study even more pronounced peculiarities of resonant absorption, reflection, fluorescence as well as magneto-optical processes for the case when L is of the order or less than resonant optical radiation wavelength λ [8-15]. Particularly, the contrast of VSOP resonance is very good for $L = \lambda$, which allows direct recording in a single transmission spectrum of the nanocell. In addition, also several new striking features were revealed in the nanocell spectra, which are absent in micrometric-thin cells.

The experimental results of the absorption and resonance fluorescence spectra in the nanocells show that the main parameter determining the nature of interaction is not the thickness L itself, but the ratio L/λ. In particular, it has been revealed that for low pump intensity (< 1 mW/cm^2), the linewidth of the absorption spectrum of hyperfine transition exhibits oscillating behavior, and has minimum value when $L = (2n + 1)\,\lambda/2$ (n is an integer). The most narrow linewidth of the absorption spectrum is achieved when $L = \lambda/2$, and is 3-4 times less than the Doppler linewidth of the absorption spectrum obtained in ordinary cm-size cells. In contrast, the maximum value of the absorption linewidth is achieved when $L = n\lambda$ and is close to the Doppler linewidth of the absorption spectrum in ordinary cm-size cells [10]. This is "collapse and revival of Dicke-type coherent narrowing" effect (CRDN) [9-15], known earlier for microwave domain. Also, the oscillating behavior of the absorption magnitude depending on the ratio of L/λ was demonstrated. The signature of CRDN effect has been detected for the thicknesses up to $L = 7/2\lambda$ [12,14]. In [16] it was shown that also in the case of a two-photon absorption, there is clearly a length when the spectral width of the transmission line shape is minimal, the effect being a two-photon analog of CRDN.

For relatively high pump intensity (> 1 mW/cm^2), the VSOP resonances of a reduced absorption, with the linewidth close to the natural one, and centered on the hyperfine transitions, occur when $L = n\lambda$ [10,13,14,17]. The most narrow and high contrast VSOP resonance is observable when $L = \lambda$. There are several distinct benefits to use these VSOP resonances as a frequency reference for atomic transitions of alkali atoms in comparison with widely used saturated absorption technique, and the most significant one is complete elimination of crossover (CO) resonances when $L = \lambda$.

Important to note that in contrast to the transmission spectra, the linewidth of the fluorescence spectra increases monotonically, without oscillations with L, remaining below the Doppler broadening even when the thickness of vapor column is increased up to 3-4 μm [8,10,12,18]. The narrowest linewidth of the fluorescence spectra is obtained when L is in the range $\lambda/4 < L \leq \lambda/2$ (~ 1.3 times narrower than the absorption linewidth for the same thickness). Another distinction is that VSOP resonances of reduced fluorescence appear at much higher intensity (> 20 mW/cm^2).

Sub-Doppler nature of the transmission and fluorescence spectra of nanocells gives a possibility to directly study the behavior of hyperfine atomic transitions of Rb and Cs in external magnetic fields [19-21]. The VSOP resonances in the transmission spectrum of the nanocell with $L = \lambda$ are split in an external magnetic field B into several components, with frequency positions and probabilities depending on B-field. This can be straightforwardly used for investigation of individual hyperfine transition between Zeeman sublevels in magnetic field ranged between 0.1 mT and 0.25 T.

The experimentally observed CRDN effect, as well as appearance of VSOP resonances in the absorption and fluorescence spectra of nanocells are rather well described by developed theoretical models [11,13,14].

Furthermore, the use of a nanocell with thickness $L < 100$ nm is shown to be efficient for studies of atom – dielectric surface interaction. As a manifestation of this interaction, an expected strong red frequency shift has been observed in the transmission and fluorescence spectra, of the Cs and Rb atomic vapor [15,22] (in some cases also possible to observe a blue shift).

The recent interest to the electromagnetically induced transparency (EIT) phenomenon is caused by the possibility of forming narrow optical resonances, particularly, in atomic vapor of alkali metals [23-28]. As the size of the vapor cell is reduced, the lifetime of the ground-state coherence becomes shorter because of quenching collisions of the atoms with the cell windows: $\Gamma_{21} = (2\pi t)^{-1}$, where $t = L/u$ (u is the thermal velocity). Hence, in the case of $L \sim 1$ μm ($2\Gamma_{21} > 100$ MHz), one could expect that the EIT effect should be completely washed out. However, in [29,30] the EIT resonance was successfully observed in a nanocell under special conditions, along with a number of new features and distinctions from the case of ordinary cells. The reason is again linked with the contribution of atoms flying nearly parallel to the cell windows, which is enhanced thanks to their longer interaction time with laser field. Due to such atomic velocity-selectivity, when the frequency of the coupling laser is resonant with the atomic transition, the linewidth of the EIT resonance can be more than 10 times narrower than that expected from the inversed window-to-window flight time of the atoms; the influence of atom-wall collisions can be directly detected when the coupling laser frequency is shifted with respect to the atomic transition frequency.

Thus, the nanocell containing atomic vapor of metal is a promising and efficient tool for studies of atom-light, atom-atom, atom-surface and atom-external magnetic filed interactions.

II. Experimental

II.1. The Design of Nano- and Micrometric Thin Cells

The first design of the nanometric, or so called extremely thin cell, which consists of windows and a vertical side-arm (a metal reservoir) was presented in [8]. Later, this design has been somewhat modified; the typical example of the last version is presented in Fig.1. We use the window wafers, 25 ÷ 30 mm in diameter (or 15 × 30 mm rectangular) and 2.5 mm thick, fabricated either from commercial sapphire or garnet ($Y_3Al_5O_{12}$) crystal, which are both chemically resistant to the chemically aggressive vapors of alkali metals. In order to minimize the birefringence effect, the sapphire plates are cut across the c-axis, while the garnet is free of birefringence. The wafers have to be perfectly polished before the use. In order to form the wedge gap in the vertical direction (this is needed to have a variable thickness of atomic vapor column), an Al_2O_3 strip, 3000 nm thick, 10 mm long, and 1 mm wide, is deposited on the surface of one of the windows in its lower part (in some cases, a platinum strip is used as a spacer: Pt is also resistant to highly corrosive hot Rb atomic vapor). Prior to the deposition, a 2 mm hole is drilled in the bottom of the windows, into which a 50 mm long tube of the same diameter, made of commercial sapphire, is inserted. Then, the entire construction is assembled and glued in a vacuum furnace. After the gluing, a glass extension is sealed to the sapphire tube, and the nanocell is filled with a natural mixture of the ^{85}Rb and ^{87}Rb (or Cs, or Cs-Rb mixture, etc.), as is done for an ordinary glass cell. The amount of rubidium is chosen so that the sapphire tube, with an inner diameter of about 1 mm, is almost completely filled. After sealing off of the nanocell, the glass tube has a length of 5 – 6 mm. As was shown in [31], the metal vapor pressure is determined by the temperature of the upper edge of the metallic column.

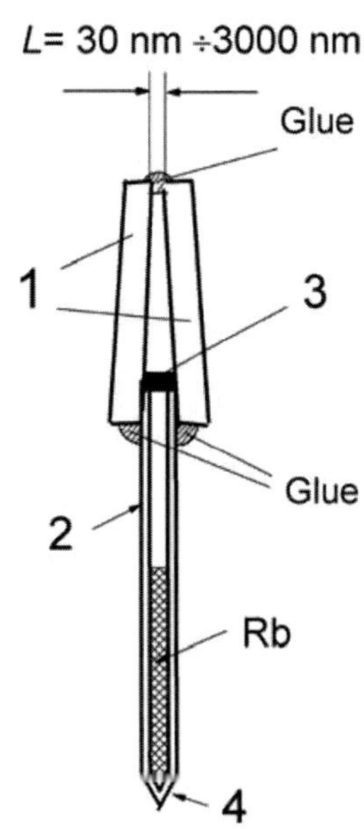

Figure 1. The design of the nano-cell. *1*- the windows of the cell; *2*- the sapphire side-arm; *3*- spacer; *4*- the glass extension.

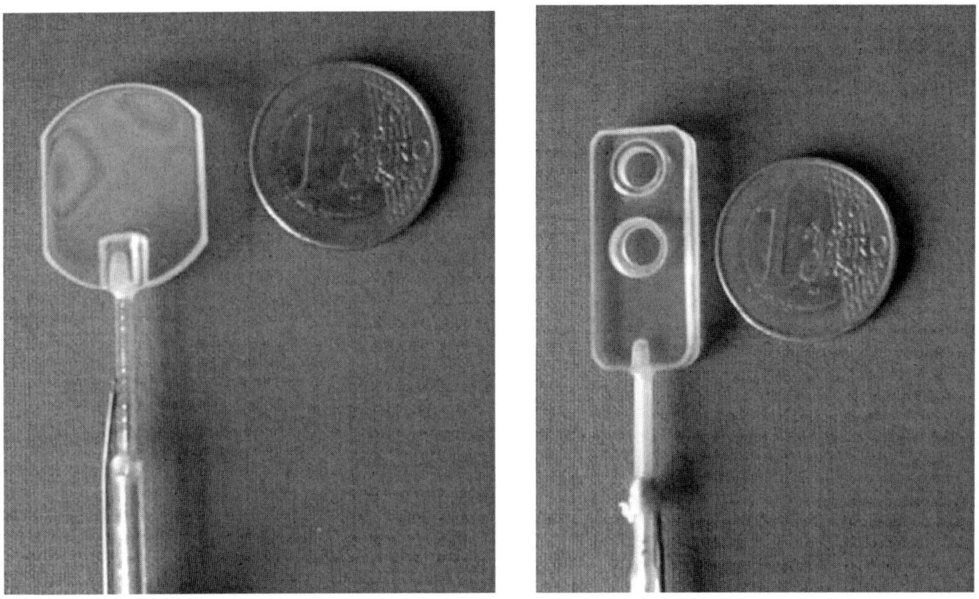

Figure 2. (a) a nanocell with smoothly variable thickness from 50 nm at the top to 6 μm near the bottom, i.e. wedged in the vertical direction; (b) a multi-region cell containing 3 regions of different thickness L: i) 2 mm; ii) 600 μm; iii) wedged region of 0.8 ÷10 μm. The1 Euro coin is shown for the size comparison.

Adjusting the heater construction, one can achieve the following limiting thermal conditions: the temperature at the upper edge of metallic Rb column is approximately 500 °C; the temperature of the glass extension (which is outside of the heater) is about 80 °C. At such a low temperature, metallic Rb only slightly reacts with glass. As for the interaction of hot Rb vapor with commercial sapphire or garnet, no chemical interaction occurs up to the temperatures of about 1000 °C [32]. In Fig.2a we present the photograph of the nanocell with smoothly variable thickness wedged in the vertical direction from 50 nm at the top to 6 μm near the bottom. Note, that since the gap thickness L is of the order of visible light wavelength, interferometric pattern is clearly seen in Fig.2. The wedged gap in this case was formed using a platinum spacer strip of 6 μm thickness (see above). For temperature control and monitoring, a thermocouple is tied close to the boundary of the Rb metallic column inside the side-arm in order to ensure thermal contact with the sapphire tube of the side-arm (a part of the thermocouple is seen in Fig.2).

The design of micrometric thin cells is similar to one presented in Fig.1, i.e. it consists of the windows and vertical side-arm filled with metal. The perfectly polished window wafers are fabricated from commercial sapphire or garnet. The gap of micrometric thickness L (10 ÷ 1000) μm between windows is formed by Pt spacer strips of appropriate thickness. Particularly, in order to form the wedged-shaped (vertically) thickness L of the gap, the spacer at the bottom of the windows had 90 μm thickness, while spacer at the top was 50 μm-thick. Then, the entire construction is assembled and glued in a vacuum furnace. All other operations are similar to ones described above. In Fig.2b we show the photograph of so-called multi-region cell (MRC), which contains 3 regions of different thickness of the atomic vapor column: i) 2 mm; ii) 600 μm; iii) a region with L smoothly varying from 0.8 μm at the top to

10 μm at the bottom. MRC is a convenient tool for various experiments, since it is possible to combine different processes using a single cell.

II.2. The Nanocell with Controllable Thickness *L* inside a Vacuum Chamber

Another way to smoothly vary *L* is as follows. We have developed a nanocell, which has thin windows. The thickness of very well polished garnet (or sapphire) window plates is less than 0.5 mm. The initial thickness of the gap between the windows *L* is determined by the ring-shaped aluminum oxide (Al_2O_3) spacer evaporation-deposited at the edge of one of the windows. With this spacer the initial distance *L* between windows ~ 1500 nm (at a room temperature). When the nanocell is filled with an alkali metal and then sealed-off, the atmosphere pressure reduces the gap thickness by pressing upon thin windows, reducing *L* to ~ 140 nm. Afterwards, the nanocell is placed inside a special glass vacuum chamber, in which the air pressure can be controlled in the range of 0 ÷ 1 atmosphere.

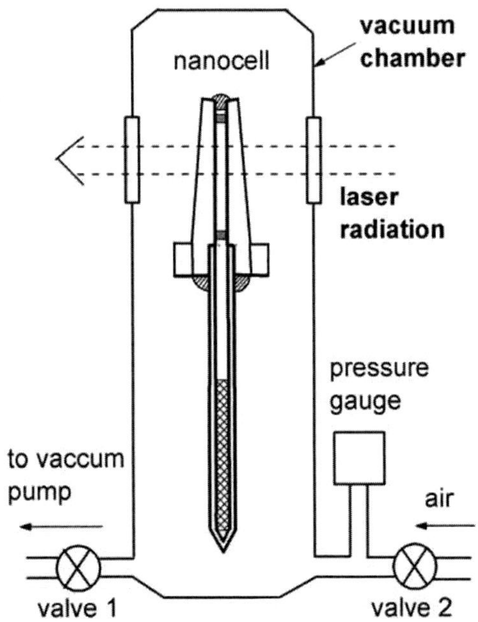

Figure 3. The design of the glass vacuum chamber with a nanocell inside.

The design of the glass vacuum chamber is shown in Fig.3: valve 1 is for pumping out the air from the chamber; valve 2 is to let in the air with the fixed pressure inside the chamber, the pressure is measured by a usual gauge. We change the pressure inside the chamber with a step of ~ 1 Torr, thus changing the thickness of the gap. When the air is completely pumped out from the vacuum chamber, *L* increases up to its initial value ~ 1500 nm. Thus, by changing the pressure inside the chamber, it is possible to control *L* accurately and smoothly in the range from 140 nm to 1700 nm with a nanometer-range step.

Fig.4 presents the gap thickness *L* as a function of the pressure inside the chamber. At the operating temperature of 120 °C, due to additional deformation of the garnet windows, the whole tuning range of *L* is slightly shifted (170 ÷ 1750 nm). Note that the advantage of this type of thickness tuning compared with the wedged gap is that the position where the laser radiation enters the nanocell is invariable, while in the case of a nanocell with wedged gap this position is different for different values of L. The pressure-controlled thickness system is found to be robust and reliable: a Rb nanocell with 0.5 mm-thick garnet windows successfully operates in our laboratory with a glass vacuum chamber without any damage over 6 years.

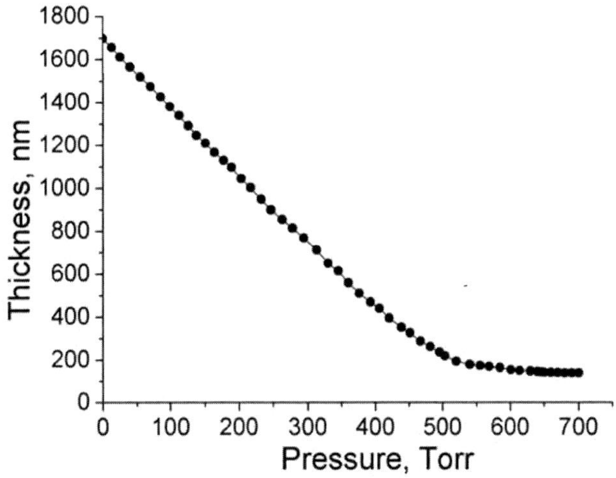

Figure 4. Dependence of the gap thickness L on the pressure inside the chamber.

II.3. Measurement of the Thickness L

Interferometric method described in [9] is used to determine the wedge-shaped thickness L of the nanocell gap. The idea of the thickness measurement is as follows. Since the thickness of the vapor gap (i.e. the distance between the inner surfaces of the nanocell) varies just by micrometers or less across the whole window aperture of ~ 3 cm, the two inner surfaces of the nanocell windows are highly parallel to each other within the laser beam aperture of ~ 0.1 mm. Hence, behavior of the transmission and reflection of the nanocell is similar to that of a low finesse Fabry-Pérot etalon, and interferometric formulas describing the thickness dependence of reflected or transmitted power can be used. As it is seen from Fig.5, there are 3 reflection beams (in order to spatially separate these beams, the window wafers are made slightly wedged): R_1 is the reflection from the first surface of the nanocell, R_2 is the reflection by the gap (i.e., joint reflection by the two inner surfaces of the windows), and R_3 is the reflection from the outer surface of the rear window. In Fig.6 we present the ratio R_2/R_1 as a function of the gap thickness L, for the laser wavelength λ = 780 nm (assuming that the windows are made of garnet: the refraction index of YAG is n = 1.80, for sapphire n = 1.76).

As it is seen, for the case $L = n\lambda/2$ (n is integer), the reflection $R_2 = 0$ (particularly, for L = 390 nm, 780 nm, 1170 nm, etc.), which is very convenient for the experimental adjustment. The curve for the thickness measurement has an oscillating behavior, which imposes ambiguity for the determination of L. This inconvenience was overcome by performing the measurements presented in Fig.6 for several wavelengths (thus, 3 different laser wavelengths have been used in [9]). However, as we will show below, the determination of $L = \lambda/2$, the thickness of particular interest, is easy enough simply from the transmission spectrum (see below), and having this reference, the ambiguity can be avoided for larger L, even for single wavelength interference case. In the particular case of Fig.1, the thickness determined by the described interferometric technique was found to range from 30 nm to 3000 nm.

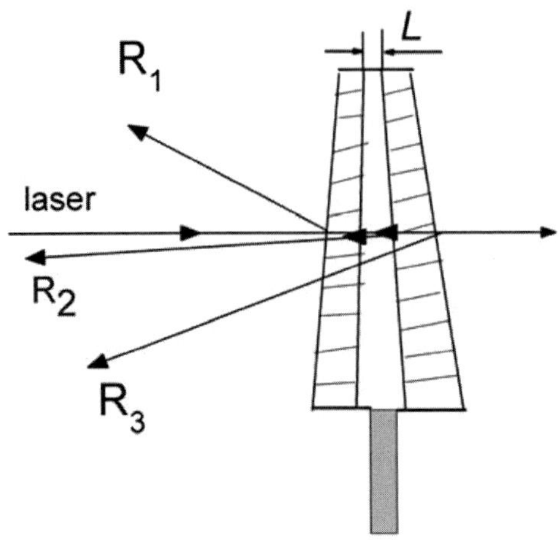

Figure 5. Three reflected beams R_1, R_2 and R_3 originated by four surfaces of the nanocell.

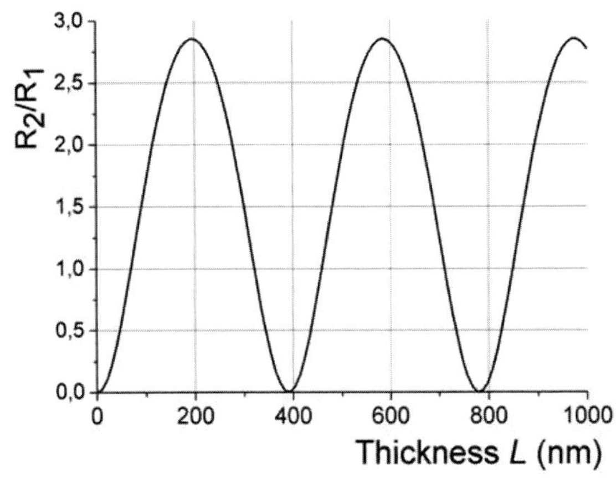

Figure 6. The ratio R_2/R_1 as a function of the gap thickness L, for $\lambda = 780$ nm.

II.4. Experimental Arrangement for Spectroscopic Measurements

The basic experimental arrangement for transmission measurements is presented in Fig.7. Different types of single-frequency tunable *cw* diode lasers (1) were used to study resonant absorption and fluorescence for D_1 and D_2 lines of Rb and Cs: extended cavity diode laser (ECDL) with radiation wavelengths at $\lambda = 780$ nm and 794 nm, with 10 - 50 mW power and spectral linewidth less than 1 MHz; single-frequency ~ 6 MHz- linewidth tunable distributed feedback (DFB) diode lasers at $\lambda = 894$ nm; single-frequency ~ 5 MHz- linewidth tunable *cw* solitary (free running) diode lasers at λ = 780 nm, 794 nm, 852 nm. The types of diode lasers utilized for the particular experimental studies will be given in the text below.

Faraday isolators were used in order to avoid optical feedback. In most cases, the diode laser radiation was focused on the nanocell nearly perpendicular to its window surface (this is very important [1,18]), with a spot size of $\varnothing \sim 0.5$ mm.

The nanocell is placed in the oven made from non-magnetic materials, which has 4 openings, by 2 for laser beam transmission and for registration of fluorescence in perpendicular direction to the laser beam. The temperature of the nanocell is kept at ~ 110 - 120 °C on the side arm (this defines the vapor pressure), and 20-30 °C higher at the windows. This regime corresponds to the number density of Rb atoms $N \sim 10^{13}$ cm^{-3}.

In order to obtain the transmission or/and fluorescence spectra, the laser radiation frequency is linearly scanned in an appropriate spectral region around the studied transition group of Rb or Cs. Part of radiation (~ 50%) is branched to an auxiliary reference setup (2) for comparison with the spectra obtained in the main nanocell. Depending on task, (2) could be: i) resonant transmission or fluorescence spectra obtained with a usual room temperature 6 cm-long cell containing either Rb or Cs; ii) saturated absorption (SA) spectra based on a room temperature 6 cm-long cell containing either Rb or Cs; iii) transmission spectra of hyperfine atomic transitions in an auxiliary nanocell, which is preferable in many cases (see below). The spectra were recorded by photodetectors (4) followed by a digital two channel storage oscilloscope, Tektronix TDS3032B or similar. Note that the basic experimental arrangement shown in Fig.7 was used for investigations presented in Chapter III. For the results presented in the Chapter IV and Chapter V, there are some modifications of the experimental arrangement, which are discussing directly in the corresponding chapter.

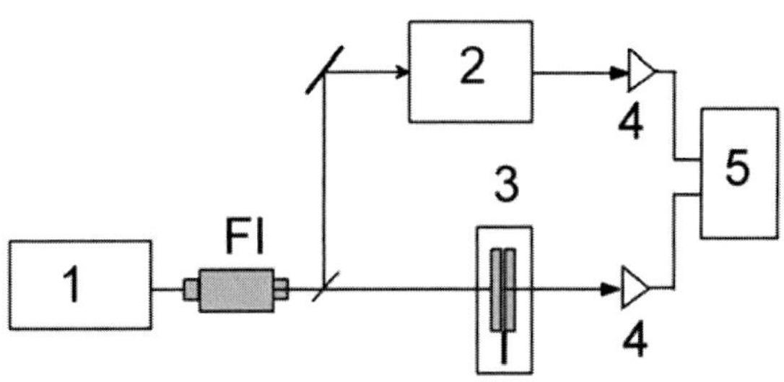

Figure 7. Sketch of the experimental setup. 1-different types of cw diode lasers with λ = 780, 794, 852, and 894 nm; FI- Faraday isolator, 2- frequency reference based either on saturated absorption technique or on an auxiliary nanocell, 3- thin cell inside the oven, 4- photodiodes, 5- digital oscilloscope.

III. Resonant Absorption and Fluorescence Processes in Nanocells

III.1. Key Features of Resonant Absorption in Nanocells

In a series of seminal papers Dicke showed that the microwave absorption of a gas in a vapor cell smaller than the wavelength (λ = 2.5 cm) can have a sub-Doppler spectrum due to frequent collisions of atoms with the walls of the cell [33]. This counterintuitive and unexpected effect can be explained under two assumptions: first, the gas has to be dilute, so that only atom–surface collisions to be considered; second, the walls have to be so close that atoms with a large velocity component towards the walls will hit the walls before they complete one interaction cycle with the incoming light (Rabi cycle). Thus atoms with a large longitudinal velocity give only a small contribution to the absorption signal, which is now dominated either by slow atoms or by atoms flying parallel to the windows. For the radiation propagating perpendicular to the nanocell's windows, the Doppler broadening is strongly reduced. In order to explain the fact that minimal linewidth of the absorption is achieving for $L = \lambda/2$, also the coherent effect has to be taken into account [34]. Note that there is another, well-known type of Dicke-narrowing caused by buffer gas collisions (DNB) [35], which differs from that presented here. The nanocells with atomic vapor of alkali metals allow studying Dicke-narrowing effect in optical domain.

There are several factors that could influence the profile (shape) and magnitude of the transmission line in fluorescence and reflection spectra of atomic vapor column confined in a nanocell and make them sensitive to the ratio L/λ: i) the transient behavior of the atomic excitation (i.e. build-up of the absorption coefficient) leads to a Dicke-type sub-Doppler narrowing, with an enhancement of the slow atom contribution, and the lineshape exhibits a maximal narrow contribution when $L = (2n+1)\lambda/2$ (i.e. with λ pseudo-periodicity), and this narrowing is practically absent (is collapsed) when $L = 2n(\lambda/2)$: this is so called CRDN, or Dicke coherent narrowing and its collapse [9]; ii) due to Fabry-Pérot nature of the nanocell (with $\lambda/2$ periodicity), both the pump and the re-emitted fields form a partially standing wave inside the nanocell, the intensities of transmitted and reflected radiations depend on L/λ, which affects the transmission, fluorescence and reflection spectra [11,13]. In addition, the pump intensity can strictly influence the shape and magnitude of the transmission, fluorescence and reflection spectra; also there is a strong the van der Waals (vdW) interaction between an atom and dielectric windows of a nanocell when $L/\lambda \ll 1$ [22].

One can conclude that because of CRDN and VSOP formation conditions, the most striking features of resonant processes in the nanocells occur at thicknesses $L=(2n+1)\lambda/2$ and $L=n\lambda$. Moreover, the non-resonant reflection vanishes for these thicknesses, and one can neglect the mixing of transmitted and reflected signals associated with intrinsic Fabry-Pérot nature of a nanocell, which is an essential ease. We present below experimental results for these two cases of particular interest and importance.

III.2. Low and High Laser Intensity Cases

Within last several years, the CRDN effect was studied on D_1 and D_2 lines of Rb and Cs (λ = 794, 780, 852, 894 nm), and in all cases the expected prominent L/λ-dependent features have been observed [9,10,12]. Here we mainly present the new results, which have not been published elsewhere, although the observed peculiarities of recorded spectra are similar to that obtained earlier.

Due to simplicity of atomic energy levels, it is more convenient to observe CRDN effect on the D_1 line of either Rb or Cs. Resonance transmission spectrum of $F_g=1 \rightarrow F_e=1,2$ transitions of ^{87}Rb D_1 line (the subscripts g and e refer to lower and upper states, respectively) for the nanocell with $L = \lambda/2$ and that for an ordinary cell (TOC) with $L = 5$ cm are shown in Fig.8 (the diagram of the levels is presented in Fig.20). The laser beam (ECDL, $\lambda = 794$ nm,

γ_L = 1 MHz) is focused by F = 35 cm lens into the nanocell at nearly normal incidence. The laser intensity is < 0.1 mW/cm^2. As it is seen, there is a strong linewidth narrowing in the case of $L = \lambda/2$. The full width at half maximum (FWHM) for the TOC is ~ 550 MHz (well fitted by Gaussian function), while for $L = \lambda/2$ nanocell the FWHM is 150 MHz (well fitted by "pseudo-Voigt2" function). Thus, the ratio FWHM(TOC) / FWHM(L=λ/2) is ≈ 3.7.

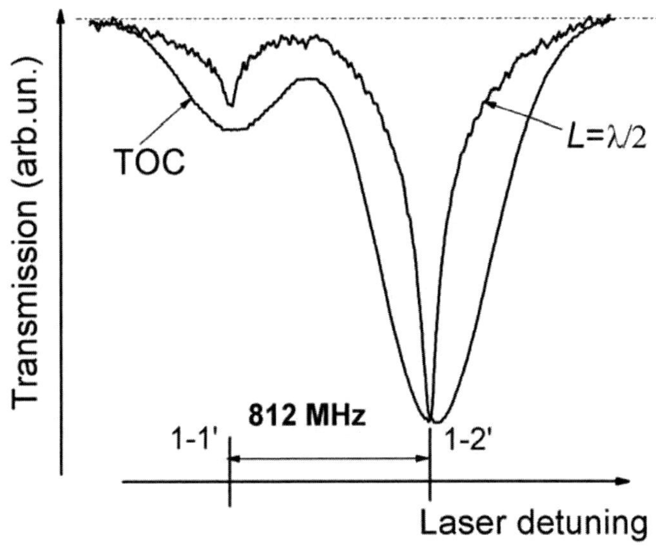

Figure 8. Transmission spectra for the nanocell with $L = \lambda/2$ and for the ordinary cell (TOC) with L = 5 cm , for the ^{87}Rb, $F_g=1 \rightarrow F_e=1,2$ transitions and laser intensity < 0.1 mW/cm^2.

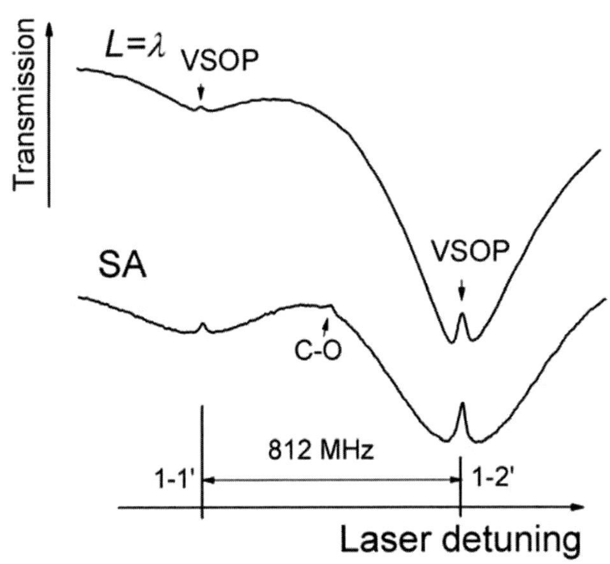

Figure 9. Transmission spectrum for $L = \lambda$ (upper curve) with simultaneously recorded SA spectrum (lower curve), for the ^{87}Rb, $F_g=1 \rightarrow F_e=1,2$ transitions and laser intensity 8 mW/cm^2.

In Fig.9 the transmission spectrum on the same transitions is presented for the thickness $L = \lambda$ (upper curve). Laser intensity is ~ 8 mW/cm^2. Obviously, the linewidth of the spectrum is much larger than in the case of $L = \lambda/2$, however the VSOP resonances appear exactly at the frequency positions of atomic transitions (the VSOP resonances are absent when the laser intensity is less than 0.1 mW/cm^2). The lower curve shows simultaneously recorded SA spectrum. One can see that the crossover (CO) resonance, which is recorded in the SA spectrum, is absent in the nanocell spectrum for $L = \lambda$. This is a very important advantage of the nanocell transmission spectrum. Although there is a reflected beam from the outer rear surface of the nanocell, which propagates backwards with respect to the main beam, it does not cause formation of CO resonance, since the interaction time t_L for atoms with a longitudinal velocity $u_z = 2\pi\varepsilon/k$ (where ε is half of the frequency separation of corresponding excited levels) is rather short at $L = \lambda$ for developing an efficient optical pumping needed for buildup of CO resonance. The FWHM of the peak of reduced absorption in the SA spectrum presented in Fig.9 is ~ 15 MHz, and somewhat wider for the case of $L = \lambda$.

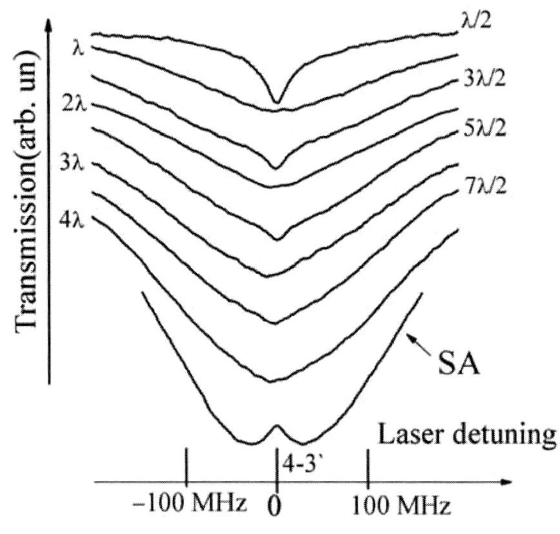

Figure 10. Transmission spectra of Cs D_1 line, for L varying from $\lambda/2$ to 4λ with the step of $\lambda/2$; the lower curve is SA spectrum.

Fig.10. shows the experimental transmission spectra when the thickness of the vapor column is varied from $\lambda/2$ to 4λ with the step of $\lambda/2$ at the laser intensity ~ 0.1 mW/cm^2. In this experiment, the ∅ 0.7 mm beam of a single-frequency tunable distributed feedback diode laser (laser, DFB) with λ = 894 nm and spectral width γ_L ~ 6 MHz was directed onto the nanocell at nearly-normal angle. The laser frequency was scanned in the range of ~ 1 GHz covering F_g=4→F_e=3 hf transition of Cs D_1 line. The CRDN effect is well seen: the minimum value of spectral width is attained at the vapor column thicknesses $L = (2n+1)\lambda/2$.

Note that for $L = \lambda/2$ the width is minimal: 75–80 MHz, whereas for $L = 3\lambda/2$, $5\lambda/2$, $7\lambda/2$, and $9\lambda/2$ the spectral narrowing is imposed on broad Doppler pedestal increasing with L (at the further increase, the spectrum will obviously approach the Doppler-broadened absorption spectrum of the usual cell). On the other hand, as is seen in Fig.10, for $L = n\lambda$ the spectral width is maximal (~ 300 MHz for $L = \lambda$). So, there is a fourfold narrowing of the transmission spectrum for $L = \lambda/2$ as compared to $L = \lambda$ (this experiment has been carried out jointly with the group of A. Weis, Fribourg University, Switzerland).

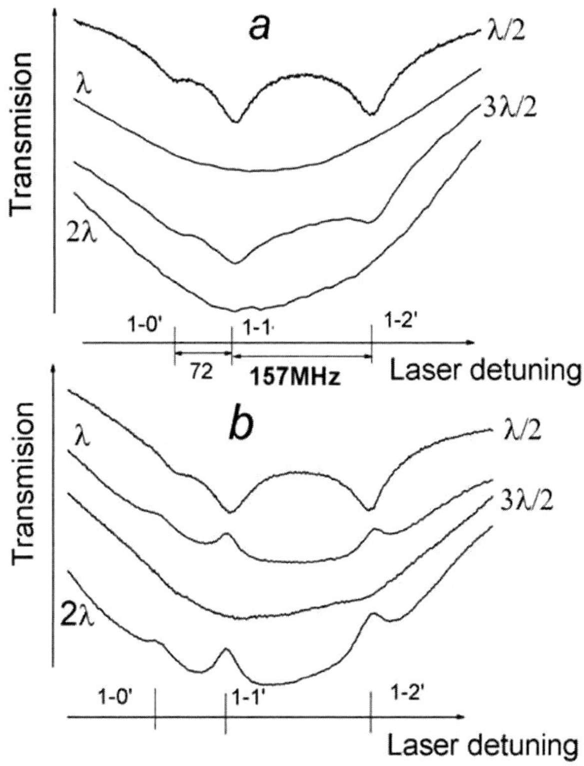

Figure 11. Transmission spectra of ^{85}Rb D_2 line, for L varying from $\lambda/2$ to 2λ with the step of $\lambda/2$. The laser intensity is: a) 0.1 mW/cm^2; b) 5 mW/cm^2.

In the case of D_2 lines of Rb and Cs the transmission spectra are more complicated as compared with D_1 lines because of several Doppler-overlapped excited hyperfine levels. Nevertheless, the use of a nanocell allows separation of these transitions in transmission spectra (Fig.11). For this case the laser beam (λ = 780 nm, γ_L = 5 MHz) is focused by F = 35 cm lens into the nanocell at nearly normal incidence. The vertically wedge-shaped thickness L of the nanocell gap permits smooth varying of L in the range from 100 to 2800 nm. The operation temperature is $T_W \sim 140$ °C at the window and $T_{SA} \sim 110$ °C at the side-arm. In Fig.11 we present the nanocell transmission spectra for $F_g=1 \rightarrow F_e=0,1,2$ transitions of ^{87}Rb D_2 line, with the probe intensity ~ 0.1 mW/cm^2 (a) and ~ 5 mW/cm^2 (b). The thicknesses L was varied in the range of $L = \lambda/2 - 2\lambda$ with the step of $\lambda/2$. The effect of Dicke type coherent narrowing is well observable in Fig.11a: minimal values of spectral linewidth are achieved at the thickness $L = \lambda/2$ and $3\lambda/2$ (at $L = \lambda/2$ the linewidth is sharply minimal, whereas at $L = 3\lambda/2$, the narrowing is observed on a wide pedestal), and maximal values are recorded at $L = \lambda$ and 2λ. The peaks of CRDN in transmission with some modification of spectra compared with low intensity case are still observable in Fig.11b at $L = \lambda/2$ and $3\lambda/2$. On the other hand, the laser intensity in this case is sufficient to form sub-Doppler VSOP resonances at $L = \lambda$ and 2λ, which are well seen in the transmission spectra.

In [11] a theoretical model is presented, which adequately takes into account the peculiarities of real nanocell at low intensity, and in [13,14] this model is generalized to include also dependence on radiation intensity as well.

It is important to note that for the case of $L = \lambda/2$, the atomic vapor density can be risen up to 10^{16} - 10^{17} cm^{-3} (this depends on atomic transition probability), and the absorption will be still below 100% due to small value of L ("optically thin" absorption regime). This is simply determined by the product $N \times L$ (N is atomic vapor number density), which a determining parameter in absorption. This enables to study the broadening and shift of atomic transitions at very high atomic densities from transmission spectra recorded with the help of nanocell with $L = \lambda/2$: the self-broadening and frequency shifts become pronounced at atomic vapor densities > 10^{14} cm^{-3}, when on-resonance D line absorption of atomic vapor in ordinary cm-size cells is in deep "optically thick" regime.

We have studied also nanocell absorption spectra for higher-excited atomic vapor. The frequency of laser diode radiation with ~ 1 MHz linewidth was scanned in the vicinity of $5^2S_{1/2} - 6^2P_{3/2}$ blue transitions (λ = 420.2 nm) of ^{85}Rb and ^{87}Rb isotopes. A strong red shift of all the atomic transitions was detected at the temperature $T > 300$ °C ($N > 10^{16}$ cm^{-3}) [36] (the experiment was carried out jointly with G. Pikhler's group, Institute of Physics, Zagreb).

III.3. Frequency Reference for Atomic Transitions Based on a Nanocell with $L=\lambda$

If hyperfine splitting of excited level does not exceed the Doppler width (such as for D_2 line of Rb and Cs), crossover resonances appear in SA spectra, which can distort recorded positions of real transitions, as well as complicate their identification. The transmission spectra on $F_g=2 \rightarrow F_e=1,2,3$ atomic transitions of ^{85}Rb D_2 line for the nanocell with $L = \lambda$ are presented in Fig.12. The transmission spectra on $F_g=3 \rightarrow F_e=2,3,4$ transitions of ^{85}Rb D_2 line for $L = \lambda$ and $L = 2\lambda$ are shown in Fig.13. The lower curves in both figures show the simultaneously recorded saturated absorption (SA) spectrum. A striking peculiarity of high-intensity $L = \lambda$ absorption spectra is that the amplitude ratios of VSOP peaks are close to ratios of the corresponding atomic transition probabilities. Thus, as is seen from Fig.13, for $F_g=3 \rightarrow F_e=2,3,4$ transitions of ^{85}Rb, the amplitude ratios are $A(3\text{-}4')/A(3\text{-}3') \sim 2$, $A(3\text{-}3')/A(3\text{-}2') \sim 3$ [18], while for the SA spectrum it is not the case due to nonlinear nature of the process (compare with the lower curve, where $A(3\text{-}4') < A(3\text{-}3')$). As the thickness increases, the amplitude of the VSOP peak corresponding to the $F_g=3 \rightarrow F_e=4$ cycling transition decreases, while the amplitudes of the VSOP peaks for the non-cycling transitions $F_g=3 \rightarrow F_e=2,3$ increase.

Hence, $L = \lambda$ nanocell can be employed as a convenient frequency reference for atomic resonance transitions, having the following advantages as opposed to widely used SA technique: i) no crossover resonances, which is important for some applications (see above); ii) the amplitudes of VSOP peaks are proportional to the corresponding transition probabilities, which makes it possible to get an information not only of frequency positions but also of strength of studied transitions; iii) a single-beam transmission spectrum is needed;

iv) the required laser power is less by more than an order of magnitude. Note that in order to achieve the VSOP resonance linewidth close to that shown in Fig.12 and Fig.13 by using a sub-millimeter thin cell, a special complicated technique has to be implemented [37].

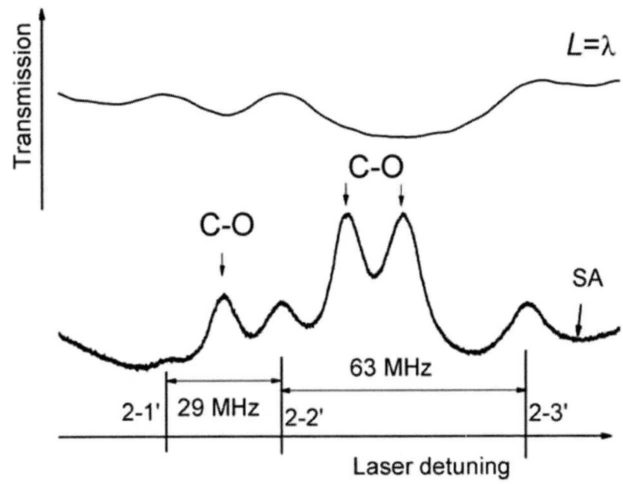

Figure 12. Transmission spectrum of ^{85}Rb D_2 line (transitions from F_g=2) for $L = \lambda$ (upper curve); the lower curve is SA spectrum.

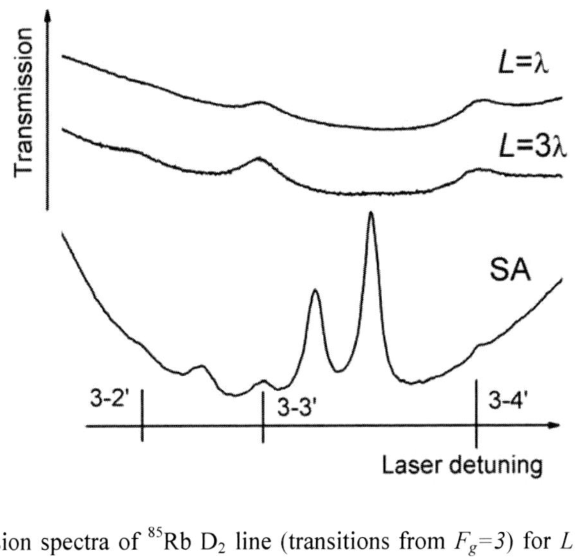

Figure 13. Transmission spectra of ^{85}Rb D_2 line (transitions from F_g=3) for $L = \lambda$ and $L = 3\lambda$ (upper curves); the lower curve is SA spectrum.

III.4. Resonant Fluorescence in Nanocells

In ordinary cm-size cells, for low absorption and low laser intensity regime, the behaviors of the resonant absorption and fluorescence spectra are very similar to each other [35]. In the case of a nanocell, the following five distinctions have been revealed: i) the fluorescence spectrum linewidth is always narrower (~ 1.5 times for $L = \lambda/2$, and ~ 3 times for $L = \lambda$), than the absorption width for the same transition; this general result is related with longer time needed to buildup the fluorescence, which implies absorption and emission cycle, thus imposing more strict atomic velocity restrictions (the atom has to emit photon before being quenched on the cell window); ii) surprisingly, the CRDN effect, which is present in transmission spectra, is not observable in the fluorescence; in contrast to transmission, the increase of L results in monotonic increase of fluorescence linewidth, without oscillating behavior versus L; iii) the fluorescence linewidth remains below the Doppler broadening even when the thickness of vapor column is increased to $L = 3 \div 4$ μm, while the absorption linewidth for these values of L reaches the Doppler broadened width of cm-size cells; iv) the VSOP resonance peaks in nanocell transmission at $L = n\lambda$ are observable for radiation intensities of several mW/cm^2, while the intensity required to observe VSOP resonance dips in the fluorescence spectrum is more than by one order of magnitude larger; v) the VSOP resonance peaks in transmission spectrum for $L = n\lambda$ appear on cycling and open transitions simultaneously (for I_L ~ a few mW/cm^2), meanwhile the VSOP resonance dips in fluorescence first appear on open transitions, and then, with further increase of L (or I_L), also on cycling transition (for $F_g < F_e$) [34].

Although sensitive detection, as well as higher laser intensity is needed to record resonant fluorescence emitted by a nanocell, the first gained advantage is narrower sub-Doppler linewidth, and the second is that the fluorescence signal appears on a nearly zero background.

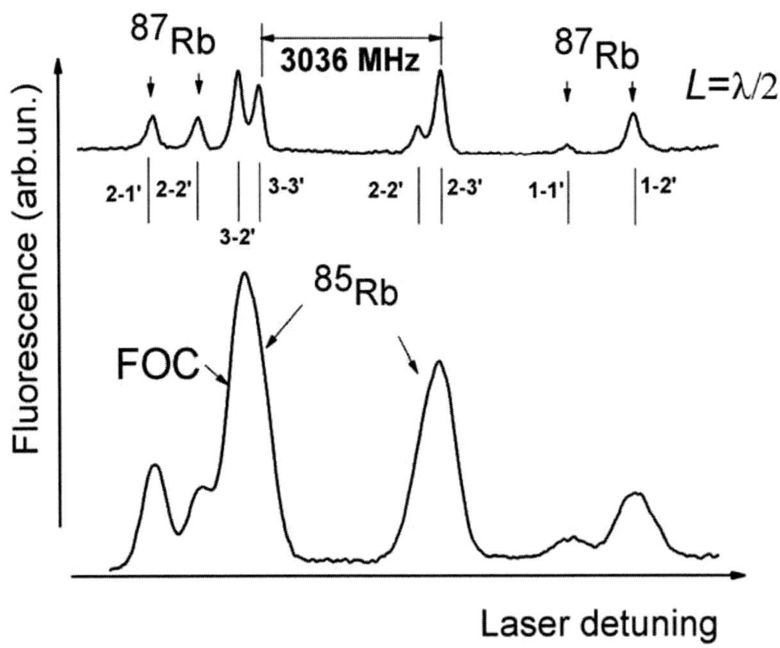

Figure 14. Fluorescence spectrum on Rb D_1 line emitted by a nanocell with $L = \lambda/2$ (upper curve); the lower curve is fluorescence emitted by an ordinary-length cell (FOC).

In particular experiment, the laser beam (ECDL, $\lambda = 794$ nm, $\gamma_L = 1$ MHz) is sent onto the nanocell at nearly normal incidence, laser intensity is ~ 10 mW/cm^2, the operation temperatures are T_{SA} ~ 110 °C, and T_W ~ 140 °C. The fluorescence spectrum of Rb D_1 line for $L = \lambda/2$ is shown in Fig.14 (upper trace). The lower trace is the fluorescence emitted by an ordinary cell (FOC). As it is seen, all the hyperfine transitions (labeled correspondingly) are resolved in the upper trace. Under optimal conditions (i.e. low I_L and sensitive detection), the FWHM of the fluorescence spectrum for an individual transition can be reduced down to 70 - 80 MHz (in the case of FOC it is ~ 500 MHz).

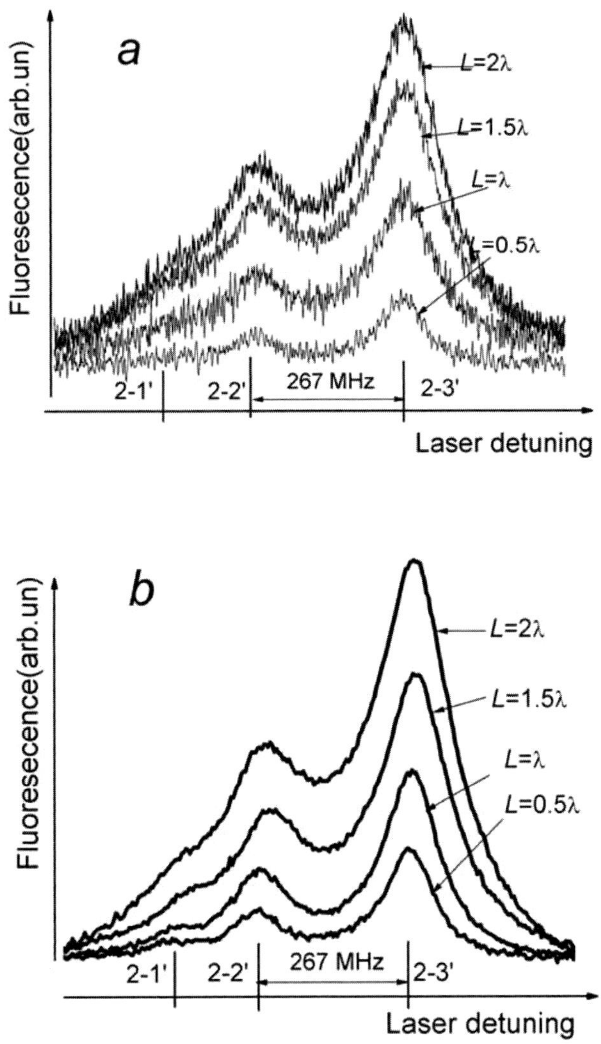

Figure 15. Fluorescence spectra of the nanocell on ^{87}Rb D_2 line, with L varying from $\lambda/2$ to 2λ with the step of $\lambda/2$. The laser intensity is $I_L = 0.4$ mW/cm^2 (a), and 20 mW/cm^2 (b).

The line shape and magnitude of the resonant fluorescence on F_g=2→F_e=1,2,3 transitions of ^{87}Rb D_2 line, obtained for thicknesses $L = \lambda/2 - 2\lambda$ with a $\lambda/2$ step, are shown in Fig.15. A free running laser diode (λ = 780 nm, linewidth 18 MHz) was used. The cases (a) and (b) correspond to laser intensity I_L = 0.4 mW/cm^2 (to detect the fluorescence for such a low intensity, a special oven was designed to place detector closer to the nanocell), and I_L = 20 mW/cm^2, respectively. Obviously, the CRDN effect revealed in transmission spectra (Figs.10, 11) is not observable in the fluorescence. Instead, the linewidth of the fluorescence spectra increases monotonically with L, without oscillations of the linewidth, remaining below the Doppler broadening even for $L = 2\lambda$. Note that the VSOP resonances observable in nanocell transmission spectra at $L = \lambda$, 2λ, 3λ etc. at laser intensity of a few mW/cm^2 (see Figs.12, 13), are absent in fluorescence, even for $I_L >> 1$ mW/cm^2 (Fig.15b). The narrowest linewidth of fluorescence is achieved when L is in the range $\lambda/4 < L < \lambda/2$, particularly for $L = \lambda/2$ presented in Fig.15a, and is 80 MHz FWHM. Optimization of experimental parameters (in particular, narrower laser linewidth) should result in even smaller spectral width.

IV. Magneto-optical Processes in Nanocells

IV.1. Fluorescence of a Nanocell with $L = \lambda/2$ in External Magnetic Field, HLZT Method

It is well known that energy levels of atoms placed in an external magnetic field undergo frequency shifts and changes in their transition probabilities. Below it is demonstrated that using fluorescence spectra from a nanocell with thickness $L = \lambda/2$, it is possible to efficiently study the above mentioned changes (so called "half-λ Zeeman technique" (HLZT)). As we wrote above, in spite of some drawbacks (signal weakness, relatively high required excitation intensity), the fluorescence emitted by a nanocell with $L = \lambda/2$ has an advantages of narrow sub-Doppler linewidth (80 ÷ 100 MHz) and zero detection off-resonance background.

The experimental setup is similar to that presented in Fig.7. The radiation beam (∅ 1 mm) of λ = 794 nm, 6 MHz- linewidth single-frequency *cw* laser diode was directed at normal incidence onto the nanocell with $L = \lambda/2$. The needed temperature regime of the nanocell ($T_{SA} \sim$ 110 – 120 ºC, $T_W \sim$ 140 – 150 ºC, corresponding $N \sim 6\times10^{12} - 1.5\times10^{13}$ cm^{-3}) was provided by a special oven with 4 openings: 2 inlets for laser beam passage, and 2 orthogonal inlets for side fluorescence detection. This geometry allows simultaneous recording of fluorescence and transmission spectra. The fluorescence signal was detected by a photodiode with an aperture of 1 cm^2 which was placed at 90° angle to the laser propagation direction. The photodiode signal was recorded by Tektronix TDS 3032B digital storage oscilloscope. The nanocell was placed in 3 pairs of mutually perpendicular Helmholtz coils providing possibility to apply homogeneous magnetic field in arbitrary direction. A Glan prism was used to purify initial linear radiation polarization of the laser radiation; to produce a circular polarization, a $\lambda/4$ plate was utilized. Geometrical configuration of the experiment is shown in Fig.16. By the help of a beam splitter, 50% of the pump beam was directed on the auxiliary nanocell with Rb, which was used as a frequency and amplitude reference for $B = 0$ (light curves in the Figs.17, 18 below).

The fluorescence spectrum of ^{87}Rb D_1 line for the case of longitudinal magnetic field B = 22.5 mT, with σ^+ laser radiation excitation of the $F_g=1 \rightarrow F_e=1,2$ hyperfine transitions is presented in Fig.17. It is easy to identify all the five split peaks. The labels denote corresponding transitions between magnetic sublevels of $F_g=1 \rightarrow F_e=1,2$ shown in the inset of Fig.17: $F_g=1$, $m_F=-1,0 \rightarrow F_e=1$, $m_F=0,+1$ (labeled *5* and *4*, respectively), and $F_g=1$, $m_F=-1,0,+1 \rightarrow F_e=2$, $m_F=0,+1,+2$ (labeled *3*, *2* and *1*, respectively). Note, that the fluorescence amplitude of the individual transition is proportional to its probability. Particularly, for small B values the theoretical ratio of transition probabilities corresponding to resonances numbered *1*, *2*, *3* is 6:3:1, which is close to what is seen in Fig.17.

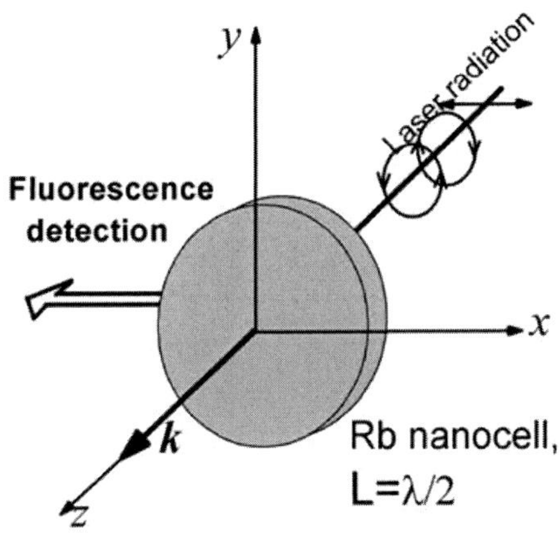

Figure 16. Mutual orientations of laser excitation, B-field, and detection of fluorescence emitted by a nanocell with $L = \lambda/2$.

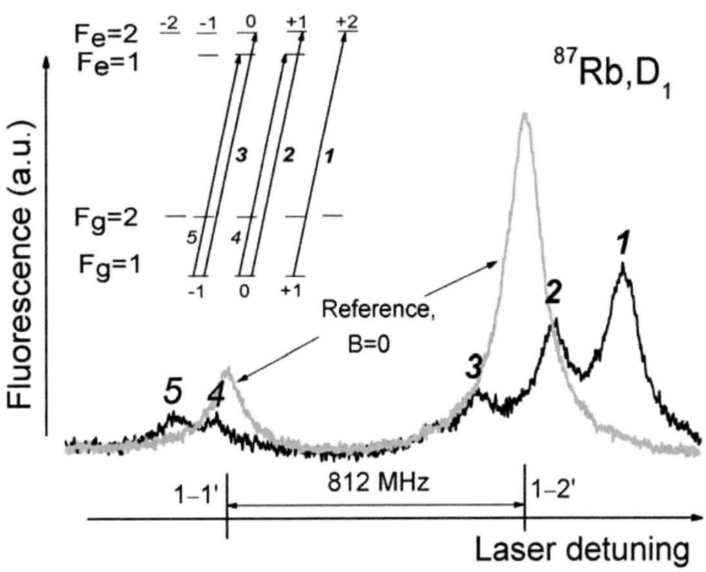

Figure 17. Spectrum of fluorescence emitted by a nanocell with $L = \lambda/2$ in external magnetic field (black line), the five peaks are well resolved (see the text); the gray line is the spectrum of a reference nanocell for $B = 0$.

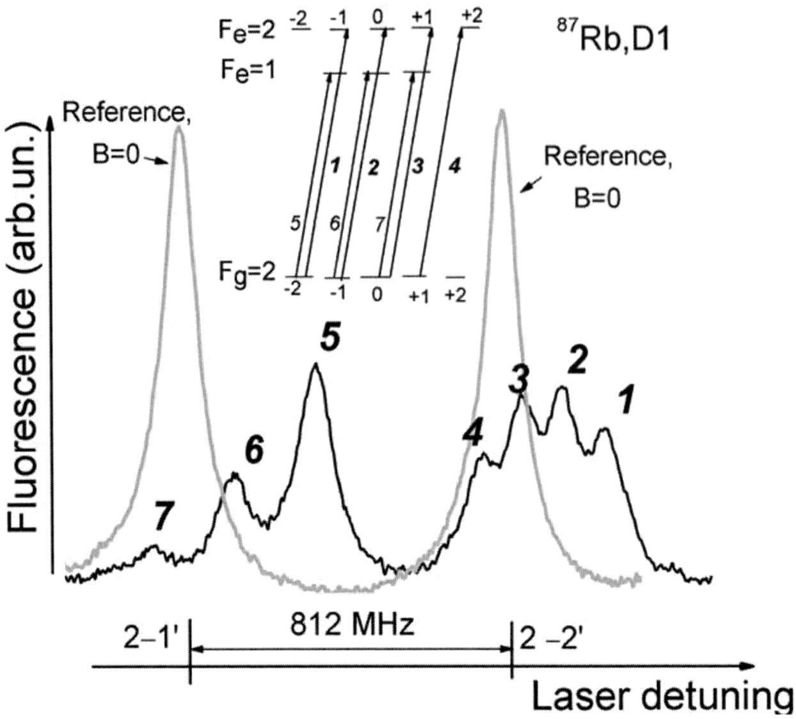

Figure 18. Fluorescence spectrum emitted by a nanocell with $L = \lambda/2$ (black line) in external magnetic field, the seven peaks are well identified (see the text); the gray line is the reference spectrum of an auxiliary nanocell for $B = 0$.

The fluorescence spectra on ^{87}Rb $F_g=2 \rightarrow F_e=1,2$ transitions for the case of longitudinal magnetic field ($B \parallel k$), B = 22.5 mT with σ^- excitation is shown in Fig.18. It is easy to identify all the seven peaks, which correspond to Zeeman transitions presented in the inset of the figure. For small values of B, the theoretical ratio of transition probabilities corresponding to the resonances numbered *5, 6, 7* is 6:3:1, and for resonances numbered *4, 3, 2, 1* is 2:3:3:2, in conformity with the experiment. Thus, the HLZT allows one to study not only the frequency shift of any individual Zeeman transition, but also the modification in its transition probability in the region of 1 mT - 25 mT.

IV.2. The Nanocell Transmission for $L = \lambda$ in External Magnetic Field, LZT Method

It was revealed that the use of VSOP resonances forming in transmission spectra of a nanocell with thickness $L = \lambda$ (we call this technique "L=λ Zeeman technique" (LZT)), could have several advantages as compared with the HLZT. The particular advantages of the new LZT method is that the linewidth of VSOP resonances used in LZT is at least ~ 5 times narrower, and this means that the resolution of this technique is much higher. Also, laser power required for LZT is ~ 100 times lower than that needed for HLZT. Finally, recording of resonant transmission spectra does not require sensitive detectors with a large aperture as in the case of fluorescence spectra.

Implementation of LZT allowed us to study experimentally unknown peculiarities of modification of atomic transition probabilities and frequency shift of individual atomic Zeeman transitions of D lines of ^{87}Rb, ^{85}Rb atoms in an external magnetic field B. As it was mentioned, the VSOP peaks start to appear at laser intensity of ~ 1 mW/cm^2 in the transmission spectrum of a nanocell with atomic vapor column of $L = \lambda$ thickness, with λ being the resonant wavelength of laser radiation (794 nm or 780 nm for Rb D_1 or D_2 line). At $B > 0$, the VSOP resonance is split to several Zeeman components, the number of which depends on quantum numbers F of the lower and upper levels. The LZT allows one to study not only frequency shift of any individual hf transition, but also the modification in transition probability in the tested region of 0.1 mT - 0.25 T, since both the amplitudes of VSOP peaks and their frequency positions depend unambiguously on B value (LZT is expected to be valid up to several T).

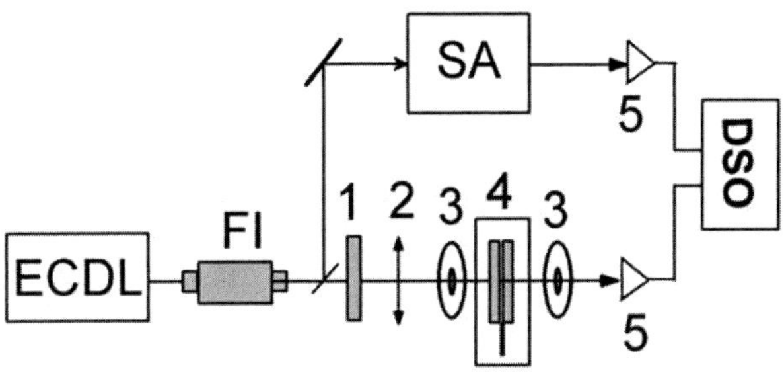

Figure 19. Experimental setup for LZT. FI- Faraday isolator, *1*- λ/4 plate, *2*- lens (F = 35 cm), *3*- ring magnets, *4*- nanocell in the oven, *5*- photodetectors, DSO-digital storage oscilloscope.

Experimental realization of LZT is simple enough (see Fig.19). The circularly polarized beam of extended cavity diode laser (ECDL, λ = 794 nm, $P_L \sim 5$ mW, $\gamma_L < 1$ MHz) resonant with ^{87}Rb D_1 transition frequency, after passing through Faraday isolator was focused (∅ 0.5 mm) onto Rb nanocell with the thickness of vapor column $L = \lambda$, at an angle close to normal. A part of the laser radiation was branched to a cm-size Rb cell to form a $B = 0$ SA spectrum serving as frequency reference. The small longitudinal magnetic field ($B < 25$ mT) was applied to the nanocell by a system of Helmholtz coils (not shown in Fig.19). The B-field strength was measured by a calibrated Hall gauge.

Another advantage of LZT is possibility to apply much stronger magnetic field using widely available strong permanent ring magnets (PRM). In spite of strong inhomogeneity of a B-field (in our case it can reach 15 mT/mm), due to small thickness of the nanocell (794 nm) the variation of B inside atomic vapor column is a few μT, i.e. by several orders less than the applied B value. The allowed transitions between magnetic sublevels of hf states for ^{87}Rb D_1 line in the case of σ^+ (left circular) polarized excitation are depicted in Fig.20 (LZT works well for σ^- excitation, too).

It is important to note that the absence of CO resonances in transmission spectra is an important advantage of a nanocell method (see below). The strong magnetic field was produced by two ∅ 30 mm PRMs with ∅ 3 mm hole for radiation passage, placed at variable distance 30 ÷ 50 mm from each other, with the nanocell oven between (see Fig.19). To control the magnetic field strength, one of the magnets was mounted on a micrometric translation stage for axial (longitudinal) displacement.

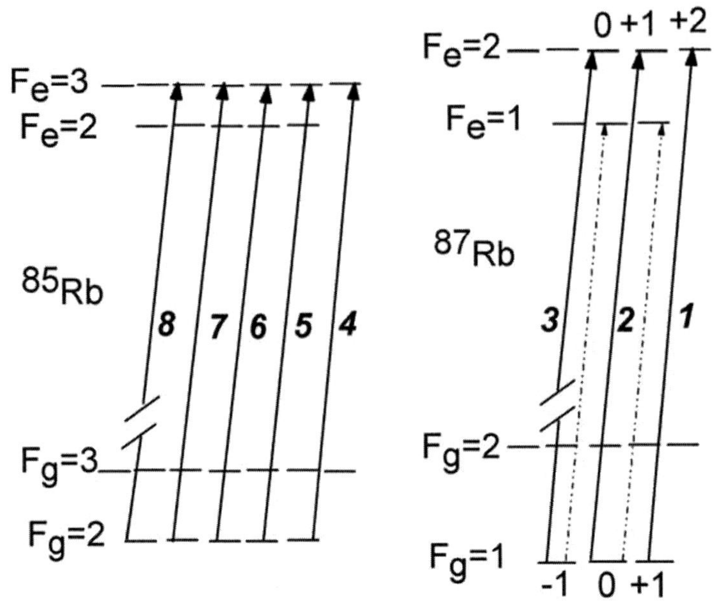

Figure 20. Energy level hfs diagram of D_1 line of ^{85}Rb and ^{87}Rb in magnetic field and possible atomic Zeeman transitions for σ^+-polarized exciting laser radiation.

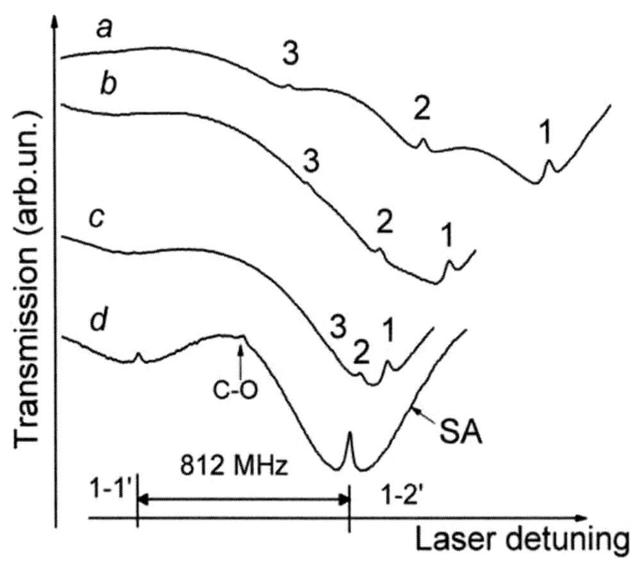

Figure 21. The transmission spectra on $F_g=1\rightarrow F_e=1,2$ transitions of ^{87}Rb D_1 line for $B = 59$ mT (a), 31 mT (b), and 11.5 mT (c); the lower trace is SA spectrum.

Fig.21 shows the transmission spectra of the nanocell for small $B = 59$ mT (a), 31 mT (b), and 11.5 mT (c). Curve (d) shows saturation absorption spectrum for $B = 0$, and is used as a frequency reference. As it is seen, all the individual Zeeman transitions are clearly resolved. The two transitions F_g=1→F_e=1 (shown in Fig.20 by dotted lines) are detectable for B < 12 mT, while at higher B their probabilities strictly reduce. The nanocell transmission spectra for larger B values are presented in Fig.22: B = 0.24 T (a), 0.154 T (b), 0.117 (c). The lower curve is again SA spectrum. Note that at such a high B values, the Zeeman components of $F_g=1\rightarrow F_e=2$ transition of ^{87}Rb D_1 line get mixed in spectra with the Zeeman components of $F_g=2\rightarrow F_e=3$ transition of ^{85}Rb D_1 line, in spite of large initial B = 0 frequency separation (> 2 GHz).

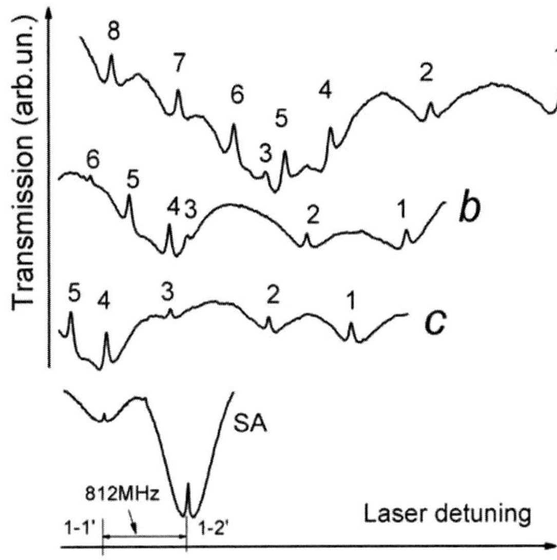

Figure 22. The transmission spectra on ^{87}Rb and ^{85}Rb D_1 line for B = 0.24 T (a), 0.154 T (b), and 0.117 T (c); the lower trace is SA spectrum.

We also implemented LZT to study atomic transitions $F_g=1\rightarrow F_e=0,1,2,3$ of ^{87}Rb D_2 line (σ^+ excitation was used, but LZT works perfectly well also for σ^- excitation). In this experiment, the beam from ECLD (λ = 780 nm, spectral width < 1 MHz) of ~ 0.7 mm diameter was directed onto the nanocell with the thickness $L = \lambda$.

The relevant energy levels of ^{87}Rb D_2 line in magnetic field and possible atomic transitions for an σ^+ polarized exciting laser radiation are depicted in the inset of Fig.23. Particularly, it is revealed that in relatively weak magnetic field (~ 10 mT), also those atomic transitions are recorded, for which new selection rules with respect to the quantum number F take place: the probabilities of ^{87}Rb D_2 $F_g=1 \rightarrow F_e=3$ transition components (VSOP peaks numbered *7, 8, 9*) increase with the increase of B, and for ~ 21.7 mT the probability of the transition numbered *7* equalize with the probability of the strongest (for B = 0) transition component of $F_g=1 \rightarrow F_e=2$ numbered *6* (see Fig.23). At higher magnetic field up to 0.2 T, the probability of transition *7* is the largest and only for B > 0.2 T transition probability of *6* starts to exceed again that of *7*. This is well seen in Fig.24, where the transmission spectrum for B = 0.21 T in the case of σ^+ excitation is presented (numbers *6, 7, 8* denote the corresponding transitions shown in the inset of Fig.23). The two pointing arrows show the positions of Zeeman components of ^{85}Rb $F_g=2\rightarrow F_e=4$ transition.

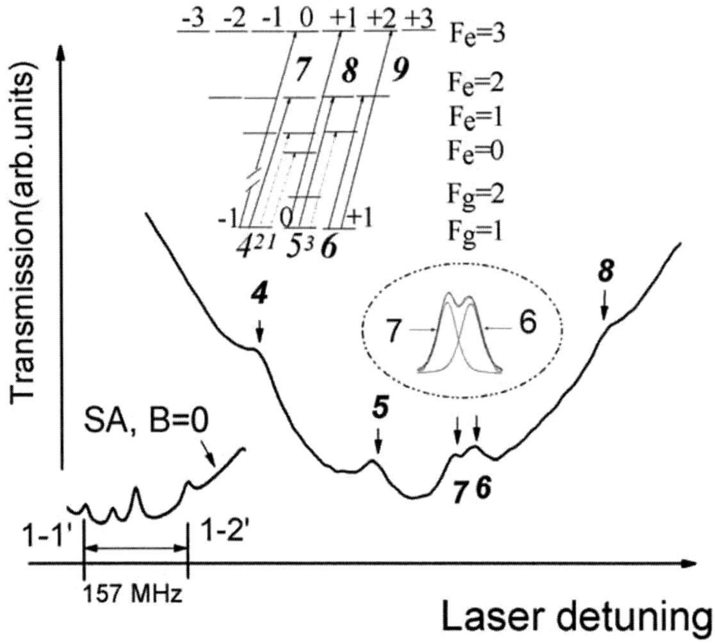

Figure 23. The nanocell transmission spectrum for the transitions $F_g=1\rightarrow F_e=0,1,2,3$ of ^{87}Rb D_2 line. $B = 21.7$ mT, the amplitude of the VSOP peak *7* ($\Delta F=2$) is equal to that of the peak *6*, and the amplitudes of the VSOP peak *8* (also $\Delta F=2$) is half the amplitude of the peaks *6* and *7*.

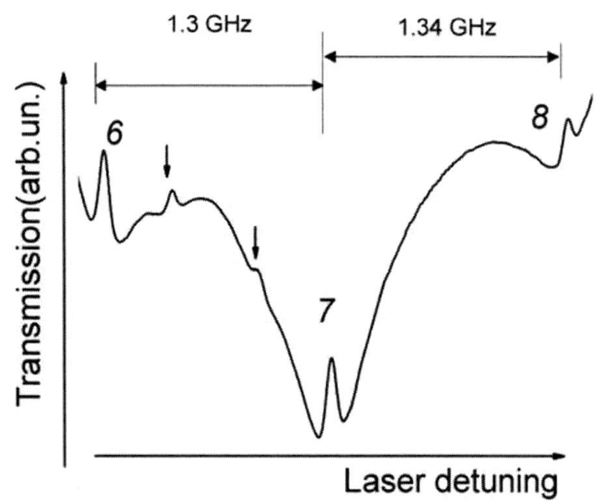

Figure 24. The nanocell transmission spectrum for $B = 0.21$ T; the amplitude of the VSOP peak *7* ($\Delta F = 2$) is somewhat smaller than that of the peak *6*; the two arrows show the positions of ^{85}Rb D_2 line $F_g=2\rightarrow F_e=4$ Zeeman transitions.

It is important to note that transition *7* is strongly shifted (by 5.6 GHz) from $F_g=1 \rightarrow F_e=2$. Thus, a tunable frequency reference based on the nanocell and ring magnets can be developed, since changing the magnetic field (by simple displacement of a ring magnet), it is possible to shift the reference frequency position by several GHz.

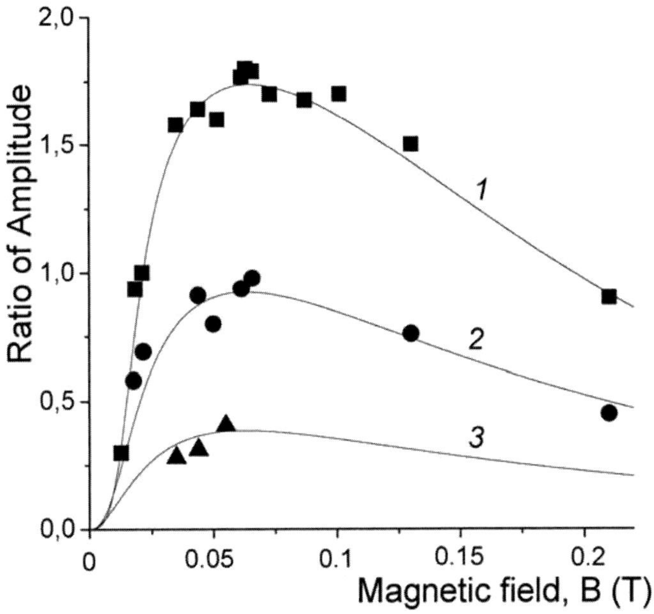

Figure 25. Dependence of the VSOP amplitude ratios (i.e., ratios of transition probabilities) on *B* for Zeeman components of ^{87}Rb D_2 line atomic transitions with σ^+ excitation: (1)- ratio A(*7*)/A(*6*), (2)- ratio A(*8*)/A(*6*), (3)- ratio A(*9*)/A(*6*).

Fig.25 gives the *B*-field dependence of VSOP amplitude (transition probability) ratio A(*7*)/A(*6*), A(*8*)/A(*6*), and A(*9*)/A(*6*) (curves *1*, *2*, *3*). The solid lines represent theoretical calculations [21]. Obviously, there is a good agreement between the experiment and theory.

V. Electromagnetically Induced Transparency Effect in Micro- and Nanometric Thin Cells

V.1. Peculiarities of EIT in Thin Vapor Cells

There has been considerable interest in recent years about the fascinating properties of coherent population trapping (CPT) and the related electromagnetically induced transparency (EIT). This scientific concern is caused also by a number of important applications [24-26]. EIT resonances can occur in Λ-system with two long-lived states and one excited state coupled by two laser fields. In order to achieve narrow EIT resonance, the two lasers have to be coherently coupled, and for this purpose modulation techniques are implemented [26]. However, there are some situations, for example as it is in the case of EIT in the "ladder-system", where the frequency separation between the two lasers is very large [38,39], or as it is for the EIT experiment in indium vapor [40], and in these cases the only way is to use two entirely different lasers.

The linewidth of EIT resonance for small laser intensity is $\gamma_{EIT} \approx 2\Gamma_{21} + \Omega^2/\gamma_N$ [23,25], where Γ_{21} is the relaxation rate between the ground hyperfine levels, with Ω being a Rabi frequency, γ_N -natural linewidth . For the case of ETC, the value of Γ_{21} is highly affected by atom-wall (cell window) collisions. It is known that a unique collision with the dielectric surface of an uncoated vapor cell is sufficient to thermalize the ground hyperfine levels, with depolarization probability 0.5 – 1. Note, that in order to preserve the atomic coherence in wall collisions and to detect ultranarrow linewidths with the help of coherent processes like EIT, nonlinear magneto-optical rotation (NMOR) and others, paraffin-coated walls are used [38]. As the size of the vapor cell is reduced, the lifetime of the ground state coherence becomes shorter because of collisions of the atoms with the cell windows: $\Gamma_{21} = (2\pi t)^{-1}$, where $t = L/u$ (L is distance between the windows, and u is the most probable thermal velocity). Also, the EIT resonance contrast (defined as the ratio of EIT depth to the height of shoulders of the EIT window) strongly depends on Γ_{21}. Hence, in the case of $L \sim 1$ μm ($2\Gamma_{21} > 100$ MHz), one could expect that the EIT effect will be completely vanished. In order to prevent atom–window collisions for the case of sub-millimeter thin cells, buffer gas has been successfully used in [27].

Nevertheless, in [26] it was predicted that even in the case of pure atomic vapor, narrow EIT resonance can be observed for the cell with $L \sim 10$ μm. The explanation is that the contribution of atoms flying nearly parallel to the cell windows is enhanced thanks to their longer interaction time with the laser field. Thus, these atoms yield stronger contribution to the EIT resonance. Nonlinear Hanle resonance in atomic vapor when the thickness $L = \lambda$, where λ is 852 nm was reported in [42], and for the case of micrometric thin vapor column was presented in [43]. Note that in [44] the EIT effect has been studied for selective reflection (SR) process on Rb D_1 line in a several cm-long cell, and though for this case the vapor column is limited from one side by a cell window, the second boundary is diffused, thus the thickness is not clearly determined. As it was demonstrated in [7,9], there is a dramatic difference of the SR spectra obtained with an ordinary cell and with a thin vapor layer.

V.2. Experimental Results and Discussion on EIT in Micro- and Nanometric Cells

The experimental arrangement is sketched in Fig.26. Both beams (∅ 3 mm) of two separate single-frequency diode lasers, "coupling" and "probe", with $\lambda \approx 780$ nm (the linewidth is ~ 5 MHz) are well superposed and directed onto the thin cell at near-normal incidence with the help of the first Glan prism (the coupling and probe beams have orthogonal linear polarizations). The second Glan prism allows to separately detect transmission spectra of probe and coupling lasers. Thin cell *4*, of either micrometric or nanometric thickness, is placed inside the three pairs of mutually perpendicular Helmholtz coils *2* providing a possibility to cancel laboratory magnetic field, as well as to apply the needed homogeneous magnetic field.

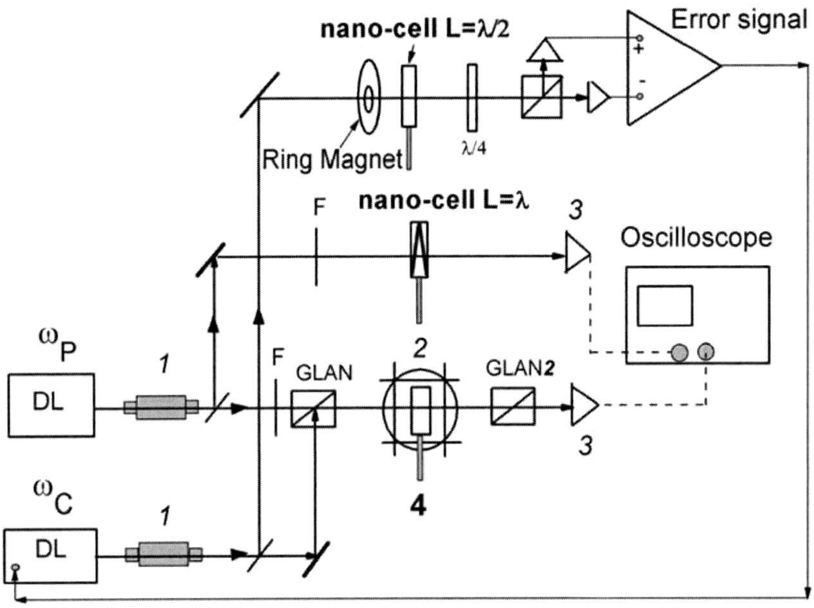

Figure 26. Sketch of the experimental setup. *1*- Faraday isolators, *2*- Helmholtz coils, *3*- photodiodes, *4*- micrometric or nanometric thin cell, *F*- beam attenuators.

The powers of coupling and probe lasers were in the range of 0.5 - 1 mW and 0.01 - 0.1 mW, respectively. A modified Dichroic-Atomic-Vapor-Laser-Locking (DAVLL) method, realized in a separate nanocell with the thickness $L = \lambda/2$, is used for stabilization of the coupling laser frequency; this technique is similar to that published in [45]. In [45], the selective reflection spectra of a circularly-polarized laser beam from a nanometric Rb vapor cell with L varied around $\lambda/2$ ($\lambda = 780$ nm, D_2 line) have been studied. The spectra revealed striking change of signal shape occurring in the vicinity of $L = \lambda/2$. Application of a longitudinal B-field strongly modifies the spectral line shapes and causes a sign-dependent overall frequency shift. Based on the results of these studies, a new method for tunable locking of diode laser frequency, which does not require frequency modulation nor complex electronics, has been developed.

The frequency reference spectra have been formed with the help of an auxiliary nanocell with thickness $L = \lambda$. In some cases, SA reference has been implemented. The optical EIT and reference radiations were simultaneously detected by photodiodes *3*, the signals of the photodiodes were amplified and recorded by a two-channel digital storage oscilloscope Tektronix TDS 3032B.

In Fig.27 it is shown the fluorescence spectrum (upper curve) and transmission spectrum (middle curve) of the probe laser in $L \sim 10$ μm cell when the coupling laser (~ 0.6 mW) is resonant with $F_g=3 \rightarrow F_e=3$ transition, while the probe laser (~ 0.4 mW) frequency is scanned across the $F_g=2 \rightarrow F_e=1,2,3$ transitions (see Fig.28). The lower curve is the reference saturated absorption spectrum. In the upper and middle traces, the EIT resonance is seen together with six sub-Doppler VSOP resonances marked by the numbered arrows. The EIT resonance has a linewidth of ~ 6 MHz.

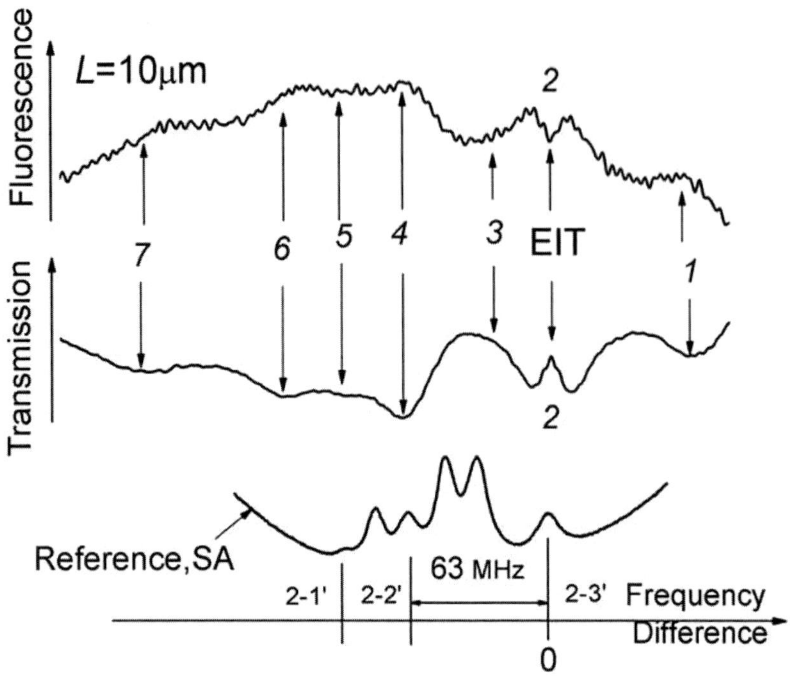

Figure 27. Fluorescence (upper trace) and transmission (middle trace) spectra of L ~ 10 μm cell. The coupling laser is resonant with F_g=3→F_e=3 transition, the probe is scanned across F_g=2→F_e=1,2,3. The EIT resonance and six VSOP resonances are seen both in the fluorescence and transmission spectra. The lower curve is reference SA spectrum.

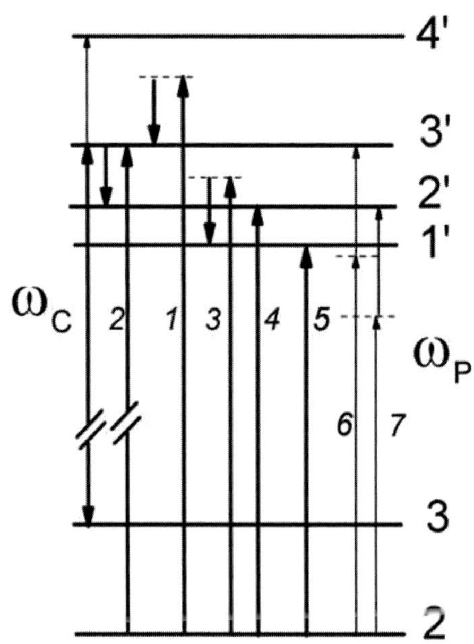

Figure 28. Diagram of relevant levels of ^{85}Rb D_2 line, which participate in EIT and VSOP resonance formation. The coupling laser is resonant with F_g=3→F_e=3 transition.

The linewidth of VSOP resonances is 3 times larger. The VSOP resonances labeled *2, 4, 5* are formed by atoms flying parallel to the thin cell windows; the resonances labeled *1, 3* are formed by the atoms flying with $u \approx 50$ m/s in the direction of laser radiation propagation k; the resonances labeled *6, 7* (they are present only in the coupling laser spectrum [46]) are formed by the atoms flying with $u \approx 95$ m/s in the direction opposite to the laser radiation propagation k (see the diagram in Fig.28).

As it mentioned above, the polarizations of coupling and probe lasers are perpendicular to each other, thus with the help of the second Glan prism (see Fig.26) it is possible to separate and record either probe or coupling laser transmission spectrum. Below we demonstrate the dramatic difference in spectra when the coupling and probe lasers start either from the same hf level (the case of V-type system), or from two different levels (the case of Λ-type system).

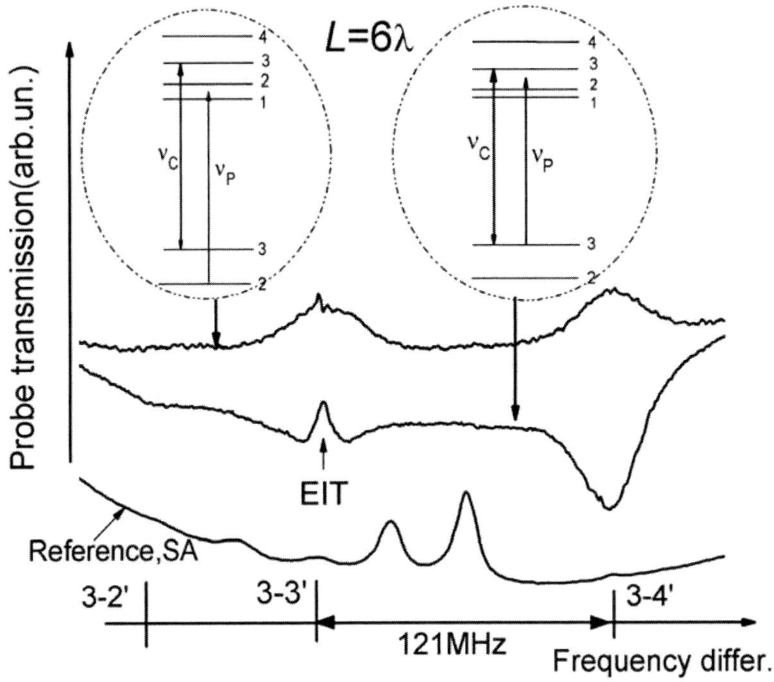

Figure 29. The probe transmission spectra when the coupling laser and probe laser start from the same ground level (upper curve, the left inset diagram), and from different levels (middle curve, the right inset diagram). The lower curve is reference SA spectrum.

Fig.29 (upper curve) shows the transmission spectrum of the probe laser when the coupling laser (~ 0.2 mW) is resonant with $F_g=3 \rightarrow F_e=3$ transition, while the frequency of probe laser (~ 0.02 mW) is scanned across $F_g=3 \rightarrow F_e=2,3,4$ transitions (see the diagram presented in the left inset of Fig.29). Here all the VSOP peaks demonstrate reduction of absorption, since the coupling laser transfers a number of atoms from $F_g=3$ level to $F_g=2$, which causes reduction of the probe absorption for $F_g=3 \rightarrow F_e=2,3,4$ transitions (a small signature of beating of the coupling and probe frequencies is seen on the VSOP located at $F_g=3 \rightarrow F_e=3$ transition). The middle curve in Fig.29 shows the transmission spectrum of the probe laser when the coupling laser is resonant with $F_g=2 \rightarrow F_e=3$ transition (see the diagram presented in the right inset).

Here we see an EIT resonance, which has much narrower linewidth (5 times) than that of the corresponding VSOP peak in the upper curve ($F_g=3 \rightarrow F_e=3$ transition). This is explained by coherent nature of the EIT process (it is well known that the linewidth of the EIT resonance can be much narrower than the natural one [23,25]), while the VSOP peak originates by a non-coherent process, and its linewidth can not be narrower than the natural one. As it is expected, the VSOP peak at $F_g=3 \rightarrow F_e=4$ transition (middle curve) exhibits increase of absorption, since the coupling laser transfers a fraction of atomic population from $F_g=2$ level to $F_g=3$, which results in increase of the probe absorption for $F_g=3 \rightarrow F_e=4$ as well as $F_g=3 \rightarrow F_e=2,3$ transitions.

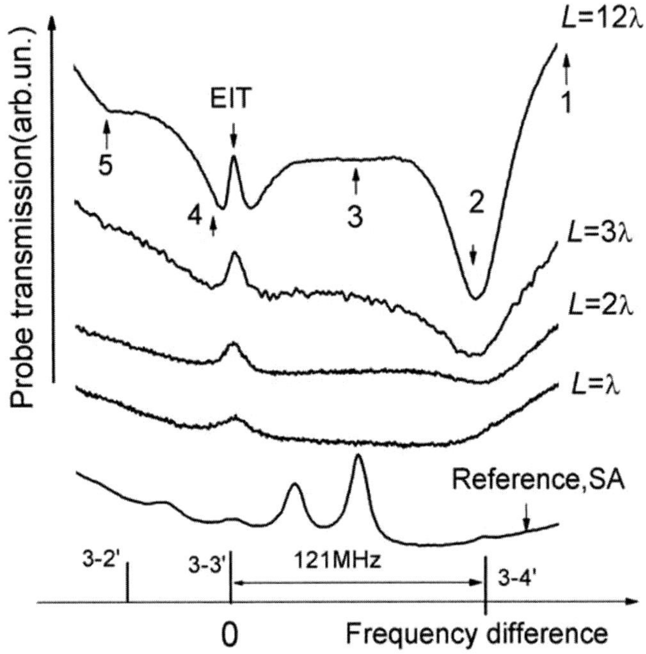

Figure 30. The probe transmission spectrum for L = 12λ; 3λ; 2λ and λ. The coupling laser is resonant with $F_g=2 \rightarrow F_e=3$ transition, the probe laser is scanned across $F_g=3 \rightarrow F_e=2,3,4$ transitions. The lower curve is SA spectrum.

One of the most striking results on EIT in thin cells is that when the coupling laser is resonant with the atomic transition, the reduction of thickness L does not cause a rapid increase of EIT resonance linewidth (γ_{EIT}) as might be expected intuitively. Fig.30 presents probe transmission spectra at different L values, when the coupling laser (~ 0.1 mW) is resonant with $F_g=2 \rightarrow F_e=3$ transition, while the probe laser (~ 0.01 mW) is scanned across the $F_g=3 \rightarrow F_e=2,3,4$ transitions (see the diagram in Fig.31). As we see, when the thickness is consecutively reduced (L = 12λ, 3λ, 2λ, λ), the FWHM of γ_{EIT} increases slowly (5.5 MHz, 7.5 MHz, 8.5 MHz, and 10 MHz, correspondingly). Taking into account our previous experimental results on EIT in mm-thick cells, the γ_{EIT} dependence as a function of thickness L can be presented as $\gamma_{\text{EIT}} \sim L^{-0.25}$ (this is valid only for the case when the coupling laser frequency is resonant with an atomic transition). The explanation of this weak dependence is that for the case when the coupling laser frequency is resonant with an atomic transition, the contribution of atoms flying nearly parallel to the cell windows is enhanced thanks to their longer interaction time with laser field, and they yield a dominant contribution to the EIT resonance formation.

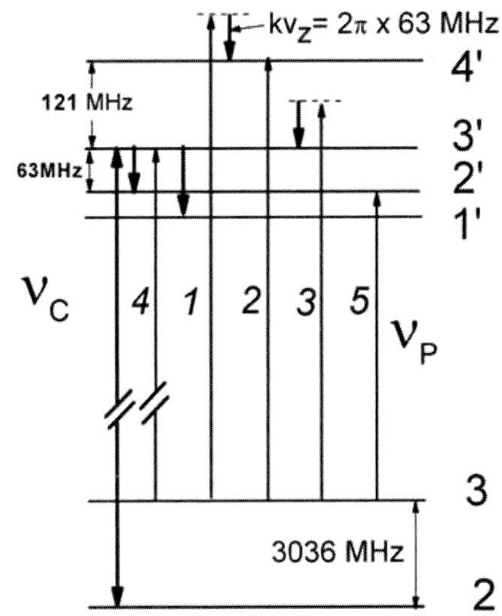

Figure 31. The diagram of levels and transitions of ^{85}Rb D_2 line, related to formation of the EIT and VSOP resonances (see Figure 30); the coupling laser is resonant with F_g=2→F_e=3 transition.

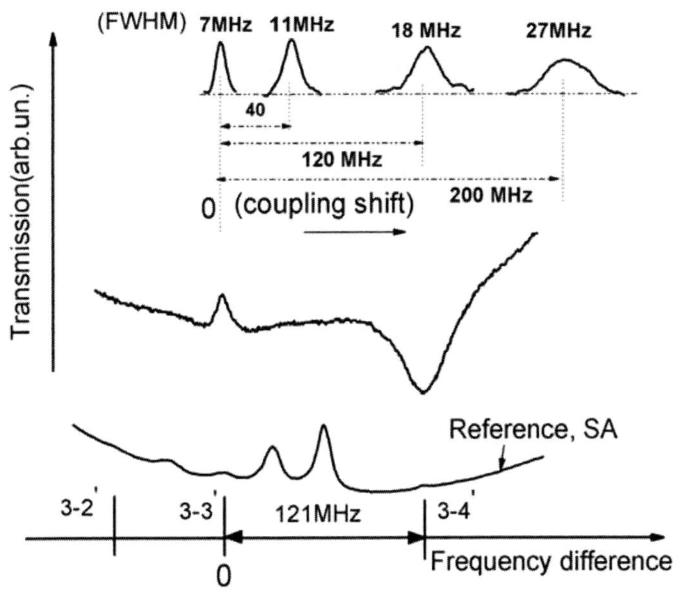

Figure 32. Fragments of the transmission spectrum of the probe laser for L = 6λ, when the coupling laser frequency is shifted by Δ with the respect of F_g=2→F_e=3 transition: Δ = 0, 40, 120, and 200 MHz (upper curve). The middle curve shows the whole spectrum when Δ = 0, including the VSOP resonance at the F_g=3→F_e=4 transition. The lower curve is SA spectrum.

Also, the EIT contrast for this case has a weak dependence on L. Stronger dependence is observed for the VSOP resonance contrast. The resonances labeled *1, 2, 3, 4, 5* (see Fig.31) are well seen in the upper curve of Fig.30, when $L = 12\lambda$ (the VSOP resonance labeled *1* is out of the presented curve, but it exists in the spectrum). With reduction of thickness L, the VSOP amplitudes rapidly reduce, caused by reduction of the interaction time below the value needed for buildup of optical pumping.

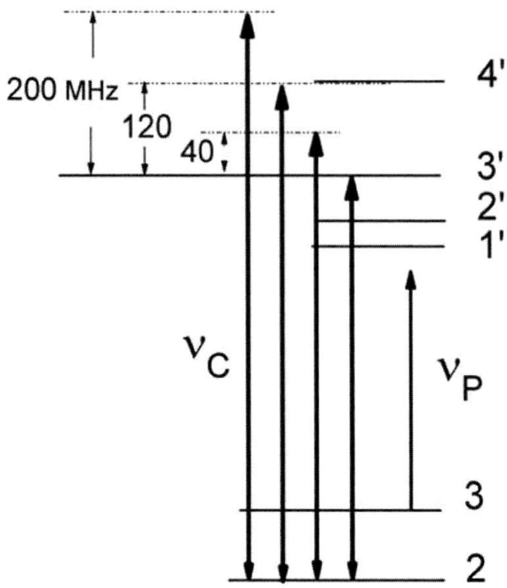

Figure 33. The diagram of relevant levels of ^{85}Rb D_2 line involved in EIT effect. The coupling laser is detuned by $\Delta = 0$, 40 MHz, 120 MHz, and 200 MHz with the respect to $F_g=2 \rightarrow F_e=3$ transition.

The situation with the EIT linewidth is dramatically different when the coupling laser frequency is strongly detuned with respect to the corresponding atomic transition resonance. The upper picture in Fig.32 shows fragments of the probe transmission spectra for the cell with $L = 6\lambda = 4.68$ μm (only the part that contains the EIT resonance), when the coupling laser frequency is shifted by Δ with the respect of $F_g=2 \rightarrow F_e=3$ transition (see the diagram in Fig.33), while the probe laser is scanned across $F_g=3 \rightarrow F_e=2,3,4$ transitions. In the case of zero detuning of the coupling laser, the FWHM of γ_{EIT} is 7 MHz, while for $\Delta = 40$ MHz, 120 MHz, and 200 MHz, the FWHM of γ_{EIT} is 11 MHz, 18 MHz, and 27 MHz, correspondingly. The physical explanation of this fact is that when the coupling laser is detuned from the corresponding transition by Δ, only those atoms having the velocity projection $u_z = 2\pi\Delta/k$ participate in formation of the EIT, and for these atoms the time of flight between the cell windows $\tau_L = L/u_z$ becomes smaller as Δ increases (for the atoms flying nearly parallel to nanocell windows $\tau_D = D/u$, and $D >> L$). Large longitudinal velocity projection causes frequent collisions of atoms with the nanocell windows, resulting in increase of Γ_{21} and, consequently, in broadening of the EIT resonance linewidth and reduction of its contrast. Broadening rate (the ratio of broadening in MHz to the frequency shift in MHz) for the thickness $L = 6\lambda$ and blue detuning of the coupling laser, is $K_B(6\lambda) \approx 0.1 \pm 0.01$ (for the red detuning of coupling laser, $K_R \approx 1.4 \times K_B$).

We have mentioned above that the linewidth of EIT resonance in the case of low laser intensity is $\gamma_{EIT} \approx 2\Gamma_{21} + \Omega^2/\gamma_N$. For large frequency detuning, the condition $\Delta >> \Omega^2/\gamma_N$ is fulfilled and the linewidth is determined mainly by collisions, i.e. by the value $\Gamma_{21} = 1/(2\pi\tau) = u_z/(2\pi L)$. Since $u_z = 2\pi\Delta/k$, one expects a linear dependence of γ_{EIT} as a function of Δ. Note that different atoms fly different longitudinal distances d inside the nanocell before colliding with its window. These distances are ranged from $d \approx 0$ (i.e., interaction starts close to the window) to $d \approx L$ (full-length interaction), and it is reasonable to use an average distance $d \approx L/2$ instead of L. Finally, we get the following expression to estimate Γ_{21} in the case of a large detuning: $\Gamma_{21} \approx u_z/(\pi L)$, where $ku_z = 2\pi\Delta$.

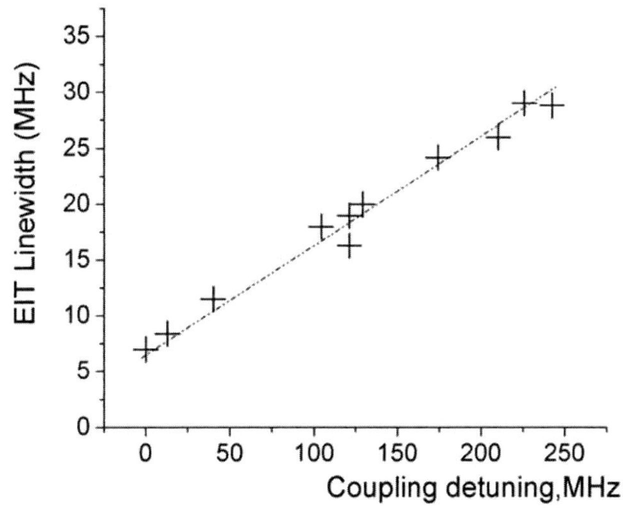

Figure 34. The EIT linewidth (MHz) for an $L = 6\lambda$ cell as a function of the coupling laser detuning from the $F_g{=}2 \rightarrow F_e{=}3$ transition of ^{85}Rb D_2 line. Crosses show the values measured in the experiment.

Dependence of the EIT resonance width γ_{EIT} on coupling laser detuning Δ is presented in Fig.34 for the cell with $L = 6\lambda$. In particular case of Fig.32, for $\Delta = 200$ MHz, $u_z = 156$ m/s, and $2\Gamma_{21} \approx 21$ MHz. Taking into account the γ_{EIT} value 7 MHz for $\Delta = 0$, we get $\gamma_{EIT} \approx 7$ MHz + 21 MHz = 28 MHz. This value is in good agreement with the experimental one (27 MHz). It is important to note that it was experimentally proved that for a fixed value of Δ, K_B is proportional to $1/L$, particularly $K_B(6\lambda) = K_B(\lambda)/6$. Hence, for any thickness in the range $\lambda \leq L \leq 20\lambda$ (for the larger thicknesses the validity of these estimations is not yet clarified), one can simply estimate the expected value of the γ_{EIT}, knowing the value of $K_B(n\lambda)$ for any n and the blue detuning Δ.

The probe transmission spectra were calculated for $L = \lambda$ in [30]. Fig.35 shows the most important fragments of the calculated spectra, which contain the EIT resonance, for different values of a blue detuning Δ of the couple field with respect to ^{87}Rb $F_g{=}2 \rightarrow F_e{=}2$ transition frequency. The probe is scanned across $F_g{=}1 \rightarrow F_e{=}0,1,2$. In the model, we have chosen the blue detuning in order to have an unperturbed dependence from the detuning, since in the case of red detuning, the transition $F_g{=}1 \rightarrow F_e{=}1$ will have an influence on EIT resonance. The following parameters have been used: cell thickness $L = \lambda = 780$ nm, laser linewidth $\gamma_L = 1$ MHz, Rabi frequency of the coupling laser $\Omega_c = 1.5\Gamma_N$, Rabi frequency of the probe laser $\Omega_p =$

$0.35\Gamma_N$ (the probe laser intensity is chosen relatively high in order to study also the VSOP resonance behavior versus detuning), Γ_N = 6 MHz. As can be seen, the narrowest EIT resonance is observed when the coupling laser is resonant with the $F_g{=}2{\rightarrow}F_e{=}2$ transition (i.e. Δ = 0). The increase of the coupling field detuning leads to broadening of EIT resonance (from 12 to 36 MHz) with simultaneous rapid worsening of EIT contrast.

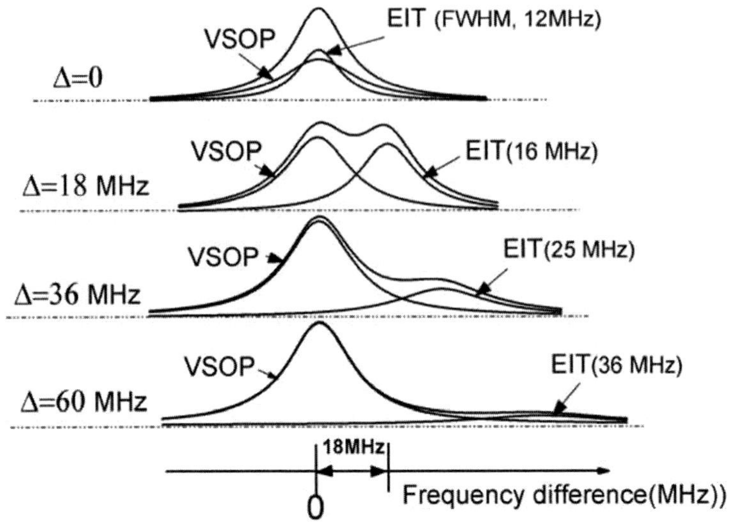

Figure 35. Calculated probe field transmission spectra as a function of blue detuning Δ of the couple field frequency from the ^{87}Rb $F_g{=}2{\rightarrow}F_e{=}2$ transition. The cell thickness is $L = \lambda$ = 780 nm, γ_L = 1 MHz, $\Omega_c = 1.5\Gamma_N$, $\Omega_p = 0.35\ \Gamma_N$, Γ_N = 6 MHz. γ_{EIT} = 12, 16, 25, and 36 MHz for Δ = 0, 18, 36, and 60 MHz, respectively. A rapid increase of γ_{EIT} and EIT contrast worsening versus Δ is clearly seen.

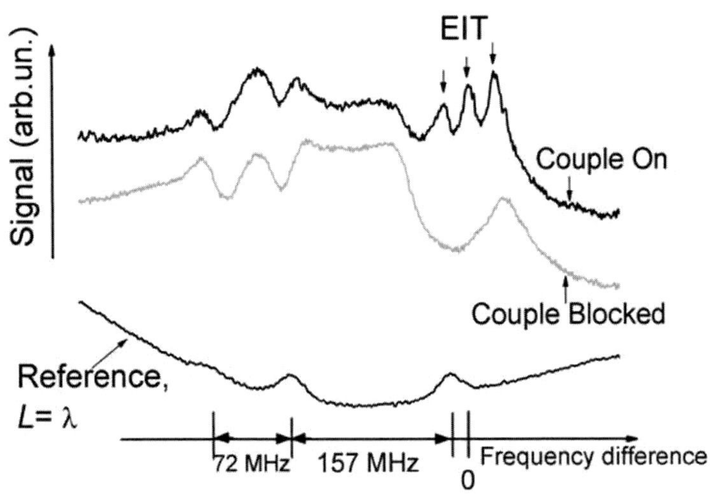

Figure 36. The probe laser transmission spectrum of $L = 2\lambda$ nanocell in a longitudinal magnetic field B = 15 G. The coupling laser is: on (upper curve); off (middle curve). The lower curve is the reference spectrum of $L = \lambda$ nanocell for B = 0. The three EIT resonances are well seen on the upper trace.

Noteworthy, the width of VSOP resonances remains practically unchanged with an increase of the coupling field detuning, since for the case of a nanocell, the VSOP resonances are located exactly at the atomic transitions. For the four cases presented in Fig.35, the width of VSOP resonances is equal to 18 MHz, and their amplitudes increase with Δ, since in this case the coupling laser transfers atoms from $F_g=1$ to $F_g=2$, thus causing the increase of probe absorption from the $F_g=2$ level.

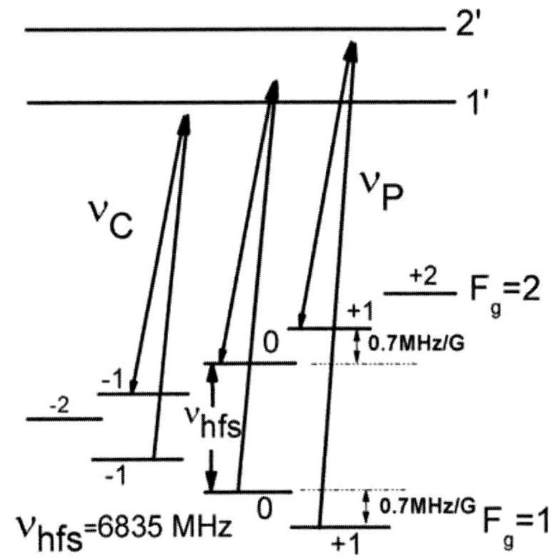

Figure 37. The diagram showing three Λ-systems formed in an external magnetic field on ^{87}Rb D_2 line; the frequency separation between the EIT resonances is 21 MHz.

It is interesting to note, that in the case of an ordinary cm-size length cell (COL) the EIT resonance width and contrast are practically not influenced by the coupling laser frequency shift. Moreover, in some conditions an additional narrowing of the EIT resonance occurs for large frequency shift Δ (since $L >> D$ for COL).

In [29] it was shown that one can determine the magnitude of a magnetic field by measuring the splitting of the EIT resonances in a nanocell. Fig.36 shows the transmitted probe laser spectrum when a longitudinal magnetic field B = 15 G is applied to the Rb cell with $L = 2\lambda$. In order to increase the recording contrast of the probe spectrum, the polarizer (Glan 2 in Fig.26) located before the photodetector was adjusted to a non-zero angle. Note that in this case also the Faraday rotation is influent on the probe transmission spectrum. The coupling laser is blue detuned by 18 MHz from $F_g=2 \rightarrow F_e=2$ transition of ^{87}Rb D_2 line; the probe is scanned across $F_g=1 \rightarrow F_e=0,1,2$ transitions. Three well resolved EIT resonances with a linewidth ≤ 10 MHz are marked by arrows on the upper trace of Fig.36. As it is shown in Fig.37, three Λ-systems are formed in this case, and the frequency separation between the EIT resonances is ≈ 1.4 MHz/G × 15 G = 21 MHz. The middle grey trace shows the probe spectrum, when the coupling laser is off (the curve is slightly shifted vertically for convenience). The substructure of the probe spectrum is caused by VSOP resonances. The lower curve is reference transmission spectrum of $L = \lambda$ nanocell for $B = 0$.

The use of a COL yields much higher *B*-field sensitivity due to narrower EIT resonances [25]. Nevertheless, the nanocell may provide a sub-micrometer spatial resolution, which is advantageous for the measurement of highly-inhomogeneous magnetic fields.

A theoretical model, which describes well the peculiarities of EIT and VSOP resonances for 10 μm-thin cell, has been developed by Yu. Malakyan and D. Petrosyan [26]. Later, the model applicable for nanocells has been developed by Y. Pashayan-Leroy and C. Leroy (University of Bourgogne, France) [30].

VI. Conclusion

Section II. Different types of micro- and nanometric thin cells containing atomic vapor of alkali metals are developed for studies of resonant absorption, fluorescence, reflection as well as magneto-optical processes in the case when atomic vapor thickness L is comparable with the laser radiation wavelength λ, in the range from 30 nm up to several mm. The designed cells allow smooth variation of thickness, either by lateral translation of wedge-shaped gap or by external pressure control in a special chamber. The operation temperature limit of micro- and nanometric thin cells is 500 °C.

Section III. Resonant absorption and fluorescence processes are studied in the cells of nanometric thickness (atomic vapor column L varies in the range from 100 to 3000 nm). Several unusual peculiarities are revealed. Particularly, it was found that the main parameter determining the spectral lineshape of resonant absorption and reflection is the ratio L/λ (λ is the laser radiation wavelength equal to an atomic transition wavelength). The phenomena of Dicke-type coherent narrowing (DCN) and collapse of DCN were revealed and studied jointly with the group of M. Ducloy and D. Bloch from Paris-Nord University [9]. The minima of resonant absorption spectral width take place for $L = (2n+1)\lambda/2$. The smallest width ~ 100 MHz, i.e. about 4 times narrower than Doppler-broadened line in the "usual" cell, occurs at $L = \lambda/2$, whereas for $L = 3\lambda/2$, $5\lambda/2$, $7\lambda/2$, and $9\lambda/2$ the reducing spectral narrowing is superimposed on an increasing broad Doppler pedestal. The maxima of the spectral width take place for $L = n\lambda$.

At least five distinctions between the absorption and fluorescence processes are revealed: i) the spectral linewidth of the fluorescence is always narrower (~ 1.5 times for $L = \lambda/2$, and ~ 3 times for $L = \lambda$) than that of the transmission for the same transition; ii) the DCN effects do not influence the fluorescence, and in contrast to absorption spectra, the increase of L results in monotonic increase of the fluorescence linewidth; iii) the fluorescence linewidth remains below the Doppler broadening even for $L = 3\div4$ μm, while the transmission linewidth for these L values becomes equal to Doppler width of a cm-size cell; iv) the VSOP resonances in nanocell transmission spectra appear at $L = n\lambda$ for radiation intensities ~ 1 mW/cm^2, meanwhile the intensity required to observe VSOP resonances in fluorescence spectra is more than one order of magnitude higher; v) the VSOP resonance dips at $L = n\lambda$ appear in the transmission spectrum on cycling and open transitions simultaneously, meanwhile, the VSOP dips in fluorescence spectrum first appear on open transition and, with further increase of L (or radiation intensity), also on cycling transition (for $F_g < F_e$) (this was jointly studied with S. Cartaleva group, Institute of Electronics, Sofia).

It is demonstrated that the atomic frequency reference based on a nanocell has several advantages in comparison with a widely used saturated absorption technique: i) the absence of crossover resonances, which is very important for some applications; ii) the ratio of amplitudes of VSOP peaks for a nanocell with $L = \lambda$ is close to the ratio of corresponding transition probabilities; iii) a single-beam transmission spectrum is needed for a nanocell; iv) the laser power required for a nanocell spectral reference is as low as ~ 0.1 mW.

Section IV. The "λ-Zeeman technique" (LZT) is shown to be efficient for studies of individual transitions between the Zeeman sublevels of hyperfine levels in an external magnetic field of 0.1 mT ÷ 0.25 T. LZT is based on atomic nanocell with thickness $L = \lambda$, where λ is the resonant wavelength (794 or 780 nm for D_1 or D_2 line of Rb). Narrow VSOP resonances in the transmission spectrum of the nanocell are split into several components in a magnetic field; their frequency positions and probabilities depend on the B-field. Possible applications, such as magnetometers with nanometric spatial resolution and tunable atomic frequency reference are described. Implementation of LZT allowed us to reveal appearance of "forbidden" transitions in recorded spectra, for which new selection rules for the quantum number F take place. Thus, Zeeman components of $F_g=1 \rightarrow F_e=3$ transition of ^{87}Rb D_2 line appear at $B > 10$ mT, moreover, for 20 mT $< B <$ 0.2 T the probability of one of this "forbidden" components exceeds that of the strongest transition at $B = 0$. Also, the LZT allowed us to reveal that the atomic transitions of ^{85}Rb and ^{87}Rb D_1 line, which are separated by a few GHz at $B = 0$, start to merge at $B > 0.2$ T (the theoretical model which fairly describes the experiment was developed by the group of M. Auzinsh, University of Latvia, Riga [19-21]).

Another method for studies of atomic spectra in magnetic field, "λ/2-Zeeman technique" (HLZT), is based on resonant fluorescence spectrum emitted by a nanocell with $L = \lambda/2$. The results of implementation of HLZT to investigate individual hyperfine transitions between Zeeman sublevels in an external magnetic field of 1 mT ÷ 25 mT is presented.

Section V. We have studied peculiarities of the EIT effect in micrometric and nanometric thin cell filled with pure atomic vapor of alkali metal. The behavior of EIT resonance, as well as VSOP resonances, which are the satellites of EIT for the case of L ~ 50 μm ÷ 1 mm are, in general, similar to that obtained earlier with ordinary cm-size cells (COL). The distinctions start to develop at L ~ 5 ÷ 50 μm. However, the most remarkable differences of EIT and VSOP resonance formation as compared with COL are revealed for the thickness $L <$ 3 ÷ 5 μm. Particularly, in the case when the coupling laser frequency is resonant with corresponding atomic transition, the dependence of EIT resonance linewidth γ_{EIT} on L is rather weak: $\gamma_{EIT} \sim L^{-0.25}$. This is caused by predominant contribution to the EIT resonance by the atoms flying nearly parallel to the cell windows with zero velocity projection $u_z = 0$ in the direction of laser beam propagation, so that the interaction time of the laser field with these atoms is $\tau_D = D/u$.

A strong dependence of EIT resonance linewidth ($\gamma_{EIT} \sim 1/L$) exhibits for a fixed detuning of the coupling laser with respect to the corresponding transition. Particularly, when the coupling laser frequency is resonant with corresponding transition for the case of $L = \lambda$ thickness, $\gamma_{EIT} \leq 10$ MHz, the EIT resonance has a relatively good contrast and can be easily detected, while when the coupling laser is detuned by Δ ~ 70 - 80 MHz for the same L, the EIT resonance linewidth rapidly increases to γ_{EIT} ~ 90 - 100 MHz, and the contrast reduces practically to zero. The explanation is that when the coupling laser is detuned from the corresponding transition by Δ value, only the atoms having the velocity projection $u_z = 2\pi\Delta/k$ participate in formation of the EIT resonance, and for these atoms the time of flight between the cell windows is very short: $\tau_L = L/u_z$. This causes frequent collisions of atoms with the nanocell windows, which leads to dramatic increase of Γ_{21} and, consequently, to the broadening of EIT resonance linewidth and sharp reduction of its contrast. It is important to note, that measuring the EIT linewidth in the case when the coupling laser is detuned, it is possible to determine the dephasing rate of coherence between the ground states Γ_{21}, and possibly also the atom-wall collision rate. In the case of an ordinary cm-size cell the above mentioned effect is completely absent.

Acknowledgements

The authors express their gratitude to A.S. Sarkisyan for his valuable participation in fabrication of nanocells. We are grateful to M. Ducloy, A. Weis, Yu. Malakyan, S. Cartaleva, M. Auzinsh, D. Bloch, Y. Pashayan-Leroy, T. Varzhapetyan, C. Leroy, P. Moroshkin, G. Pikhler, and A. Akulshin for useful and fruitful discussions. We thank A. Sargsyan for valuable experimental assistance. This work was done with the financial support provided by INTAS South-Caucasus Grant 06-1000017-9001 and by SCOPES Grant IB7320-110684/1.

References

[1] S. Briaudeau, D. Bloch, M. Ducloy, *Europhys. Lett.* **35**(5), 337 (1996).

[2] S. Briaudeau, S. Saltiel, G. Nienhuis, D. Bloch, M. Ducloy, *Phys. Rev. A* **57**, R3169 (1998).

[3] S. Briaudeau, D. Bloch, M. Ducloy, *Phys. Rev. A* **59**, 3723 (1999).

[4] A.Ch. Izmailov, *Opt. and Spectr.* **74**(1), 25 (1993); A.Ch. Izmailov, *Laser Phys. Lett.* **3**(3), 132 (2006).

[5] D.S. Glassner, B. ai, R.J. Knize, *Opt. Lett.* **19**, 2071 (1994); B. ai, D.S. Glassner, R.J. Knize, J.P. Partanen, *App. Phys. Lett.* **64**(8), 951 (1994).

[6] T.A. Vartanyan, D.L. Lin, *Phys. Rev. A* **38**, 5197 (1995); T. Vartanyan, *Rus. Physics-Uspekhi* **44**, 1203 (2001).

[7] B. Zambon, G. Nienhuis, *Opt. Commun.* **143**, 308 (1997).

[8] D. Sarkisyan, D. Bloch, A. Papoyan, M. Ducloy, *Opt. Commun.* **200**, 201 (2001).

[9] G. Dutier, A. Yarovitski, S. Saltiel, A. Papoyan, D. Sarkisyan, D. Bloch, M. Ducloy, *Europhys. Lett.* **63**, 35 (2003).

[10] D. Sarkisyan, T. Varzhapetyan, A. Sarkisyan, Yu. Malakyan, A. Papoyan, A. Lezama, D. Bloch, M. Ducloy, *Phys. Rev. A* **69**, 065802 (2004).

[11] G. Dutier, S. Saltiel, D. Bloch, M. Ducloy, *J. Opt. Soc. Am. B* **20**, 793 (2003).

[12] T. Varzhapetyan, D. Sarkisyan, L. Petrov, C. Andreeva, D. Slavov, S. Saltiel, A. Markovski, S. Cartaleva, *Proc. SPIE* **5830**, 196 (2005); D. Sarkisyan, T. Varzhapetyan, A. Papoyan, D. Bloch, M. Ducloy, *Proc. SPIE* **6257**, 625701 (2006); D. Sarkisyan, A. Sargsyan, A. Papoyan, Y. Pashayan-Leroy, *Proc. SPIE* **6604**, 660405 (2007).

[13] G. Nikogosyan, D. Sarkisyan, Yu. Malakyan, *J. Opt. Technol.* **71**, 602 (2004).

[14] A.D. Sargsyan, D.H. Sarkisyan, Y. Pashayan-Leroy, C. Leroy, P. Moroshkin, A. Weis, *J. Contemp. Phys*. **43**, 7 (2008).

[15] I. Hamdi, P. Todorov, A. Yarovitski, G. Dutier, I. Maurin, S. Saltiel, Y. Li, A. Lezama, T. Varzhapetyan, D. Sarkisyan, M.-P. Gorza, M. Fichet, D. Bloch, M. Ducloy, *Laser Phys*. **15**, 987 (2005).

[16] G. Dutier, P. Todorov, I. Hamdi, I. Maurin, S. Saltiel, D. Bloch, M. Ducloy, *Phys. Rev. A* **72**, 040501(R) (2005).

[17] A. Sargsyan, D. Sarkisyan, A. Papoyan, Y. Pashayan-Leroy, P. Moroshkin, A. Weis, A. Khanbekyan, E. Mariotti, L. Moi, *Laser Phys*. **18**, 1 (2008).

[18] D. Sarkisyan, T. Becker, A. Papoyan, P. Thoumany, H. Walther, *Appl. Phys. B: Las. Opt*. **76**, 625 (2003).

[19] A. Papoyan, D. Sarkisyan, K. Blush, M. Auzinsh, D. Bloch, M. Ducloy, *Laser Phys*. **13**, 1467 (2003); D. Sarkisyan, A. Papoyan, T. Varzhapetyan, J. Alnis, K. Blush, M. Auzinsh, *J. Opt. A: Pure Appl. Opt*. **6,** S142 (2004).

[20] D. Sarkisyan, A. Papoyan, T. Varzhapetyan, K. Blush, M. Auzinsh, *Opt. Spectr*. **96**, 328 (2004); D. Sarkisyan, A. Papoyan, T. Varzhapetyan, K. Blush, M. Auzinsh, *J. Opt. Soc. Am. B* **22**, 88 (2005).

[21] T. Varzhapetyan, H. Hakhumyan, D. Sarkisyan, V. Babushkin, A. Atvars, M. Auzinsh, *J. Contemp. Phys*. **42**, 223 (2007).

[22] M. Fichet, G. Dutier, A. Yarovitsky, P. Todorov, I. Hamdi, I.Maurin, S. Saltiel, D. Sarkisyan, M.P. Gorza, D. Bloch, M. Ducloy, *Europhys. Lett*. **77**, 540001 (2007).

[23] E. Arimondo, in: *Progress in Optics*, edited by E. Wolf (Elsevier, Amsterdam) 35, 257 (1996).

[24] S. Harris, *Physics Today* **50**(7), 36–42 (1997).

[25] R. Wynands, A. Nagel, *Appl. Phys. B: Lasers Opt*. **68**, 1 (1999) and references therein.

[26] D. Petrosyan, Yu. Malakyan, *Phys. Rev. A* **61**, 053820 (2000).

[27] S. Knappe, L. Hollberg, J. Kitching, *Opt. Lett*. **29**, 388 (2004).

[28] K. Fukuda, A. Toriyama, A. Izmailov, M.Tachikawa, *Appl. Phys. B* **80**, 503 (2005).

[29] A. Sargsyan, D. Sarkisyan, A. Papoyan, *Phys. Rev. A* **73**, 033803 (2006).

[30] Y. Pashayan-Leroy, C. Leroy, A. Sargsyan, A. Papoyan, D. Sarkisyan, *J. Opt. Soc. Am. B* **24**, 1829 (2007); A. Sargsyan, *Proc. SPIE* **6227**, 622719 (2007).

[31] D. Sarkisyan, U. Hinze, L. Meyer, B. Wellegehausen, *Appl. Phys. B* **70**, 351 (2000).

[32] D. Sarkisyan, A. Sarkisyan, A. Yalanusyan, *Appl. Phys. B* **66**, 241 (1998).

[33] R.H. Romer, R.H. Dicke, *Phys. Rev*. **99**, 532 (1955).

[34] C. Andreeva, S. Cartaleva, L. Petrov, S.M. Saltiel, D. Sarkisyan, T. Varzhapetyan, D. Bloch, M. Ducloy, *Phys. Rev. A* **76**, 013837 (2007).

[35] W. Demtroder, *Laser Spectroscopy: Basic Concepts & Instr*., Springer, Heidelberg, 1996.

[36] T. Varzhapetyan, A. Nersisyan, D. Sarkisyan, S. Vdovic, G. Pichler, Proc. "Laser Physics-2005", *Yerevan - Ashtarak*, **135** (2006).

[37] A. Mikata, U. Tanaka, Sh. Urabe, *Appl. Opt*. **47**, 640 (2008).

[38] J. Gea-Banacloche, Y. Li, Sh. Jin, M. Xiao, *Phys. Rev. A* **51**, 576 (1995).

[39] A.K. Mohapatra, T.R. Jackson, C.S. Adams, *Phys. Rev. Lett*. **98**, 113003 (2007).

[40] U. Rasbach, J. Wang, R. dela Torre, V. Leung, B. Klaoter, D. Meschede, T. Varzhapetyan, D. Sarkisyan, *Phys. Rev. A* **70**, 033810 (2004).

[41] D. Budker, V. Yaschuk, M. Zolotarev, *Phys. Rev. Lett*. **81**, 5788 (1998).

[42] C. Andreeva, A. Atvars, M. Auzinsh, K. Blush, S. Cartaleva, L. Petrov, D. Slavov, *Phys. Rev. A* **76**, 063804 (2007).

[43] H. Failache, L. Lenci, A. Lezama, D. Bloch, M. Ducloy, *Phys. Rev. A* **76**, 053826 (2007).

[44] B. Gross, N. Papageorgiou, V. Sautenkov, A. Weis, *Phys. Rev. A* **55**, 2973 (1997).

[45] E.A. Gazazyan, A.V. Papoyan, D. Sarkisyan, A. Weis, *Laser Phys. Lett.* **4**, 801 (2007).

[46] A. Sargsyan, D. Sarkisyan, D. Staedter, A.M. Akulshin, *Opt. Spectr.* **101**, 762 (2006).

In: New Trends in Quantum Coherence...
Editor: Rafael Drampyan, pp. 123-148
ISBN 978-1-60741-025-6

Chapter 4

TIME-DOMAIN NONLINEAR OPTICAL SPECTROSCOPY OF DENSE ATOMIC VAPORS

V.O. Lorenz[1,2,*] ***and S.T. Cundiff***[1]
[1]JILA, University of Colorado and National Institute of Standards and Technology, Boulder, CO 80309-0440, USA
[2]Department of Physics, University of Colorado, Boulder, CO 80309-0390, USA

Abstract

Time-domain experiments are ideal for studies of fast dynamics because the fastest dynamics dominate at short times and with high signal-to-noise ratio. Applying techniques of nonlinear optical spectroscopy in the time domain can separate short timescales from long ones, providing a different perspective from frequency-domain studies of optical lineshapes. We discuss two- and three-pulse transient four-wave mixing techniques and present results from experimental studies of collisional dynamics in dense atomic vapors. Dense alkali vapors are interesting for their many-body interactions but have proven difficult to study because of technical limitations. We present a method of fabricating high-temperature, high-density alkali vapor cells for spectroscopy experiments.

1. Introduction

Studies of fast dynamics in systems such as atomic vapors and molecular liquids can provide insight into many-body interactions and the way bonds stretch and molecules conform [1, 2, 3]. In these systems, interactions occur on short timescales and thus their frequency-domain manifestation is most dominant in the wings of the absorption lineshape [4]. Frequency-domain experiments are limited in their ability to probe the fastest dynamics by the inherently low signal-to-noise ratio and influence of other resonances in the wings of absorption lineshapes. This is in contrast to time-domain experiments, where the fastest dynamics dominate at short times and with high signal-to-noise ratio. While lineshapes are integrated over many collision durations, time-domain techniques can separate short

*Present address: University of Oxford, Clarendon Laboratory, Parks Road, Oxford OX13PU, UK

timescales from long ones, providing a different perspective from frequency-domain studies of fast dynamics.

Here we present time-domain studies of a dense potassium vapor, in which the dephasing caused by collisions occurs on picosecond timescales. These collision durations are long compared to the optical pulses used in the experiments, allowing the study of the effects of memory on the dephasing process. When memory effects must be taken into account, the dynamics are called non-Markovian. Such dynamics have been studied for molecular systems, but the dephasing caused by interactions of the molecules with the solvent often occur at femtosecond time scales, causing ultrafast dephasing, which obscures the non-Markovian dynamics at short times [5, 6, 7, 8]. In molecular experiments, it is also common to have dephasing times on the order of the pulse duration. In contrast, in a dense atomic vapor the pulse duration, collision duration and dephasing time are all well-separated, making dense atomic vapors ideal systems in which to apply ultrafast spectroscopy to the study of non-Markovian dynamics. In this chapter we describe the novel high-temperature, high-density alkali vapor cell used in the experiments. We show that the dynamics revealed by the experiments are modeled well using the non-Markovian theory commonly applied to more complex systems [9]. Further modeling using molecular dynamics simulations reveal the effects of long-range resonant interactions in dense vapors. We begin with a brief overview of the historical background and a qualitative description of the non-Markovian regime.

Historically, the effects of fast dynamics on frequency-domain lineshapes attracted interest about a century after the first studies of spectral emission lines in the mid-1800's. After decades of improving the technique of white-light spectroscopy, in the 1920's pressure-broadened lineshapes were resolved for a number of atomic vapors and theoretical treatments began to elucidate the underlying physics [10]. Initially, theories that approximated collisions between atoms and perturbers as infinitely fast in what is known as the impact approximation were sufficient to describe the majority of experiments, with good agreement for linewidths at low densities.

Studies took on new emphasis with the improved intensities and spectral resolution provided by the advent of the laser in the 1960's. Initially, theoretical treatments employing the impact approximation were still sufficient to explain most experimental results [11, 12]. Starting in the 1970's, predictions were made for signatures of the finite duration of collisions to be manifested in frequency-domain lineshapes as a deviation from Lorentzian form, both near line center and at detunings from resonance on the order of the inverse duration of a collision [13]. These were quickly verified experimentally [14, 15]. A non-impact theory of the technique of frequency-domain four-wave mixing (FWM) was presented in 1986, and FWM experiments revealed non-impact behavior in the wings of helium-broadened sodium vapor lineshapes in 1989 [16, 17].

The application of time-domain spectroscopy with pulsed lasers began in the 1970's in conjunction with the development of the theory of transient nonlinear optics [18]. Initial applications of ultrafast spectroscopy were made largely in condensed phase systems, with only a few ultrafast experiments in dense atomic vapors [19]. The theory for ultrafast spectroscopy was initially based on the optical Bloch equations in the Markov approximation, which like the impact approximation assumes infinitely fast interactions [20, 21]. As pulse durations approached the sub-picosecond scale in the 1980's, studies began to reveal ultrafast relaxation phenomena and theoretical treatments were developed that modified the optical Bloch equations to include the finite duration of interactions, or non-Markovian dephasing, with a unified approach to transient nonlinear optical phenomena using Liouville operators published in 1983 [9, 22]. Currently, ultrafast nonlinear spectroscopy is commonly applied to such intricate systems as biological light-harvesting complexes to gain insight into their structure and dynamics [23]. In molecular systems, studies of ultrafast interactions of atoms in molecules and between molecules and solvents have provided detailed inter- and intramolecular information [2, 3].

The irreversible loss of order due to interactions in quantum systems that is the subject of non-Markovian studies is known generally as quantum dissipation [24]. We consider an atomic vapor as an example, where phase-destroying collisions are the cause of quantum dissipation. Figure 1 is a pictorial representation of the resonance frequency of a two-level atom as it undergoes a collision, which also can be thought of as the transient formation of a dimer. The effect of the collision is to cause the resonance frequency to fluctuate by an amount that is in principle determined by the interatomic potential energy surfaces of the dimer.

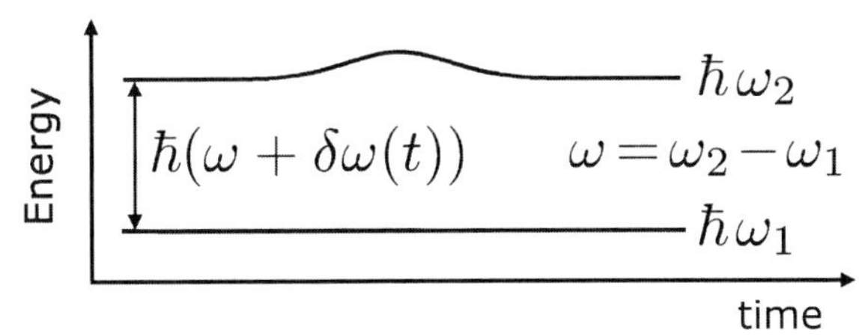

Figure 1. A schematic of the ground state and excited state energy levels of an atom undergoing a collision. During the collision the transition frequency fluctuates by a time-dependent amount $\delta\omega(t)$.

If we now consider the quantum state evolution of an excited atom undergoing a collision, represented by the superposition between ground and excited state shown in Fig. 2, collisions cause the phase of the superposition to be different for atoms that have undergone a collision compared to those that have not. This loss of phase coherence is known as dephasing. Using the superpositions of an unperturbed and perturbed atom, we can qualitatively define the correlation time t_c as the time it takes for a perturbed atom to become out of phase with an unperturbed atom, shown in Fig. 2. The correlation time is a measure of the phase memory of the system, representing how long an atom retains approximately the same phase as an unperturbed atom.

In physics literature, dynamics that occur on timescales less than the correlation time are called non-Markovian dynamics, while those that occur in timescales much greater than the correlation time, where the time evolution is determined solely by the instantaneous state of the system, are called Markovian dynamics. For non-Markovian dynamics the phase coherence is retained such that the evolution of the superpositions of the atoms is reversible. Conversely, for Markovian dynamics the phase coherence of the system is lost over the timescale of interest due to the randomization of the phases of the superpositions due to collisions, making the evolution irreversible. Stated from an experimental perspective, non-Markovian dynamics arise in experiment whenever the finite duration of collisions must be taken into account, while dynamics are Markovian whenever collisions can be considered to occur infinitely fast.

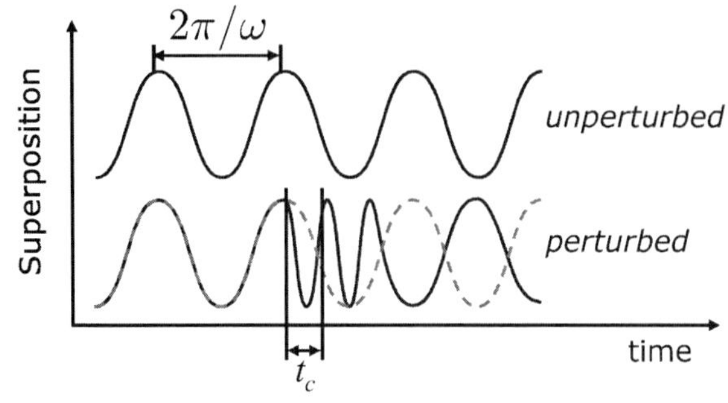

Figure 2. The superposition between a ground state and excited state as a function of time for an unperturbed (top) and perturbed (bottom) atom. The correlation time t_c is the time it takes for a perturbed atom to become out of phase with an unperturbed atom.

2. Theoretical Framework

In our studies of non-Markovian dynamics we employ an ultrafast spectroscopic technique called Transient Four-Wave Mixing (TFWM). This technique uses a sequence of ultrashort pulses to create a kind of stop-motion picture of atomic interactions. In a three-pulse TFWM experiment, the first pulse, with wavevector $\mathbf{k}_1$, excites a coherent superposition between the ground and excited states. During the ensuing delay τ, dephasing occurs due to collisions. Then the second pulse, $\mathbf{k}_2$, arrives and converts the remaining coherence to excited-state or ground-state population. During the second delay T, loss of population and spectral redistribution occurs due to decay and collisions, and the system's ability to rephase decreases. The third and final pulse, $\mathbf{k}_3$, forms a third-order coherence that radiates as the signal. The time-integrated signal is measured using a slow detector by scanning the first delay τ for various fixed values of the second delay T.

The above description of the transient four-wave mixing process uses vocabulary from the density matrix formalism. In this context, varying the time delays between pulses allows the characterization of the intermediate density matrix states. To be more explicit, varying the first delay probes the off-diagonal density matrix elements corresponding to coherence between states i and j, ρ_{ij}, and varying the second delay probes the density matrix elements representing population, ρ_{ii} and, in the case of a three-level system, the Raman coherence ρ_{ij}.

The usual theoretical framework applied to TFWM experiments expands the density matrix equations of motion to third order in the field. The third-order density matrix $\rho^{(3)}$ is related to the polarization $[\mathbf{P}^{(3)}]_{ij}$ via the equation

$$[\mathbf{P}^{(3)}]_{ij} = N\mathrm{Tr}(\rho^{(3)}\mu)_{ij}, \tag{1}$$

where N is the number density and μ the dipole moment operator. The intensity of the time-integrated signal is

$$S(\tau,T) = \int_{-\infty}^{\infty} |\mathbf{P}^{(3)}(\mathbf{r},t)|^2 dt. \tag{2}$$

In TFWM experiments it is common for the non-Markovian behavior of the coherence to be fast enough to be neglected and to account for dephasing phenomenologically by adding dephasing rates to the density matrix equations of motion. The q-th order equation of motion for a two-level system then takes the form

$$\dot{\rho}_{ij}^{(q)} = -(i\omega_{ij} + \gamma_{ij})\rho_{ij}^{(q)} - \frac{i}{\hbar}[-\hat{\mu} \cdot \mathbf{E}(t), \hat{\rho}^{(q-1)}]_{ij}, \tag{3}$$

where ω_{ij} is the transition frequency and γ_{ij} is the population relaxation rate T_1^{-1} if the state is diagonal or the dephasing rate T_2^{-1} if the state is off-diagonal. Although the population decay term can be justified by quantizing the field [20], adding phenomenological dephasing terms is equivalent to making the Markov approximation for the coherence decay process.

Deriving the material response in the Markov approximation can be done more transparently by adding a term to the density matrix equation of motion to represent frequency fluctuations $\delta\omega(t)$:

$$\dot{\rho}_{ij}^{(q)} = -(i\omega_{ij} + i\delta\omega_{ij}(t) + \gamma_{ij}^{pop})\rho_{ij}^{(q)} - \frac{i}{\hbar}[-\hat{\mu} \cdot \mathbf{E}(t), \hat{\rho}^{(q-1)}]_{ij}, \tag{4}$$

where γ_{ij}^{pop} now only represents the population relaxation rate T_1^{-1} ($\gamma_{ij}^{pop} = 0$ for $i \neq j$). This equation is easily solved for $\rho_{ij}^{(q)}(t)$ by using an integrating factor. Because variations in $\delta\omega_{ij}(t)$ are random for each atom, we take an ensemble average of the density matrix, which only affects an exponential term containing the fluctuations. To solve, this term is Taylor-expanded, with each term representing an order of correlation of frequency fluctuations:

$$\begin{aligned}\langle \exp[-i\int_0^t \delta\omega_{ij}(t')dt']\rangle &= 1 + i\int_0^t dt' \langle \delta\omega_{ij}(t')\rangle \\ &-\frac{1}{2}\int_0^t dt' \int_0^t dt'' \langle \delta\omega_{ij}(t')\delta\omega_{ij}(t'')\rangle + \cdots \\ &+\frac{(-i)^{2n}}{(2n)!}\int_0^t dt_1 \ldots \int_0^t dt_{2n} \langle \delta\omega_{ij}(t_1)\ldots\delta\omega_{ij}(t_{2n})\rangle + \cdots,\end{aligned} \tag{5}$$

where the angle brackets denote an ensemble average. The odd terms are zero because the fluctuations are as often positive as negative and thus have zero mean. The Markov approximation assumes that variations in ω are rapid compared to other changes such as population decay rates and dephasing rates such that the fluctuation at one instant in time can be considered completely uncorrelated with that of any different time. This means the two-time correlation function can be approximated as a delta function with an overall constant representative of the amplitude of the fluctuation:

$$\langle \delta\omega_{ij}(t)\delta\omega_{ij}(t')\rangle = \gamma_{ij}^{ph}\delta(t - t'). \tag{6}$$

Inserting this into the equation of motion results in the same solution as found using phenomenological decay rates, with $\gamma_{ij}^{ph} = T_2^{-1}$ for $i \neq j$, $\gamma_{ij}^{ph} = 0$ for $i = j$.

To model non-Markovian dephasing, the two-time correlation function of frequency fluctuations is assumed to have a form with some finite temporal width. It is common practice to apply a stochastic model in which the excited state frequency fluctuations are Gauss-Markov processes. In this model the fluctuation of the i to j transition $\delta\omega_{ij}$ obeys a correlation function given by

$$\langle \delta\omega_{ij}(t_1)\delta\omega_{ij}(t_2)\rangle = \Delta_{ij}^2 \exp(-\Lambda_{ij}|t_2 - t_1|), \tag{7a}$$

with

$$\langle \delta\omega_{ij}(t)\rangle = 0, \tag{7b}$$

where Δ_{ij} corresponds to the root-mean-squared amplitude of the frequency fluctuations and Λ_{ij}^{-1} is their time scale. Although it is possible to derive the response of the system by inserting the time-dependent fluctuations directly into the density matrix equations of motion, as done above for the Markov approximation, there is a parallel approach that simplifies the derivation by using the Liouville-space formalism [9, 22]. This approach allows us to specify the perturbation sequences of interest and only derive response functions for those sequences, instead of carrying unused wavevector combinations only to discard them later as done in the optical Bloch equations. We choose this approach for our derivations [25]. Using two- and three-level subsets of the most general four-level four-point correlation function of dipole operators, we derive the response functions of a three-level system containing two excited states b and c radiatively coupled to a ground state a but not radiatively coupled to one another. Assuming delta-function pulse durations, this allows us to solve for the response functions for two- and three-pulse TFWM in terms of the stochastic model lineshape function, given by

$$\begin{aligned} g_{ij}(t) &= \int_0^t d\tau_1 \int_0^{\tau_1} d\tau_2 \langle \delta\omega_{ij}(\tau_1 - \tau_2)\delta\omega_{ij}(0)\rangle \\ &= \frac{\Delta_{ij}^2}{\Lambda_{ij}^2}[\exp(-\Lambda_{ij}t) + \Lambda_{ij}t - 1]. \end{aligned} \tag{8}$$

The response functions for three-pulse TFWM with the stochastic model for the lineshape function are

$$\begin{aligned} {}^2R_2^m(\tau_1,\tau_2,\tau_3) &= \mu_{am}^4 \mathrm{e}^{-i\omega_{ma}(\tau_1-\tau_2-\tau_3)} \\ &\quad \times \exp\Big\{\frac{1}{2}[-g_{am}(-\tau_2) + g_{am}(-\tau_1) - g_{am}(\tau_2 - \tau_1) \\ &\quad -g_{am}(-\tau_3) + g_{am}(\tau_2 - \tau_3) - g_{am}(\tau_1 - \tau_3)]\Big\}, \end{aligned} \tag{9a}$$

$$\begin{aligned} {}^2R_3^m(\tau_1,\tau_2,\tau_3) &= \mu_{am}^4 \mathrm{e}^{-i\omega_{ma}(\tau_1-\tau_2-\tau_3)} \\ &\quad \times \exp\Big\{\frac{1}{2}[-g_{am}(-\tau_3) + g_{am}(-\tau_1) - g_{am}(\tau_3 - \tau_1) \\ &\quad -g_{am}(-\tau_2) + g_{am}(\tau_3 - \tau_2) - g_{am}(\tau_1 - \tau_2)]\Big\}, \end{aligned} \tag{9b}$$

$$\begin{aligned} {}^3R_2^n(\tau_1,\tau_2,\tau_3) &= \mu_{am}^2\mu_{an}^2 \mathrm{e}^{-i\omega_{na}(\tau_1-\tau_2)-i\omega_{nm}(\tau_2-\tau_3)-i\omega_{ma}(-\tau_3)} \\ &\quad \times \exp\Big\{-\frac{1}{2}[2g_{am}(-\tau_2) + 2g_{an}(\tau_1 - \tau_3) \\ &\quad -g_{am}(-\tau_1) + g_{am}(-\tau_3) + g_{am}(\tau_2 - \tau_1) - g_{am}(\tau_2 - \tau_3) \\ &\quad -g_{an}(-\tau_1) + g_{an}(-\tau_3) + g_{an}(\tau_2 - \tau_1) - g_{an}(\tau_2 - \tau_3) \end{aligned}$$

$$+g_{nm}(-\tau_1) - g_{nm}(-\tau_3) - g_{nm}(\tau_2 - \tau_1) + g_{nm}(\tau_2 - \tau_3)]\Big\}, \tag{9c}$$

$$
\begin{aligned}
{}^3R_3^m(\tau_1, \tau_2, \tau_3) &= \mu_{am}^2 \mu_{an}^2 \mathrm{e}^{-i\omega_{na}(\tau_1-\tau_2)-i\omega_{ma}(-\tau_3)} \\
&\times \exp\Big\{-\frac{1}{2}[\,2g_{an}(-\tau_3) + 2g_{am}(\tau_1 - \tau_2) \\
&-g_{am}(-\tau_1) + g_{am}(-\tau_2) + g_{am}(\tau_3 - \tau_1) - g_{am}(\tau_3 - \tau_2) \\
&-g_{an}(-\tau_1) + g_{an}(-\tau_2) + g_{an}(\tau_3 - \tau_1) - g_{an}(\tau_3 - \tau_2) \\
&+g_{nm}(-\tau_1) - g_{nm}(-\tau_2) - g_{nm}(\tau_3 - \tau_1) + g_{nm}(\tau_3 - \tau_2)]\Big\},
\end{aligned} \tag{9d}
$$

with $m, n = b, c$; $m \neq n$ and with the interactions with the field occuring at times $0 \leq \tau_3 \leq \tau_2 \leq \tau_1$. The R labels for each response function have been written according to the pathways they represent. The first superscript represents 2- or 3-level physics. For 2-level physics, the subscript represents an excited-state (2) or ground-state (3) population to second order. For 3-level physics, the subscript represents a Raman coherence (2) or ground state population (3) to second order. The second superscript is the final excited state that forms a coherence with the ground state at third order. The sum of the above response functions gives the polarization. Thus we have for the spectrally integrated polarization for time ordering $(-\mathbf{k}_1, \mathbf{k}_2, \mathbf{k}_3)$:

$$
\begin{aligned}
P^{(3)}(t) &= {}^2R_2^2(\tau_1, \tau_2, \tau_3) + {}^2R_2^3(\tau_1, \tau_2, \tau_3) + {}^2R_3^2(\tau_1, \tau_2, \tau_3) \\
&+{}^2R_3^3(\tau_1, \tau_2, \tau_3) + {}^3R_2^2(\tau_1, \tau_2, \tau_3) + {}^3R_2^3(\tau_1, \tau_2, \tau_3) \\
&+{}^3R_3^2(\tau_1, \tau_2, \tau_3) + {}^3R_3^3(\tau_1, \tau_2, \tau_3).
\end{aligned} \tag{10}
$$

The signal intensity, S, is calculated by time-integrating the square of the polarization:

$$S(\tau, T) = \int_{-\infty}^{\infty} |P^{(3)}(t)|^2 dt. \tag{11}$$

3. Experiment and Theory

In a medium that exhibits non-Markovian dephasing, the TFWM signal at short delays is a photon echo; thus, when TFWM is applied to the study of non-Markovian dynamics, it is referred to as photon echo spectroscopy. Photon echoes are usually associated with static inhomogeneously broadened systems, while atomic vapors are homogeneously broadened at high temperatures. However, a photon echo can still arise for short time delays between pulses. This is because at short delays the atoms can be considered as almost stationary over the duration of the delay. The atoms are at varying distances from one another, and this corresponds to an inhomogeneous distribution of dipole-dipole interaction induced frequency shifts. A photon echo arises from the combination of this inhomogeneous distribution of energy levels and the retention of phase memory such that the dephasing process can be reversed on the arrival of the third pulse. The photon-echo-like behavior dominates on timescales short compared to the correlation time of the frequency fluctuations responsible for the broadening. These timescales correspond to the non-Markovian regime. Measurements of this transient photon echo shed light on the dynamics that lead to the loss of phase memory, and thus probe interatomic [1, 25, 26, 27], intermolecular [2], and solute-solvent

[5, 8, 28] interactions. Photon echo spectroscopy has enabled significant progress in understanding complex molecular interactions in the condensed phase. In these systems, the dynamics often occur on timescales such that the signals arising from such dynamics compete with those of solvent interactions and pulse width effects [5, 29].

For a dense atomic vapor, the excitation pulses are short compared to the duration of collisions and the time between collisions is long compared to the collision duration. At the vapor densities and temperatures studied in our experiments, the dominant broadening mechanism is resonance broadening, or broadening due to collisions between atoms of the same species [30]. Thus there are distinct timescales that correspond to regimes in which the phase-coherence of the atomic superpositions is retained and in which the phase memory is lost due to collisions.

We perform two- and three- pulse TFWM as well as time-resolved TFWM in reflection from the sapphire-vapor interface of a high-temperature and high-density potassium vapor cell. The experiments are performed in reflection because of the short absorption length of the vapor at the densities used ($> 10^{17}$ cm^{-3}). We begin with a description of the vapor cell used in the experiments. Then, we discuss the two-pulse experiment and model the results using the non-Markovian theory derived above that assumes a stochastic form for the correlation of frequency fluctuations due to collisions. Going one step further, the three-pulse experiment reveals the correlations between fluctuations directly. We then derive the form of the correlation function from first principles and compare with the three-pulse results to find that long-range interactions extend the correlations between frequency fluctuations. Finally, the time-resolved version of the experiments gives us a more direct glimpse of the photon-echo-like behavior at short times.

3.1. Vapor Cell Design

In order to perform optical spectroscopy experiments in dense alkali vapors the challenge of containing these highly reactive metals at high temperatures must be addressed. In the experiments we use a novel vapor cell designed to withstand temperatures up to 800°C (vapor densities up to 10^{19} cm^{-3}). Optical access is provided through a sapphire window, which is bonded to a titanium body. The high-vacuum sapphire-to-titanium seals are achieved via diffusion bonding and maintain their integrity for an average of 100 hours above 500°C with about 10 heating / cooling cycles.

Diagrams of the cell components are given in Figs. 3 and 4 [1]. The materials titanium and sapphire are chosen based on their resistance to corrosion by potassium and their similar coefficients of thermal expansion. Grade-2 commercially pure titanium is machined for the body. Custom-made sapphire windows are used for optical access. The sapphire is cut in the c-axis orientation, with a diameter of 9 mm and a thickness of 0.5 mm. The windows have an anti-reflection coating on one face for maximum transmission at 768 nm and are wedged by 0.5 degree in order to spatially separate the front and back reflections. The face of the titanium body is polished for bonding to the sapphire window. The screws are made of Inconel.

[1] In the following, mention of commercial products is for information only; it does not imply NIST recommendation or endorsement, nor does it imply that the products mentioned are necessarily the best available for the purpose.

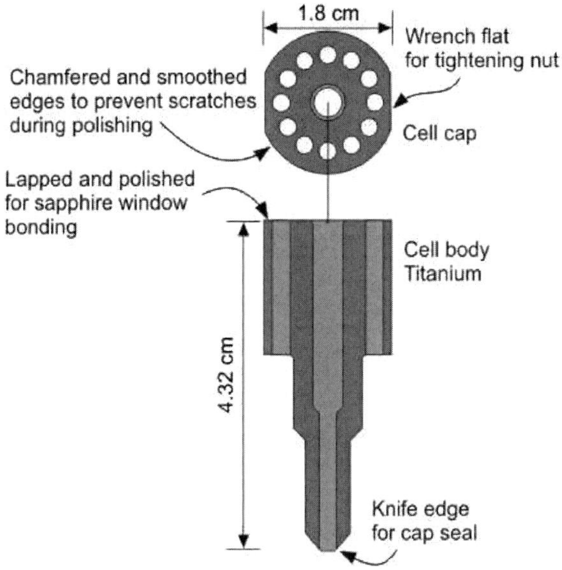

Figure 3. Schematic of potassium vapor cell components.

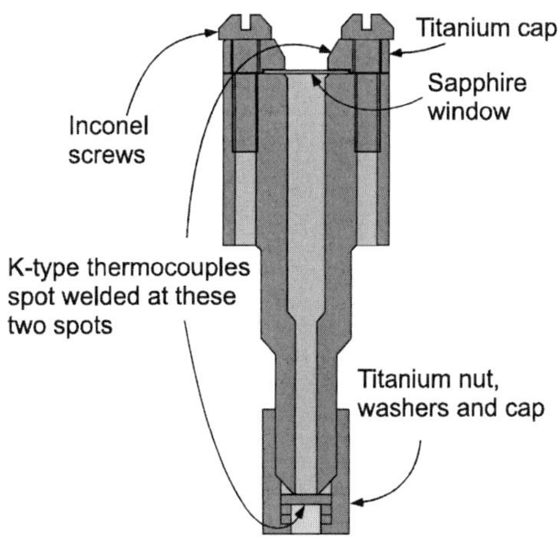

Figure 4. Schematic of an assembled potassium vapor cell.

3.1.1. Bonding Procedure

We bond the sapphire to the titanium through diffusion bonding, which can be thought of as "welding" them together: we apply pressure and heat to the softening point of titanium. First the titanium body is baked without the window to ensure that debris are not released during bonding. The sapphire window is cleaned with doubly distilled methanol and set on top of the polished end of the cell, anti-reflection coating facing out. The window is pressed against the body with a custom-made holder (not shown in figures) that applies constant pressure across the outer circumference of the window. The entire cell is then placed inside a vacuum tube with a cylindrical current-controlled heater surrounding the cell outside the tube. The tube is evacuated and the heater is ramped to reach 800°C from room temperature in about 4 hours. The temperature is kept at 800°C for 2 hours to bond the window to the cell.

3.1.2. Loading Procedure

To load the cell with potassium, we use a custom-made attachment on a vacuum system, shown in Fig. 5. The attachment is in the shape of a "T" and has three openings, one to the pump, one for the empty cell and one for the nut, washers and cap that seals the cell once it is filled with potassium. It also has a cold trap to prevent particles from entering the pump.

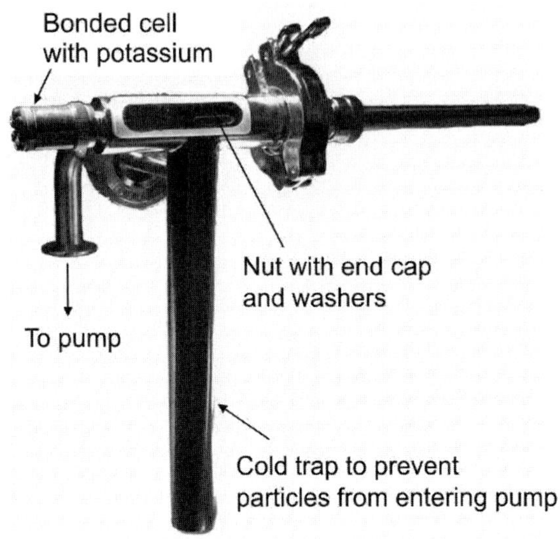

Figure 5. Vacuum attachment for loading the cell.

Once the attachment is connected to the pump, we place the nut with washers and cap together into a wrench socket connected to a steel rod that allows us to tighten the nut onto the cell inside the vacuum attachment. This is inserted into the right arm of the attachment shown in Fig. 5. Using a glove bag with an opening connected to the left arm of the attachment, we then load the potassium into the cell under an argon atmosphere to prevent the potassium from oxidizing. We use a simple stainless steel scraper to place an amount of potassium corresponding to about a 2 mm diameter sphere into the cell. Once filled, we connect the cell to the left arm of the attachment. Then we evacuate the cell to about 7×10^{-7} Torr using a turbo pump. Once under vacuum we tighten the nut to seal the cap against the knife edge of the cell and then close the valve to the turbo pump, vent to atmosphere and take out the assembled cell. Figure 6 shows the various pieces of the cell lined up for assembly and Fig. 7 shows an assembled cell.

K-type thermocouples are then spot-welded onto the titanium both at the front of the cell by the window and at the back of the cell on the cap. We slip an oven, which is custom-made for the cell, onto the back end of the cell and bind the leads of the oven and thermocouples together with a stainless steel wire. The oven consists of wire that is lead through six heaters made of poly-crystalline alumina and held together with Sauereisen. The six heaters fit into the holes drilled parallel to the cell body. The oven is designed to

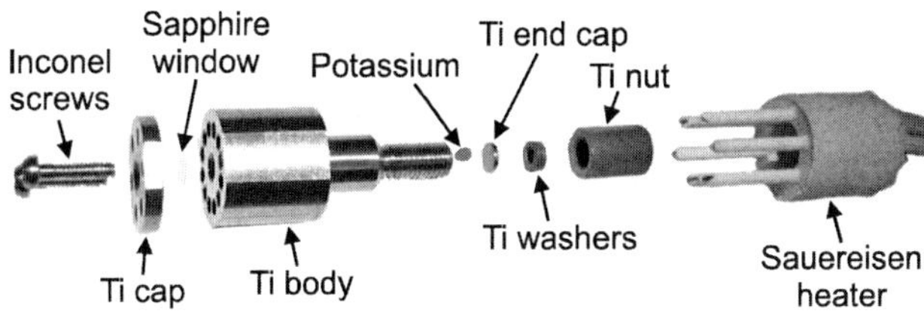

Figure 6. Pieces of the vapor cell lined up for assembly.

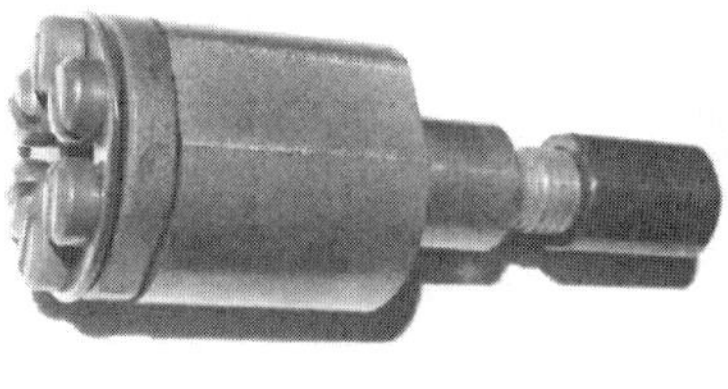

Figure 7. An assembled vapor cell.

keep the temperature at the window a few degrees higher than the temperature at the nut in order to keep the potassium from pooling at the window. Nickel foil is cut and crinkled to wrap around the cell and oven for heat shielding.

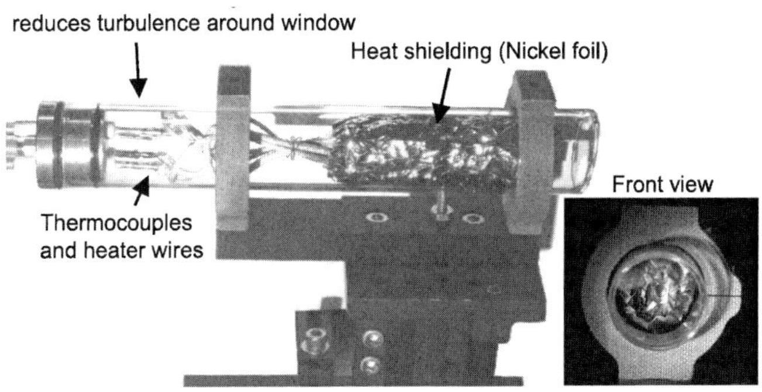

Figure 8. Potassium vapor cell inside an evacuated glass tube to reduce turbulence and for heat shielding.

The wrapped cell is then put inside a glass tube to reduce turbulence around the window and also for heat shielding, shown in Fig. 8. The glass tube is 20 cm long and 3 cm in diameter. One end of the glass tube is sealed with an angled face to prevent back-reflections, while the other end is sealed with a removable metal plug with connectors for inserting heater and thermocouple wires and tubing for evacuation of the glass tube. The vapor cell is placed inside the glass tube so that the sapphire window faces the angled face of the glass cell. The wires from the oven and thermocouples are connected through the metal plug for attachment to the temperature controller. The glass cell is pumped down to rough vacuum and the potassium cell is heated up to about 300°C to allow the potassium to pool at the nut. The density of the potassium vapor is determined from the temperature at the nut through phenomenological equations relating temperature to vapor pressure and the ideal gas equation [31]. A picture of a laser beam incident on a cell at approximately 500°C is shown in Fig. 9.

If the process goes well, the cell can last up to 130 hours above 500°C. The cells eventually fail at the titanium-sapphire interface. When the seal fails, a yellow film forms on the outside of the cell. Figure 10 shows a cell whose titanium-sapphire seal failed. The average lifetime is about 100 hours (ranging from 85 to 130 hours) with about 10 heating/cooling cycles.

Figure 9. Photo of a laser beam incident on a warm potassium vapor cell.

Figure 10. A used cell.

3.2. The Two-Pulse Photon Echo

The two-pulse photon echo (2PE) experiment and the relevant energy levels of potassium are depicted in Fig. 11. A stretched-cavity mode-locked Ti:sapphire laser produces 160 fs pulses at a repetition rate of 26 MHz. Because of their broad bandwidth, the pulses are resonant with both the D_1 and D_2 lines of potassium. The laser output is divided into two pulses and focused onto the window-vapor interface. The two pulses, with wavevectors $\mathbf{k}_a$ and $\mathbf{k}_b$, have a variable delay in time with respect to one another. The first pulse induces a coherence between the ground and excited states of the atoms with a spatial phase variation

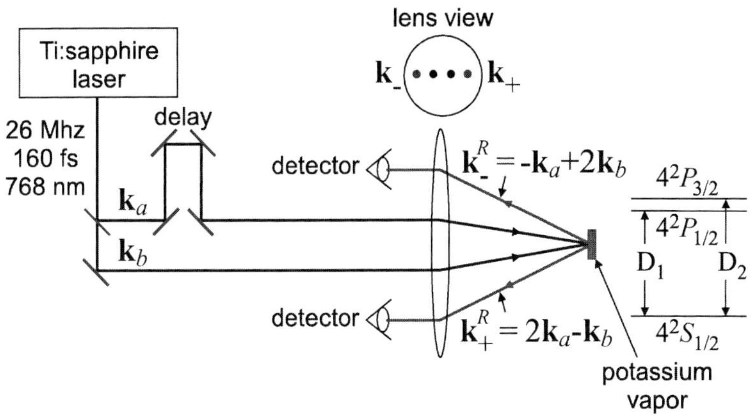

Figure 11. The time-integrated 2-pulse photon echo experimental setup and the optically active energy levels of potassium. The signals are detected in reflection (denoted by a superscript R on the wavevectors) due to the small absorption length ($\sim \lambda/2$) of the potassium vapor.

across the vapor determined by its angle of incidence and wavelength. After a delay τ, the second pulse produces an excited state population whose amplitude is greatest in areas where the coherence left by the first pulse constructively interferes with the second pulse and least where they destructively interfere. Thus the excited state population is spatially modulated, forming an amplitude grating. At the same time it creates the grating, the second pulse also scatters off the grating (i.e. creates a coherence that radiates). This background-free signal is time-integrated by a slow photodiode detector. Signals in directions $2\mathbf{k}_b - \mathbf{k}_a$ and $2\mathbf{k}_a - \mathbf{k}_b$ can be detected. For the signal in direction $2\mathbf{k}_b - \mathbf{k}_a$ positive delay corresponds to $\mathbf{k}_a$ arriving first, while for the signal in direction $2\mathbf{k}_a - \mathbf{k}_b$ positive delay corresponds to $\mathbf{k}_b$ arriving first. The scattered signal has an intensity proportional to the amplitude of the population grating. By varying the time delay between pulses, τ, the transient behavior of the induced coherence can be investigated. The temperature of the liquid potassium at the back of the vapor cell is used to determine the density from phenomenological equations for the vapor pressure and the ideal gas equation [31].

Typical experimental results for a series of vapor densities and temperatures are shown in Fig. 12, with the detected signal intensity plotted as a function of the delay between pulses. A striking feature is the oscillations, which is the quantum beating due to the simultaneous excitation of two energy levels. The modulation envelope of the beats initially increases at short delays, peaks and then exponentially decays at long delays. The non-exponential behavior is indicative of the presence of non-Markovian dynamics at short delays, while the exponential decay at long delays is representative of homogeneous dephasing. In particular, the rise of the time-integrated signal from zero delay is due to the integration up to the full area of the real-time signal, a photon echo [32]. The photon echo arises from the random atomic spacings and because the evolution of the coherent superposition between ground and excited state is still reversible over the delay. At zero delay, there is only half of a photon echo because of causality; the photon echo cannot form until after the second pulse, and so if τ is zero, only half of the echo forms. As τ is increased to

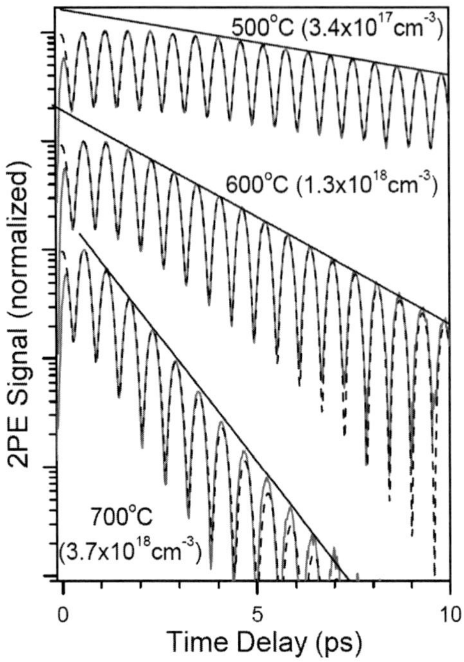

Figure 12. Experimental two-pulse photon echo signals (solid grey lines) for potassium vapor at various temperatures and densities, with curves offset for clarity and the straight lines drawn to highlight the non-exponential behavior at short times. The dashed lines are least-square fits assuming a stochastic modulation of the electronic transition frequency during collisions.

positive delays, the signal rises due to the integration over the developing photon echo. As the signal begins to peak, collisions start to take effect, randomizing the phases of the superpositions such that they cannot rephase as well to produce a photon echo. This intermediate regime occurs when τ is approximately equal to the correlation time. The exponential dephasing at long delays is due to Markovian dynamics, or pure homogeneous broadening, as the phases of the superpositions are completely randomized over the delay. The finite rise time of the signal near zero delay is not due to non-Markovian dynamics but results from the finite duration of the pulses. As temperature and density increase, the range of the non-exponential behavior representing non-Markovian dynamics decreases. This trend agrees with the expectation that the collision duration and time between collisions should decrease with increasing temperature.

Based on thermal velocities calculated assuming ideal gas behavior and considering the attractive range of a typical interatomic potential energy surface, at 700°C the collision duration is approximately 0.5 ps. This collision duration is much less than the time between collisions, which is of the order of a few ps, and much greater than the pulse duration of 160 fs. This estimated value for the collision duration agrees with the delay at which the signal peaks, which corresponds to approximately the correlation time. In condensed phase systems, the timescale of fluctuations can be on the order of 200 fs [2], with dephasing due to solvent interaction also occuring at ultrafast timescales [5]. Thus the above experimental results exhibit unusually well-separated contributions from homogeneous and inhomogeneous broadening as compared to molecules in the condensed phase, with the pulse duration shorter than the dynamics of interest.

The relative simplicity of this atomic system provides an opportunity to test the stochastic theory that is often used in more complex molecular systems, discussed in Sec. 2. [26]. The response functions for the two-pulse case are retrieved from the three-pulse response functions Eqs. 9(a-d) by setting $\tau_1 = \tau_2$ [26]. Fig. 12 contains least-square fits of the theoretical signals to the experimental data with the primary fitting variables being the parameters of the stochastic correlation function, Δ_{ij} and Λ_{ij}, for the two transitions. The final integration is performed numerically. The theoretical results match quite well with the experimental results [26]. The discrepancy near zero delay is due to the use of delta-function pulses in time, which does not allow for finite pulse width effects.

The agreement between theory and experiment implies that the stochastic model is sufficient to model the dynamics, and that taking memory in the bath (i.e. all the other atoms) into account, for example, is not necessary [33, 34]. The stochastic model is equivalent to a foreign-gas model in which the explicit transient formation of dimers is not considered, but rather it is assumed that the interactions can be averaged into the bath fluctuations. Although the agreement is encouraging, it is known that at the densities reached in this experiment resonance broadening effects and therefore long-range interactions must be considered, and a least-squares fit cannot provide new physical information about the durations of such long-range interactions or their strength. Such dynamics can be more thoroughly explored through three-pulse photon echo experiments, in which the change in photon echo signal strength is probed directly to map out the correlation function of frequency fluctuations [7, 35, 36, 37].

3.3. The Three-Pulse Photon Echo

In this section the three-pulse photon echo (3PE) experiment results are described, and it is shown that the stochastic model alone cannot account for the dynamics. Modeling using an exciton theory and molecular dynamics simulations gives insight into the effects of long-range interactions. A second 3PE experiment on potassium vapor with a buffer gas supports the theoretical interpretation of the experimental results.

3.3.1. Experiment

The experimental setup for the three-pulse photon echo is shown in Fig. 13. The 3PE technique extends the previous experimental arrangement by adding a third pulse. The first pulse excites a coherence between the ground and excited states. During the ensuing delay τ, dephasing occurs due to collisions. Then the second pulse arrives and converts the remaining coherence to excited-state or ground-state population. During the second delay T, loss of population and spectral redistribution occurs due to decay and collisions, and the system's ability to rephase and form a photon echo decreases. The third and final pulse forms a third-order coherence that radiates as the signal in directions $\mathbf{k}_- = -\mathbf{k}_a + \mathbf{k}_b + \mathbf{k}_c$ and $\mathbf{k}_+ = \mathbf{k}_a - \mathbf{k}_b + \mathbf{k}_c$. The time-integrated 3PE signal is measured as a function of the first delay τ for various fixed values of the second delay T, with positive delay corresponding to

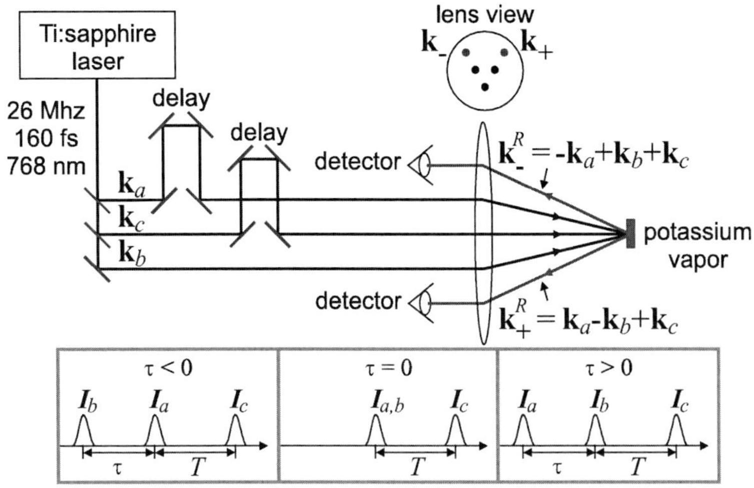

Figure 13. The time-integrated 3-pulse photon echo experimental setup and a schematic (bottom) of pulse orderings for positive, zero and negative delays.

the conjugate pulse coming first (see Fig. 13 for a schematic). The values of T are chosen at integer multiples of the beat period to maximize the signal, and thus any oscillatory behavior at the beat period as a function of T may have been aliased out [38]. We simultaneously measure $\mathbf{k}_-$ and $\mathbf{k}_+$ so that zero delay is easily determined and drifts in it compensated.

As exhibited by the data shown in Fig. 14, and considering only the $\mathbf{k}_-$ signal for simplicity, for $\tau > 0$ the modulation envelope of the signal versus τ rises, peaks, and then exponentially decays. This is the same behavior as exhibited by the two-pulse photon echo signal. Across the peak, the real-time signal changes from a photon echo to free polarization decay, corresponding to the crossover from non-Markovian to Markovian dynamics. As the second delay T is increased, the signals reveal how the system's ability to form a photon echo changes. For the case of the first delay τ being less than the correlation time, we can divide the dynamics into three regimes as a function of the second delay T. For T less than the correlation time, the non-Markovian dynamics at short times results in a real-time signal that is a photon echo, but the intensity is slightly decreased due to the phase-randomizing collisions that occur during the second delay. For T much greater than the correlation time, the changes in resonance frequency during the second delay precludes a photon echo from forming. In the time-integrated signal, this results in a shift of the peak toward $\tau = 0$. The envelope peaks are designated by arrows in Fig. 14. This shift as a function of T is referred to as the three-pulse photon echo peak shift and has proven to be a useful probe of frequency fluctuations in numerous systems [2, 3, 5, 28, 39, 40, 41, 42, 43]. To determine the echo peak shift, Gaussian fits are made to each beat of scans of τ for each T and the maxima of the beats are fit with a Gaussian to find the overall peak of the 3PE signal. Three-pulse photon echo data are taken for a series of temperatures and densities. The resulting 3PE peak shifts are shown in Fig. 15.

It has been shown that varying the delay τ between the first and second pulse for various values of the delay T between the second and third pulse allows the mapping out of the

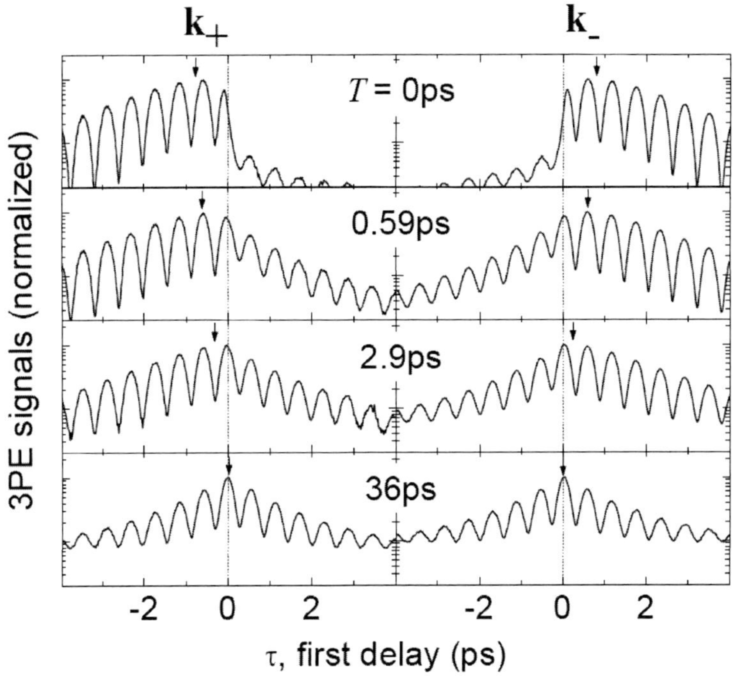

Figure 14. Experimental 3-pulse photon echo signals at 600°C ($1.3 \times 10^{18}\text{cm}^{-3}$) for $T =$ 0 ps to 36 ps. The peak of the signal, marked by an arrow in the figure, shifts to $\tau = 0$ ps, marked by a dotted line, due to the loss of rephasing capability.

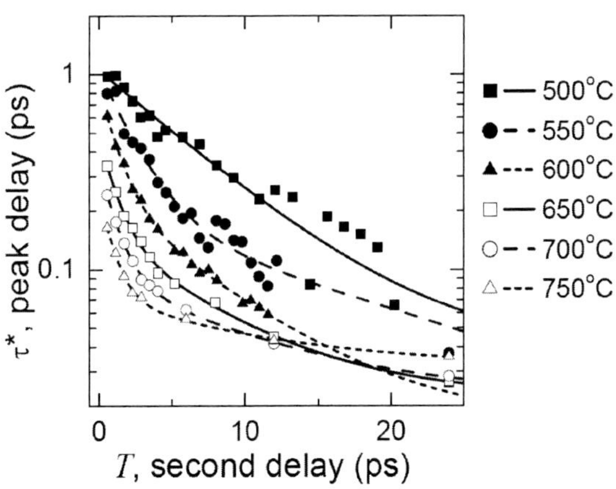

Figure 15. Three-pulse photon echo peak shifts for a series of temperatures and densities with bi-exponential fits (lines).

correlation function of frequency fluctuations for some systems [37, 44]. Physically this seems reasonable as well considering that varying the second delay probes changes that occur while the system is in the intermediate density matrix state corresponding to population. By changing that time interval we have access to the timescales of the randomization process that results in the loss of rephasing capability. That is, because the photon echo cannot arise until after the third pulse reverses the evolution of the coherent superposition created by the first pulse, varying the second-third pulse delay "exposes" the coherence to variable durations of randomizing collisions. It is just these randomizing collisions that determine the form of the correlation function. Increasing the second delay decreases the inhomogeneous system response that determines the peak location in the signal versus τ. The resulting photon echo peak shift thus reveals the form of the microscopic correlations involved in the dephasing process. In some molecular systems this relationship has been shown to take a more direct form, particularly for systems that exhibit both fast and slow components in the correlation function [29, 37, 44]. Although this is not necessarily the case for a dense atomic vapor, the experimental results yield a qualitative form.

The experimental peak delays versus T are exponential at lower temperatures and thus qualitatively support the use of an exponential correlation function in modeling two-pulse experiments, as was done for the 2PE signals above [26]. However, as temperature and density are increased, the peak shift becomes bi-exponential, which cannot be modeled with a single exponential correlation function. The biexponential fitting functions shown in Fig. 15 are of the form $\Delta_1^2 \exp(-\Lambda_1 T) + \Delta_2^2 \exp(-\Lambda_2 T)$, where the subscripts 1 and 2 refer to the fast and slow exponentials, respectively. The inverse time constants Λ_i and amplitudes Δ for the exponentials are plotted in Fig. 16 along with the parameters obtained from the theoretical fits to the two-pulse photon echo experimental results. The bi-cxponential behavior could represent the presence of two collision mechanisms, such as both short-range and long-range collisions, the latter being due to resonance effects [45].

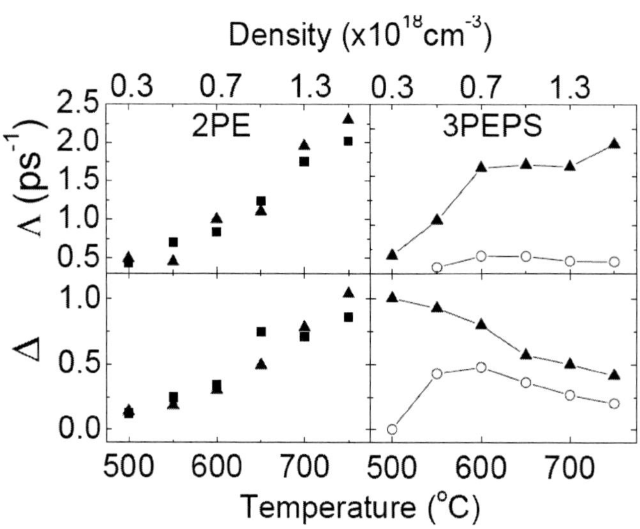

Figure 16. Left side: Stochastic theory fitting parameters Λ and Δ for the 2-pulse photon echo (2PE) experimental results for the D_1 (square) and D_2 (triangle) line versus temperature and density. Experimental 3-pulse photon echo peak shift (3PEPS) bi-exponential fitting parameters Λ and Δ for the fast (triangle) and slow (circle) exponentials.

At the densities and temperatures used in the experiment, resonant dipole collisions are the dominant broadening mechanism over Doppler broadening [45]. The stochastic model described in Sec. 2. is equivalent to a foreign-perturber model of collisions in which the interactions are short-range. This model makes the binary collision approximation, which only considers the interaction of one pair of resonant and foreign atoms at a time. For a self-broadened vapor, coupling between more than two atoms at a time must be considered due to the long-range dipole interactions between atoms of the same species. Thus, to improve the theoretical approach, we consider the effects of long-range dipole interactions on the non-Markovian dephasing process.

3.3.2. Theory

Because of the random spacing between the atoms, a self-broadened alkali vapor can be considered as a highly disordered exciton system [46]. The exciton formalism is more commonly used in condensed matter, for example in semiconductors and molecular crystals, as well as for other disordered systems for which resonant interactions are important. In semiconductors, an exciton is a bound electron-hole pair, while in molecules it is an excitation distributed across molecular constituents. For a dense atomic vapor, an exciton is similar to the molecular case, known as a Frenkel exciton. This formalism accounts for the distribution of excitations over multiple atoms. In the following we consider only first-order exciton coupling, or single excitations delocalized across multiple atoms. Two-exciton couplings, or interactions with two excitations in the system, are also present in a dense atomic vapor; however, a rigorous quantum simulation of the nonlinear optical response would require integration of the many-particle Schrödinger equation within the single- and two-exciton manifolds and is not undertaken here.

Our approach is to use the one-exciton formalism to calculate the frequency fluctuations of atoms in a molecular dynamics simulation and map the resulting correlation function into a modified stochastic model of the nonlinear response. Ultimately we find that the excitonic effects induce a slowly decaying contribution to the peak shift that resembles the slow exponential decay of the experimental peak shifts. Through further comparison of calculations using short- and long-range potential energy surfaces and a buffer gas experiment we attribute the slow decay to long-range dipole interactions.

The calculation requires four steps: first, we simulate trajectories of an ensemble of atoms; second, we determine the resonant couplings based on distances between atom pairs for each snapshot of the trajectory; third, we diagonalize the Hamiltonian to obtain the exciton energies and calculate the correlation functions; and fourth, we calculate the peak shifts within a modified stochastic model of the three-pulse–four-wave mixing response.

Using the molecular dynamics package GROMACS [47], we simulate a trajectory for 20 atoms in the gas phase using the $1^1\Sigma_g^+$ ground-state potential energy surface of the potassium dimer, K_2 [48]. Twenty runs, each 600 ps long with a resolution of 0.02 ps, are made for a series of temperatures and densities, with the box sizes scaled to include the same number of atoms for each temperature. Simulations with 20 atoms are similar to those with 40 and thus 20 atoms are used.

The Frenkel exciton Hamiltonian is

$$\hat{H}_S = \sum_m \Omega_m \hat{B}_m^\dagger \hat{B}_m + \sum_{m,n}^{m \neq n} J_{m,n} \hat{B}_m^\dagger \hat{B}_n, \tag{12}$$

where Ω_m is the excitation energy of atom m and $J_{m,n}$ is the resonant exciton coupling between atoms m and n. For each frame of the trajectory, the resonant exciton couplings $J_{m,n}$ for all atom pairs are determined from the difference between the $1^1\Sigma_u^+$ excited-state potential energy surface and the $1^1\Sigma_g^+$ ground-state surface of K_2 [48]. The other surfaces dissociating to $4\mathrm{s}+4\mathrm{p}$ are not taken into account, nor are transitions from the $1^3\Sigma_u^+$ ground state. Quartic couplings, which result in two-exciton states, are assumed to be beyond the spectral width of the laser such that they could be neglected in the simulations. This assumption would require a full simulation to justify.

For each snapshot, if the elements of the eigenvectors changed by greater than $0.3\ \mathrm{cm}^{-1}$ as compared with those of the previous snapshot, the eigenvectors are reordered to minimize the number of elements that underwent such a change, though there remain avoided crossings within eigenvectors that cannot be reconciled frame-to-frame. The fluctuating one-exciton energies obtained from diagonalizing the Hamiltonian are then used to calculate the correlation functions of frequency fluctuations, with its discretized form given by

$$C(\tau_j) = \langle \sum_{i=0}^{N} \delta\omega(t_i - \tau_j)\delta\omega(t_i) \rangle, \tag{13}$$

where $\delta\omega$ is the difference between the one-exciton energy levels and Ω_m, the brackets denote an ensemble average over all the one-exciton states, and N is the number of frames in the trajectory for calculating $C(\tau)$. In the experiment only fluctuations within the laser bandwidth ($\sim 200\ \mathrm{cm}^{-1}$) contribute to the response, and thus only energies within the laser bandwidth are used in the calculation. A survey of the range of exciton energies in the simulation reveals that most interactions stay within the laser bandwidth, representing the fact that the majority of collisions occur beyond the range of the most attractive or repulsive parts of the potential energy surfaces. The correlation functions are averaged over all runs for a series of temperatures and densities.

For direct comparison with experiment, echo peak shifts are calculated using the correlation functions calculated using the exciton Hamiltonian. Each $C(\tau)$ is fit to a biexponential, with four fitting parameters corresponding to the amplitudes $\Delta_f(\Delta_s)$ and decay rates $\Lambda_f(\Lambda_s)$ of the fast (slow) component. These parameters are used as the amplitudes and decay rates of two stochastic correlation functions. The peak shifts are calculated from these correlation functions using the response functions similar to those in Sec. 2., except for a two-level system [9, 26] and now with g_{ij} equal to the sum of the line shapes of the fast and slow correlation functions. The sum of the response functions yields the polarization [26, 27]. The signal is calculated by numerically time integrating the intensity for a sequence of time delays between pulses.

The calculated peak shifts are shown in Fig. 17 and exhibit similar changes in the ratio of slow-to-fast exponential amplitudes as a function of temperature and density as experimental peak shifts. We note that there are no adjustable parameters in the theory. To determine the origin of the biexponential behavior of the correlation function, we use a

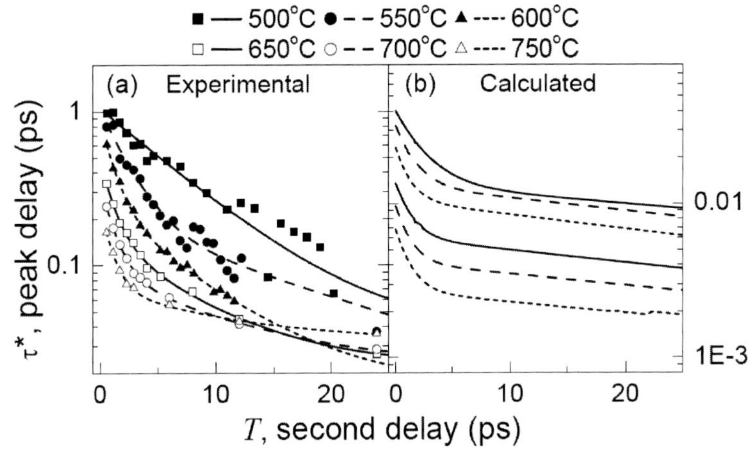

Figure 17. Experimental peak shifts (left) and theoretical peak shifts (right) from molecular dynamics simulations with 20 atoms.

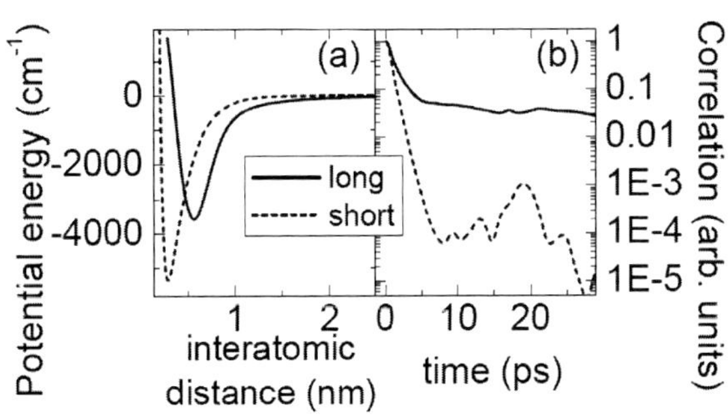

Figure 18. (a) Potential energy surfaces used to calculate correlation functions, with long- (solid) and short-range (dashed) attraction. (b) Resulting correlation functions of energy level fluctuations for long- (solid) and short-range (dashed) attraction.

shorter-range potential to calculate the one-exciton couplings, scaled to preserve the integrated potential energy change of the long-range potential, shown in Fig. 18(a). Using the shorter-range potential energy surface results in a single fast exponential decay, as shown in Fig. 18(b), revealing that the slow decay is due to long-range interactions.

3.3.3. Buffer Gas Experimental Results

To experimentally check this conclusion, we compare these results with peak shifts from a buffer-gas-broadened vapor. Echo peak shift data are taken of potassium vapor with approximately 220 Torr argon buffer gas at 600°C. As shown in Fig. 19, introducing argon has an effect similar to increasing both the density and temperature of a pure potassium vapor, increasing the decay rate and suppressing the slow exponential component relative to the fast. This supports our attribution of the slow component of the correlation function to long-range resonant interactions.

Considering the number of assumptions and the fact that there are no adjustable param-

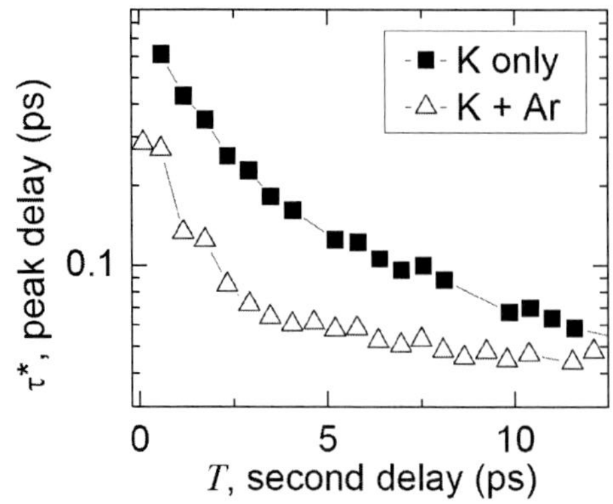

Figure 19. (color online) Experimental echo-peak shifts for potassium vapor with (triangle) and without (square) buffer gas.

eters, the agreement between theory and experiment is quite good. The major approximations are the use of only ground-state atoms in the dynamics simulation and the inclusion of only one of the dipole-allowed potential energy surfaces dissociating to 4s+4p (and also ignoring the fine structure and hyperfine splitting) in calculating the exciton Hamiltonian. Using only ground-state atoms in the dynamics simulation means we do not take into account the sensitivity of the experiment to the frequency fluctuations of excited atoms. The theoretical results do not agree as well with experiment for the lower temperatures, which may imply that the range of the potential surface is less than that represented by the surface used in the calculation. It is expected that theoretical improvements would result in qualitatively similar behavior as a function of temperature and density.

These results show that first-principle simulations using currently available potential energy surfaces are capable of modelling the dynamics. This analytical approach can be extended to other disordered excitonic systems such as molecular aggregates, semiconductors, and photosynthetic complexes.

3.4. The Time-Resolved Three-Pulse Photon Echo

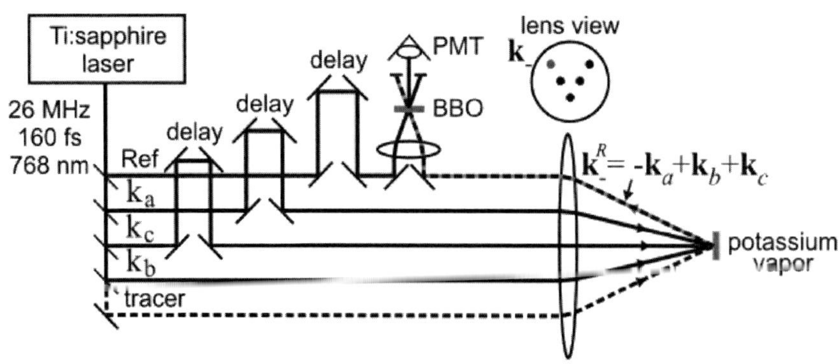

Figure 20. The time-resolved 3-pulse photon echo experimental setup.

We now consider the time-resolved version of the three-pulse photon echo experiment to explore the photon-echo-like behavior [29, 49, 50]. In the time-resolved version of 3PE, also called time-gated 3PE, a reference pulse is correlated with the emitted 3PE signal at a particular τ and T in a BBO crystal for up-conversion, resulting in a time-resolved scan of the nonlinear response. In the setup, shown in Fig. 20, there are three pulses used for four-wave mixing, one used as a tracer for the signal from the vapor, and one used as a reference for upconversion of the signal. The tracer is used for alignment and determining zero delay only and blocked during the photon echo experiment. In addition to the two delays for the 3PE signal, τ and T, there is a delay between the signal and the reference, t. Data are taken as a function of t for various fixed values of τ and T. Experimental results are shown in Fig. 21.

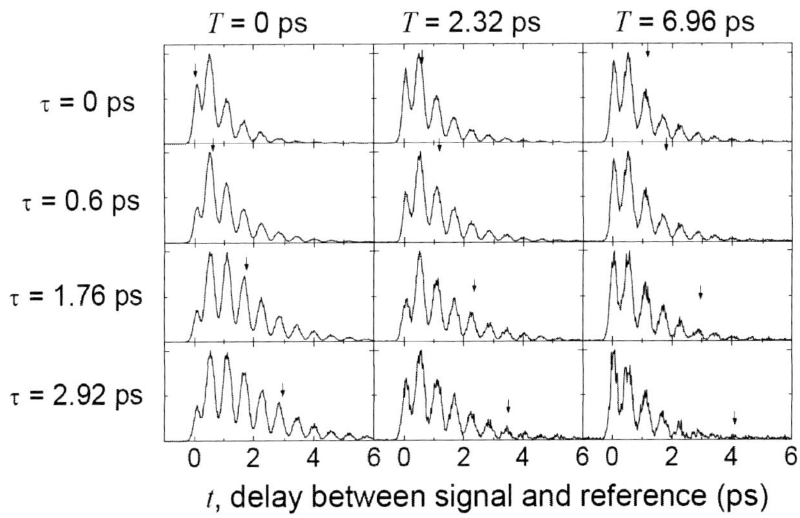

Figure 21. Experimental time-resolved 3-pulse photon echo signals at 600°C ($1.3 \times 10^{18}\text{cm}^{-3}$). Arrows point to $t = \tau$, the photon echo location for perfect rephasing. Photon-echo-like behavior decreases with increasing T.

The signals exhibits a free polarization decay and a photon echo for short delays. Although the photon echo is somewhat obscured by the free polarization decay, there are significant changes in the signal as a function of the delays. The locations of the photon echo in the case of perfect rephasing are marked by arrows in Fig. 21. Comparing the time-resolved signal across τ for small T, the signal exhibits a photon echo that moves to larger delays as τ increases. Considering the signals as T is increased, this photon-echo-like behavior goes away and the signal eventually becomes an exponential decay as a function of t for all τ. This behavior agrees with the qualitative description of the real-time response used to interpret the time-integrated results.

4. Conclusion

We have given a review of time-domain nonlinear optical experiments in a dense atomic vapor. A novel design for a vapor cell enables optical experiments in dense alkali vapors at high temperatures. Time-integrated and time-resolved four-wave mixing spectroscopy with two and three pulses is used to study ultrafast non-Markovian dynamics in a dense potassium vapor. Least-square fits based on the stochastic theory compare well with two-pulse time-integrated four-wave mixing results. The three-pulse photon echo peak shift reveals bi-exponential behavior at high temperatures and densities not modeled by the stochastic theory. A model using molecular dynamics simulations to derive correlation functions from potential energy surfaces within an exciton picture results in peak shifts similar to those in experiment. Further experiments in which argon buffer gas is introduced to the potassium vapor support the argument that the slower exponential decay arises from long-range interactions known to be present in resonance-broadened vapors. Finally, time-resolved three-pulse photon echo experiments exhibit photon-echo-like behavior.

The authors acknowledge the support of the NSF JILA Physics Frontier Center, thank T. Asnicar and H. Green for technical assistance and thank S. Mukamel and W. Zhuang for helpful discussions.

References

[1] Lorenz, V. O., Mukamel, S., Zhuang, W., and Cundiff, S. T. *Phys. Rev. Lett.* **100**, 013603 (2008).

[2] Fecko, C. J., Eaves, J. D., Loparo, J. J., Tokmakoff, A., and Geissler, P. L. *Science* **301**, 1698–1702 (2003).

[3] Stenger, J., Madsen, D., Hamm, P., Nibbering, E. T. J., and Elsaesser, T. *J. Phys. Chem. A* **106**, 2341–2350 (2002).

[4] Niemax, K. and Pichler, G. *J. Phys. B* **8**(2), 179–184 (1975).

[5] Fleming, G. R. and Cho, M. *Annu. Rev. Phys. Chem.* **47**(47), 109–134 (1996).

[6] Jimenez, R., Fleming, G. R., Kumar, P. V., and Maroncelli, M. *Nature* **369**, 471–473 (1994).

[7] Joo, T., Jia, Y., Yu, J. Y., Lang, M. J., , and Fleming, G. R. *J. Chem. Phys.* **104**(16), 6089–6108 (1996).

[8] Nibbering, E. T. J., Wiersma, D. A., and Duppen, K. *Phys. Rev. Lett.* **66**(19), 2464–2467 (1991).

[9] Mukamel, S. *Phys. Rev. A* **28**(6), 3480–3492 (1983).

[10] Margenau, H. and Watson, W. W. *Rev. Mod. Phys.* **8**, 2253 (1936).

[11] Ward, J. F. *Correlation effects in the theory of combined doppler and pressure broadening*. Thesis, University of Colorado, Boulder, (1964).

[12] Cooper, J. *Rev. Mod. Phys.* **39**, 167–177 (1967).

[13] Vleck, J. H. V. and Huber, D. L. *Rev. Mod. Phys.* **49**, 939–959 (1977).

[14] Niemax, K., Movre, M., and Pichler, G. *J. Phys. B* **12**, 3503 (1979).

[15] Walkup, R. E., Speilfiedel, A., and Pritchard, D. E. *Phys. Rev. Lett.* **45**(12), 986–989 (1980).

[16] Prior, Y. and Ben-Reuven, A. *Phys. Rev. A* **33**, 2362–2377 (1986).

[17] Asida, N., Rosenbluh, M., and Mukamel, S. *Phys. Rev. A* **40**, 41234126 (1989).

[18] Laubereau, A. and Kaiser, W. *Rev. Mod. Phys.* **50** (1978).

[19] Kinrot, O. and Prior, Y. *Phys. Rev. A* **50**(3), R1999 (1994).

[20] Sargent, M., III, Scully, M. O., Lamb, W. E., and Jr. *Laser Physics*. Perseus, Cambridge, Mass., (1974).

[21] Bloch, F. *Phys. Rev.* **70**, 460–474 (1946).

[22] Mukamel, S. *Principles of Nonlinear Optical Spectroscopy*. Oxford University Press, New York, (1995).

[23] Abramavicius, D. and Mukamel, S. *J. Phys. Chem. B* **108**, 10295–10300 (2004).

[24] Hänggi, P. and Ingold, G.-L. *Chaos* **15**(026105) (2005).

[25] Lorenz, V. O. *Ultrafast Spectroscopy of Non- Markovian Dynamics in a Dense Atomic Vapor*. Thesis, University of Colorado, (2007).

[26] Lorenz, V. O. and Cundiff, S. T. *Phys. Rev. Lett.* **95**, 163601 (2005).

[27] Lorenz, V. O. and Cundiff, S. T. *Chem. Phys.* **341**, 106–112 (2007).

[28] de Boeij, W. P., Pshenichnikov, M. S., and Wiersma, D. A. *Annu. Rev. Phys. Chem.* **49**, 99–123 (1998).

[29] de Boeij, W. P., Pshenichnikov, M. S., and Wiersma, D. A. *J. Phys. Chem.* **100**, 11806–11823 (1996).

[30] Martin, W. and Wiese, W. In *Atomic, Molecular and Optical Physics Handbook,* Drake, G. W. F., editor. Springer-Verlag, New York (2004).

[31] Nesmeyanov, A. N. *Vapor Pressure of the Chemical Elements*. Elsevier, London, (1963).

[32] Süsse, K.-E., Vogel, W., and Welsch, D.-G. *Chem. Phys. Lett.* **162**(4-5), 287–290 (1989).

[33] Aihara, M. *Phys. Rev. B* **25**(1), 53–60 (1982).

[34] Lavoine, J. P., Boeglin, A. J., Martin, P., and Villaeys, A. A. *Phys. Rev. B* **53**(17), 11535–11541 (1996).

[35] Weiner, A. M., Silvestri, S. D., and Ippen, E. P. *J. Opt. Soc. Am. B* **2**(4), 654–662 (1985).

[36] Silvestri, S. D., Weiner, A. M., Fujimoto, J. G., and Ippen, E. P. *Chem. Phys. Lett.* **112**(3), 195–199 (1984).

[37] Cho, M., Yu, J.-Y., Joo, T., Nagasawa, Y., Passino, S. A., and Fleming, G. R. *J. Phys. Chem.* **100**, 11944–11953 (1996).

[38] Huse, N., Heyne, K., Dreyer, J., Nibbering, E. T. J., and Elsaesser, T. *Phys. Rev. Lett.* **91** (2003).

[39] Jimenez, R., van Mourik, F., Yu, J. Y., and Fleming, G. R. *J. Phys. Chem. B* **101**, 7350 (1997).

[40] Hamm, P., Lim, M., and Hochstrasser, R. M. *Phys. Rev. Lett.* **81**(24), 5326–5329 (1998).

[41] Groot, M.-L., Yu, J.-Y., Agarwal, R., Norris, J. R., and Fleming, G. R. *J. Phys. Chem. B* **102**, 5923 (1998).

[42] Piryatinski, A., Lawrence, C. P., and Skinner, J. L. *J. Chem. Phys.* **118**, 9672 (2003).

[43] Carter, S. G., Chen, Z., , and Cundiff, S. T. *Phys. Rev. B* **76**, 121303 (2007).

[44] de Boeij, W. P., Pshenichnikov, M. S., and Wiersma, D. A. *Chem. Phys. Lett.* **253**, 53–60 (1996).

[45] Leegwater, J. A. and Mukamel, S. *Phys. Rev. A* **49**(1), 146–155 (1994).

[46] Mukamel, S. *Annu. Rev. Phys. Chem.* **51**, 691–729 (2000).

[47] van der Spoel, D., Lindahl, E., Hess, B., Groenhof, G., Mark, A. E., and Berendsen, H. J. C. *J. Comp. Chem.* **26**(16), 17011718 (2005).

[48] Magnier, S., Auber-Frécon, M., and Allouche, A. R. *J. Chem. Phys.* **121**(4), 1771–1781 (2004).

[49] Pshenichnikov, M. S., Duppen, K., and Wiersma, D. A. *Phys. Rev. Lett.* **74**(5), 674–677 (1995).

[50] Vöhringer, P., Arnett, D. C., Yang, T.-S., and Scherer, N. F. *Chem. Phys. Lett.* **237**, 387–398 (1995).

In: New Trends in Quantum Coherence...
Editor: Rafael Drampyan, pp. 149-169

ISBN: 978-1-60741-025-6

Chapter 5

POLARIZATION SPECTROSCOPY OF ALKALI-METAL ATOMS

Ifan G. Hughes*
Department of Physics, Durham University
Durham, DH1 3LE, UK

Abstract

A study of polarization spectroscopy of alkali-metal atoms (Rb, Cs) on the D_2 line is presented. In polarization spectroscopy a weak probe beam analyses the birefringence induced in a sample by a strong counter propagating circularly polarized pump beam. The signal is in the form of a dispersion curve. We outline a numerical treatment of optical pumping and use it to make predictions about the general form of polarization spectra in alkali-metal atoms. Theoretical spectra for the D_2 transitions in Rb and Cs are generated. Comparison of experiments conducted with diode lasers in room temperature vapours and theory demonstrates that the model accurately reproduces spectra of transitions from the upper hyperfine level of the ground state. Among these, the closed transition $F \rightarrow F' = F + 1$ dominates. The polarization spectrum provides an ideal reference, or "error", signal for frequency stabilization of a laser.

1. Introduction

In 1976 Wieman and Hänsch introduced a novel technique for generating sub-Doppler spectra by monitoring the nonlinear interaction of two laser beams in an absorbing gas via changes in light polarization [1]. As in saturation spectroscopy [2] the laser beams (pump and probe) are monochromatic and derived from the same source. Whereas in saturated absorption spectroscopy the decrease in absorption of the probe beam is caused by the pump beam having selectively depleted population from the absorbing level, in contrast the polarization spectroscopy signal arises from the change of polarization of the probe beam. The linearly polarized probe can be resolved into two equal amplitude beams of opposite circular

* E-mail address: i.g.hughes@durham.ac.uk

polarization. The pump beam optically pumps the medium resulting in different refractive indices for the two circular polarization states (birefringence) and different absorption coefficients (dichroism). The change of polarization state is detected as an increase in transmission thorough a polarizer. Another common feature shared by saturated absorption and polarization spectroscopy is that a signal is only generated when both pump and probe beams interact with the same group of atoms within a narrow velocity distribution - hence the spectra obtained are Doppler-free.

Early experimental work on polarization spectroscopy used a single detector, on which the probe beam is incident after passing through (nearly) crossed polarizers on either side of the birefringent medium [1,3]. In more recent work two detectors and a polarization beam splitting cube are used to form a balanced detector - similar to the scheme employed by Hänsch and Couillaud for stabilization of a laser frequency to a reflecting reference cavity [4]. This scheme provides dispersion-shaped resonances which can provide the "error" signal for electronic frequency stabilization ("locking") without any need for modulation. The motivation for many studies is the application of this modulation-free technique for frequency stabilization of dye, diode and solid state lasers [5-11].

Early theoretical work on modelling polarization spectroscopy [12] adopted the approximation of a single cycle of velocity selective optical pumping in a four-level system. This model partially accounted for some features in the experimental spectra, but did not succeed in describing the whole spectrum. In this chapter we describe a theoretical model of polarization spectroscopy based on a rate-equation treatment of optical pumping. Spectra are generated for the hyperfine transitions of the D_2 lines of rubidium and cesium, and show that for transitions from the upper hyperfine level of the ground state ($n\,^2S_{1/2}\left(F=\mathcal{I}+1/2\right)\rightarrow n\,^2P_{3/2}\left(F'=F+1,F,F-1\right)$, where $\mathcal{I}$ is the nuclear spin) the model gives good agreement with experiment. For transitions from the lower hyperfine level, the model breaks down, an indication that factors not included in the model play a significant role.

The layout of the remainder of the chapter is as follows. In Section 2 we model the propagation of a linearly polarized probe beam through a dichroic medium; derive the population rate equations and present results for the time-dependence of the population of the hyperfine states, and discuss how spectra are produced. Section 3 describes the experimental apparatus, and presents the results. Section 4 contains a discussion of the results, and we draw our conclusions in Section 5.

2. Theory

2.1. Propagation of a Linearly Polarized Probe through a Circular Dichroic and Birefringent Medium

It is desirable to obtain a polarization signal which has zero background. Here we consider a balanced detector, and show below that the optimal signal is obtained by orienting the plane of polarization of the probe beam at an angle $\pi/4$ with respect to an analyzer (a polarizing beam splitter). With the pump beam blocked the intensity in one arm of the polarizing beam splitter is equal to the intensity in the other arm. Subtracting these gives a zero background.

On introducing the pump beam the medium becomes birefringent and dichroic and the polarization of the probe beam rotates through a small angle.

The direction of propagation of the probe beam defines the z axis. The initial state, which is linearly polarized in the x-y plane at an angle φ with respect to the x axis, is

$$\vec{E} = E_x\,\hat{x} + E_y\,\hat{y} = E_0\left(\cos\varphi\,\hat{x} + \sin\varphi\,\hat{y}\right). \tag{1}$$

It is also convenient to rewrite this in terms of the circular polarization basis vectors,

$$\vec{E} = E_0\frac{\exp(-i\varphi)}{2}\left(\hat{x}+i\hat{y}\right) + E_0\frac{\exp(+i\varphi)}{2}\left(\hat{x}-i\hat{y}\right). \tag{2}$$

In propagating through the cell of length L the different circular components experience different absorption (dichroism) and refractive index (birefringence). The electric field of the probe beam after the cell is

$$\vec{E} = E_0\exp\left\{ik_+L - \frac{\alpha_+L}{2}\right\}\frac{\exp(-i\varphi)}{2}\left(\hat{x}+i\hat{y}\right) + E_0\exp\left\{ik_-L - \frac{\alpha_-L}{2}\right\}\frac{\exp(+i\varphi)}{2}\left(\hat{x}-i\hat{y}\right). \tag{3}$$

Here $k_\pm = \frac{\omega}{c}n_\pm$; $n_\pm$ are the refractive indices of the gas for the circular polarization components which drive $\sigma^\pm$ transitions; $\alpha_\pm$ are the corresponding absorption coefficients. It is convenient to introduce the difference between the refractive index and absorption coefficients for the different circular polarizations $\Delta n = n_+ - n_-$ and $\Delta\alpha = \alpha_+ - \alpha_-$. The angle, Φ, by which the plane of the probe beam is rotated after traversing the cell is

$$\Phi = \frac{k\Delta nL}{2} = \frac{\pi\Delta nL}{\lambda} \tag{4}$$

The probe beam is resolved into horizontal (x) and vertical (y) components by the polarizing beam splitting cube, and the difference in intensity $\left(I = \frac{1}{2}\varepsilon_0 cE^2\right)$ of the two components gives the polarization spectroscopy signal:

$$I_{\text{PS}} = I_y - I_x = -\frac{1}{2}\varepsilon_0 cE_0^2\exp\left\{-\frac{(\alpha_+ + \alpha_-)L}{2}\right\}\cos(2\varphi + 2\Phi)$$
$$I_{\text{PS}} = -I_0\cos(2\varphi + 2\Phi), \tag{5}$$

with I_0 being the intensity of the probe beam after the cell in the absence of the pump beam.

Cell-window birefringence has been neglected in the analysis, as the rotation induced by the pump beam is usually the dominant effect for the D_2 transitions for alkali-metal atoms. It is also assumed that the medium is optically thin - for a dichroic optically thick medium one hand of circular polarization will be absorbed completely, leaving a (weak) circularly polarized probe for which I_{PS} is zero.

We now make further approximations that simplify these results. The rotation angle induced by the birefringent medium is typically small. The largest polarization spectrum is thus obtained with the choice of $\varphi = \pi/4$ for the initial probe-beam polarization. In this case equation (5) reduces to

$$I_{PS} = 2I_0\Phi = I_0 \frac{2\pi \Delta n L}{\lambda}. \tag{6}$$

The refractive index and absorption coefficient of a medium are related via the Kramers-Kronig dispersion relation [13], therefore to characterize the medium it suffices to calculate either Δn or $\Delta\alpha$. In the vicinity of an isolated resonance the spectral profile of the difference in absorption is a Lorentzian of width (FWHM) Γ [2]:

$$\Delta\alpha = \frac{\Delta\alpha_0}{1 + \frac{4\Delta^2}{\Gamma^2}}. \tag{7}$$

Here the angular detuning, Δ, is defined as $\Delta = \omega_L - \omega_0$, with ω_L being the angular frequency of the laser and ω_0 the atomic transition angular frequency. $\Delta\alpha_0$ is the maximum difference in absorption at line centre. The concomitant change in the refractive index of the medium is obtained from the Kramers-Kronig relation:

$$\Delta n = \Delta\alpha_0 \frac{2c}{\omega_0\Gamma} \frac{\Delta}{1 + \frac{4\Delta^2}{\Gamma^2}}. \tag{8}$$

Combining equations (6) and (8) leads to the prediction of a dispersion-shaped polarization spectroscopy signal on a zero background; this is an ideal error signal for frequency stabilization of the laser by locking to the atomic resonance.

Cell-window birefringence and imperfections in the polarizing beam splitter can cause an offset in the pump-induced rotation angle. Empirically, one selects the angle of the probe beam to be $\varphi = \pi/4$, and then makes small modifications in order to obtain a zero-background in the absence of the pump beam.

2.2. Optical Bloch and Population Change Equations

The atom-light interaction is solved using the density matrix formulism. The optical Bloch equations for a two-level atom with ground state *a*, excited state *b*, are [14]

$$\begin{aligned}
\dot{\rho}_{bb} &= -i\frac{\Omega_\mathrm{R}}{2}\left(\tilde{\rho}_{ab}-\tilde{\rho}_{ba}\right) \quad -\Gamma\rho_{bb} \\
\dot{\rho}_{ba} &= i\Delta\tilde{\rho}_{ba}+i\frac{\Omega_\mathrm{R}}{2}\left(\rho_{bb}-\rho_{aa}\right) \quad -\frac{\Gamma}{2}\tilde{\rho}_{ba} \\
\dot{\rho}_{ab} &= -i\Delta\tilde{\rho}_{ab}-i\frac{\Omega_\mathrm{R}}{2}\left(\rho_{bb}-\rho_{aa}\right) \quad -\frac{\Gamma}{2}\tilde{\rho}_{ab} \\
\dot{\rho}_{aa} &= i\frac{\Omega_\mathrm{R}}{2}\left(\tilde{\rho}_{ab}-\tilde{\rho}_{ba}\right) \quad +\Gamma\rho_{bb}.
\end{aligned} \tag{9}$$

Here ρ_{aa} and ρ_{bb} are the ground and excited state probabilities, respectively, and the off-diagonal terms $\tilde{\rho}_{ab}$ and $\tilde{\rho}_{ba}$ are coherences. (The tilde notation indicates that the evolution of the coherence at the laser frequency has been factored out leaving "slow" variables). The Rabi frequency is denoted by Ω_R and is proportional to the product of the electric field of the laser and the transition matrix element. The effect of spontaneous emission is to reduce the excited state population at a rate $\Gamma = 1/\tau$, and the coherences at a rate $\Gamma/2$, where τ is the excited state lifetime [14]. Similar equations are derived for many-level atoms for every Zeeman sub-state.

Pressure broadening is neglected in these calculations. Gorris-Neveux *et al.* measured the Rb-Rb collisional self-broadening to be of the order of 10^{-7} Hz cm^3 [15]; therefore for a room temperature sample the pressure broadening is at least four orders of magnitude smaller than the natural broadening. Temperatures greater than 120°C are needed for the self-broadening to become comparable to the natural width.

Eliminating the coherences from equations (9) facilitates the solution of the optical Bloch equations. Numerical solutions of the optical Bloch equations containing the coherences and the rate equations for a few sample systems show that for a short time the solutions differ, however the solutions become indistinguishable after a time of $\sim 7\tau$ [9]. In section 2.4 it is shown that the typical time of flight of an atom through the laser beams is approximately two orders of magnitude longer, therefore it is an excellent approximation to solve the equations for the rate of change of populations.

To eliminate the coherences the time derivatives are set to zero in the second and third equations of (9) and the steady-state solution substituted into the first equation, obtaining

$$\dot{\rho}_{bb} = -\frac{\Omega_\mathrm{R}^2}{\Gamma}\frac{\left(\rho_{bb}-\rho_{aa}\right)}{1+\dfrac{4\Delta^2}{\Gamma^2}} \quad -\Gamma\rho_{bb}. \tag{10}$$

This leaves rate equations for populations only. It is possible to write the Rabi frequency in terms of the intensity and the saturation intensity I_{SAT}, $I/I_{\mathrm{SAT}} = 2\Omega_{\mathrm{R}}^2/\Gamma^2$. Equation (10) then becomes

$$\dot{\rho}_{bb} = -\frac{\Gamma}{2}\frac{I}{I_{\mathrm{SAT}}}\frac{\rho_{bb}}{1+\frac{4\Delta^2}{\Gamma^2}} + \frac{\Gamma}{2}\frac{I}{I_{\mathrm{SAT}}}\frac{\rho_{aa}}{1+\frac{4\Delta^2}{\Gamma^2}} - \Gamma\rho_{bb}. \tag{11}$$

The three terms on the right-hand side of (11) have obvious physical interpretation: the first and second are proportional to the light intensity, the former represents stimulated emission out of the excited state, the later absorption into the excited state. The third term accounts for spontaneous emission out of the excited state. The intensity-dependent terms have a Lorentzian line-shape, with full width at half maximum of Γ.

It is straightforward to generalize these equations for multilevel systems, with the pump beam driving σ^+ transitions. The combination of the $\Delta F = 0, \pm 1$ selection rule and three possible polarizations results in each excited state being able to decay into up to nine different ground states. The strength of the interaction between an atom and near-resonant electromagnetic radiation is characterized by the dipole matrix elements. The dipole matrix element of the transition between states $|F, m_F\rangle$ and $|F', m_{F'}\rangle$ is $\langle F, m_F | er_q | F', m_{F'}\rangle$, where the prime denotes the excited state. In order to calculate this matrix element, it is possible to factor out the angular dependence and write the matrix element as a product of Wigner 3*J* and 6*J* symbols, a phase factor and a reduced matrix element [16]. Thus,

$$\begin{aligned} \langle F, m_F | er_q | F', m_{F'}\rangle &= (-1)^{2F'+\mathcal{I}+J+J'+L+S+m_F+1} \langle L' \| e\vec{r} \| L\rangle \\ &\times \sqrt{(2F+1)(2F'+1)(2J+1)(2J'+1)(2L+1)} \\ &\times \begin{pmatrix} F & 1 & F' \\ m_F & q & -m_{F'} \end{pmatrix} \begin{Bmatrix} J & J' & 1 \\ F' & F & \mathcal{I} \end{Bmatrix} \begin{Bmatrix} L & L' & 1 \\ J' & J & S \end{Bmatrix}. \end{aligned} \tag{12}$$

Here $F, \mathcal{I}, J, L, S$ and m_F are the angular momentum quantum numbers, q the change in *z*-component of the angular momentum, with $q = 0$ for π transitions, $q = \pm 1$ for $\sigma^{\pm}$ transitions. $\mathcal{I}$, the nuclear spin, has the value 9/2, 5/2 and 3/2 for ^{133}Cs, ^{85}Rb and ^{87}Rb, respectively. S, the electron spin, has the value 1/2. The 3*J* symbol is the term contained in the large round brackets, and the 6*J* in curly brackets. The reduced matrix element $\langle L' \| e\vec{r} \| L\rangle$ can be expressed in terms of the wavelength of the transition, λ, and Γ:

$$d = \langle L = 0 \| e\vec{r} \| L' = 1\rangle = \sqrt{\frac{9\varepsilon_0 \hbar \Gamma \lambda^3}{8\pi^2}}. \tag{13}$$

For the Rb D$_2$ transition $\lambda = 780.241\,\text{nm}$ [17] and $\Gamma = 2\pi \times 6.065\,\text{MHz}$ [18], hence $\langle L' \| e\vec{r} \| L \rangle = 5.177 e a_0$ where a_0 is the Bohr radius.

For an alkali-metal atom with ground term $^2S_{1/2}$ and nuclear spin $\mathcal{I}$ there are two values for the total angular momentum F, namely $F = \mathcal{I} \pm 1/2$. There are four values for the excited state angular momentum F', namely $F = \mathcal{I} \pm 3/2,\ \mathcal{I} \pm 1/2$. The hyperfine splitting of the ground states (3.0 GHz for ^{85}Rb, 6.8 GHz for ^{87}Rb, 9.2 GHz for ^{133}Cs) exceeds the room temperature Doppler width (~0.5 GHz); in contrast, the excited-state hyperfine splitting is less than the Doppler width. Therefore, the absorption spectrum consists of two isolated Doppler-broadened absorption lines per isotope. Sub-Doppler absorption features (saturated absorption or polarization spectroscopy) are obtained by using counter-propagating pump and probe beams. Three resonances corresponding to the transitions $^2S_{1/2}(F) \rightarrow {}^2P_{3/2}(F' = F+1, F, F-1)$ are observed, as well as so-called "cross-over" resonances at frequencies halfway between each pair of conventional resonances [19].

As an example, consider ^{87}Rb ($\mathcal{I} = 3/2$), which has two hyperfine states in the ground term, $F = 2,1$. If the pump laser is tuned in the vicinity of the resonances $^2S_{1/2}(F=2) \rightarrow {}^2P_{3/2}(F' = 3,2,1)$ with the detuning, Δ, defined relative to the closed transition $F = 2 \rightarrow F' = 3$, the strongest transition will be $F = 2, m_F = 2 \rightarrow F' = 3, m_{F'} = 3$. As this is a closed two-level transition, it is possible to relate the saturation intensity I_{SAT} to the lifetime and wavelength [20]:

$$I_{\text{SAT}} = \frac{hc\Gamma\pi}{3\lambda^3}. \tag{14}$$

Using the values of λ and Γ listed above yields the value 1.6 mW/cm^2.

To simplify our notation we write the population of a state as P_{F,m_F}, which is equal to the diagonal density-matrix element $\rho_{F,m_F\,F,m_F}$. The five $F = 2, m_F$ ground states obey the rate equations

$$\frac{\mathrm{d}P_{F,m_F}}{\mathrm{d}t} = -\sum_{F'=F-1}^{F'=F+1} R_{F,m_F \rightarrow F',m_{F+1}} \frac{\Gamma}{2} \frac{I}{I_{\text{SAT}}} \frac{\left(P_{F,m_F} - P_{F',m_F+1}\right)}{1 + \frac{4\Delta'^2}{\Gamma^2}} + \sum_{m_{F'}=m_F-1}^{m_{F'}=m_F+1} \sum_{F'=F-1}^{F'=F+1} R_{F,m_F \rightarrow F',m_{F'}} \Gamma P_{F',m_{F'}} \tag{15}$$

where $\Delta' = \Delta$ if $F' = F+1$; $\Delta' = \Delta + \Delta_{32}$ if $F' = F$; and $\Delta' = \Delta + \Delta_{31}$ if $F' = F-1$. Here $\hbar\Delta_{32}$ and $\hbar\Delta_{31}$ are the excited state hyperfine intervals. The first term of equation (15) describes stimulated absorption and emission, the second term describes spontaneous emission. The line-strength ratios, which we define as

$$R_{F,m_F\to F',m_{F'}} = \left(\frac{d_{F,m_F\to F',m_{F'}}}{d_{2,2\to 3,3}}\right)^2, \tag{16}$$

are always less than one as they are normalized relative to the strongest transition.

The three $F = 1, m_F$ ground states obey the rate equations

$$\frac{\mathrm{d}P_{F,m_F}}{\mathrm{d}t} = \sum_{m_{F'}=m_F-1}^{m_{F'}=m_F+1} \sum_{F'=F-1}^{F'=F+1} R_{F,m_F\to F',m_{F'}} \Gamma P_{F',m_{F'}}. \tag{17}$$

Note that there are no stimulated terms as the large ground state splitting means that these transitions are very far off resonance; these states can only increase their population by spontaneous emission from the excited states.

Finally, the excited state population rate equations are

$$\frac{\mathrm{d}P_{F',m_{F'}}}{\mathrm{d}t} = \sum_{F=2} R_{F,m_{F'-1}\to F',m_F} \frac{\Gamma}{2}\frac{I}{I_{\mathrm{SAT}}} \frac{\left(P_{F,m_{F'-1}} - P_{F',m_{F'}}\right)}{1+\frac{4\Delta'^2}{\Gamma^2}} - \sum_{m_F=m_{F'}-1}^{m_F=m_{F'}+1} \sum_{F=F'-1}^{F=F'+1} R_{F,m_F\to F',m_{F'}} \Gamma P_{F',m_{F'}}. \tag{18}$$

Note that there are 36, 24, or 48 Zeeman levels which have to be considered for ^{85}Rb, ^{87}Rb, and ^{133}Cs atoms, respectively; eliminating the coherences from the optical Bloch equations represents a significant simplification.

2.3. Calculation of Line-Centre Absorption Difference

In section 2.1 it was shown that to predict the magnitude of the polarization spectrum it is sufficient to calculate the difference in absorption coefficients $\Delta\alpha$ for the optically pumped medium. Therefore for each resonance $F \to F'$ we calculate one parameter, the line-centre difference in absorption, $\Delta\alpha_0$.

The boundary conditions for the calculation are: (i) no excited state population initially, and (ii) that the ground state population is spread uniformly amongst the $2\times(2\mathcal{I}+1)$ different $|F,m_F\rangle$ levels. For each Zeeman state equations similar to (15) are produced, and the set of coupled equations solved numerically. The pump beam optically pumps the medium by driving the population towards the largest possible value of m_F, and the medium becomes optically anisotropic.

From these solutions, we obtain graphs of the population of each hyperfine state as a function of time. Figure (1) shows the time-evolution of these populations for a sample of ^{87}Rb atoms when $\Delta = 0$, and $\Delta = -2\pi\times 424$ MHz, corresponding to a laser on resonance with the $F = 2 \to F' = 3$ and $F = 2 \to F' = 1$ transitions respectively. When the laser is resonant with the closed $F = 2 \rightarrow F' = 3$ transition, atoms are rapidly pumped into the extreme $|2,2\rangle$ state, with over 50% of atoms in this state after $<10\,\mu s$. The population of the other states is either constant or decreasing after $\sim 1\,\mu s$.

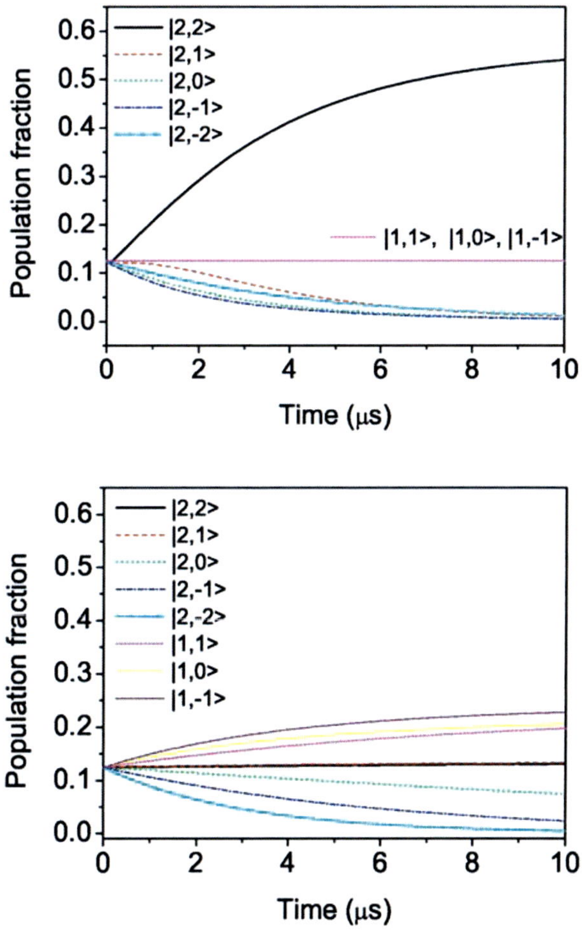

Figure 1. Optical pumping dynamics. The population of the eight ground state sub-levels of ^{87}Rb are plotted as a function of time after the pump beam is turned on. The pump drives σ^+ transitions and has an intensity of $0.1 \times \frac{I}{I_{SAT}}$ a) Laser tuned to the $F = 2 \rightarrow F' = 3$ transition. Within a few μs the medium becomes highly anisotropic as atoms accumulate in the $|2,2\rangle$ state. There is no change in the population of the $F = 1$ manifold owing to the $\Delta F = 0, \pm 1$ selection rule. b) When the laser is tuned to the $F = 2 \rightarrow F' = 1$ transition the atoms are optically pumped towards the $F = 2 \rightarrow F' = 3$ state. However, as this is an open transition, a significant fraction of population accumulates in the $F = 1$ manifold. These states do not contribute to the medium's optical anisotropy. Figure courtesy of Margaret Harris.

These graphs clearly show the different optical pumping dynamics as a function of the laser detuning. A detuning $\Delta = 0$ (Figure 1a) is optimum for achieving the most optically anisotropic medium. For $\Delta = -2\pi \times 424$ MHz (Figure 1b) the medium becomes optically anisotropic, but to a lesser extent. This is due to the open nature of the transition, which allows atoms to accumulate in the $F = 1$ manifold, where they are too far from resonance to influence the medium's anisotropy. From these population graphs it is apparent that the polarization spectroscopy signal should be dominated by the closed transition $^2S_{1/2}\left(F = \mathcal{I} + 1/2\right) \rightarrow F' = {}^2P_{3/2}\left(F' = F + 1\right)$. For this transition, dipole selection rules forbid atoms in the excited state from falling into the other ground state; consequently, the total ground state population ends up pumped into the extreme magnetic sublevels with $m_F = F$. The absorption coefficients $\alpha_{\pm}$ differ most for this extreme state; therefore, the anisotropy of the medium is maximized.

Having obtained the numerical solutions to the coupled population rate equations it is possible to calculate the time-dependent anisotropy of the medium, $\mathcal{A}(t)$. This quantity is defined as a sum over all of the ground m_F states of the difference in the absorption coefficients for the components of the probe beam driving $\sigma^{\pm}$ transitions, taking into account the ground and excited state populations:

$$\mathcal{A}(t) = \sum_{m_F=-F}^{m_F=+F} \alpha_{(F,m_F \rightarrow F',m_{F+1})}\left(P_{F,m_F} - P'_{F',m_{F+1}}\right) - \alpha_{(F,m_F \rightarrow F',m_{F-1})}\left(P_{F,m_F} - P'_{F',m_{F-1}}\right). \quad (19)$$

The populations P_{F,m_F} and $P'_{F',m_{F'}}$ are time-dependent, whereas the absorption coefficient for a given transition, $\alpha_{(F,m_F \rightarrow F',m_{F+1})}$ is constant. For each isotope and each ground state F the absorption difference is calculated for the three frequencies corresponding to the resonances $F \rightarrow \left(F' = F + 1, F, F - 1\right)$ i.e. $\Delta = 0, -\Delta_{32}$ and $-\Delta_{31}$. It is assumed that since the experiment uses counter propagating pump and probe beams only atoms with zero velocity along the axis of the beams contribute significantly. The details of the transverse motion of the atoms are outlined in the next sub-section.

2.4. Transverse Motion of Atoms

The polarization signal is proportional to the time-average of the difference in absorption coefficients. To simulate this, we average the time-dependent anisotropy, $\mathcal{A}(t)$, with a weighting function, $\mathcal{H}(t)$, which gives the distribution of times of flight transverse to the beam. For a circular beam of radius a the probability distribution, $\mathcal{F}(\ell)$, of having a path length ℓ is [9]

$$\mathcal{F}(\ell)=\frac{\ell}{2a\sqrt{4a^2-\ell^2}}. \tag{20}$$

Using the two-dimensional Maxwell velocity distribution for atoms of mass m at temperature T the probability distribution function, $\mathcal{G}(t,\ell)$, for having a transit time t for a given ℓ is:

$$\mathcal{G}(t,\ell)=\frac{2\ell^3}{u^2t^3}\exp\left\{-\frac{\ell^2}{2u^2t^2}\right\}. \tag{21}$$

Here the typical speed, u, is given by

$$u = \sqrt{\frac{2k_{\mathrm{B}}T}{m}}. \tag{22}$$

$\mathcal{H}(t)$, the distribution of times of flight, is:

$$\mathcal{H}(t)=\int_{\ell=0}^{2a}\mathcal{G}(t,\ell)\mathcal{F}(\ell)\,\mathrm{d}\ell, \tag{23}$$

which can be expressed in terms of the complex error function [21]:

$$\mathcal{H}(t)=\frac{1}{t}\left(\frac{\sqrt{\pi}}{2\eta}\left(1+2\eta^2\right)\mathrm{erfi}(\eta)\exp\left(-\eta^2\right)-1\right), \tag{24}$$

with $\eta=\dfrac{2a}{ut}$ and $\mathrm{erfi}(\eta)=\dfrac{\mathrm{erf}(i\eta)}{i}$.

The form of the distribution is shown in Figure 2 for room-temperature Rb for three typical beam radii used in polarization spectroscopy experiments. For the 1.7 mm beam the most probable time of flight is $9\,\mu\mathrm{s}$. It is instructive to compare this time with the timescale for population dynamics from Figure 1. The timescale for optical pumping is comparable to a typical time-of-flight; therefore the transient nature of the dynamics must be taken into account in the analysis. If the typical time-of-flight was significantly longer than the typical optical pumping time it would be much simpler to calculate the steady-state solution of the population dynamics. Figures 1 and 2 clearly show that this would be an oversimplification, and that the time-averaged anisotropy should be used. The average anisotropy, $\Delta\alpha$, is calculated by averaging the time-dependent anisotropy with the time of flight weighting function, i.e.:

$$\Delta\alpha = \int \mathcal{A}(t)\mathcal{H}(t)\mathrm{d}t. \tag{25}$$

This integral is calculated numerically at the three resonant frequencies of each D_2 transition to give three values of $\Delta\alpha_0$.

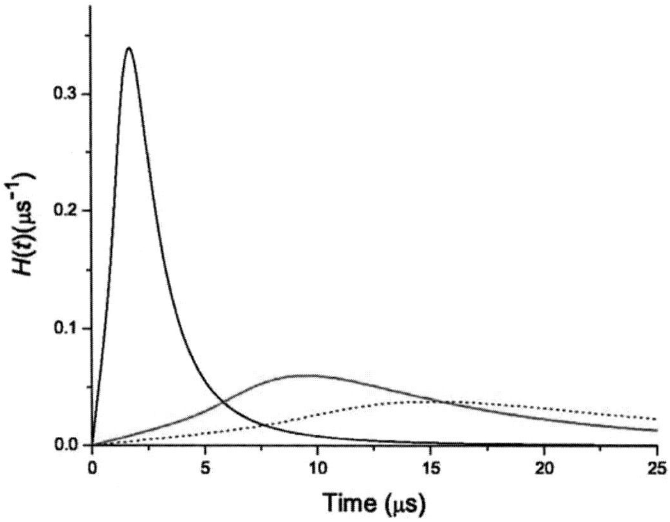

Figure 2. Probability distribution of times-of-flight as a function of time-of-flight, for beam width 0.3 mm (solid), 1.7 mm (dotted) and 2.7 mm (dashed). The sample is atomic Rb at room temperature.

2.5. Generating Theoretical Spectra

To generate the predicted spectra, the three values of $\Delta\alpha_0$ are multiplied by the appropriate lineshape, the dispersion function of equation (8). As in the work of Yoshikawa *et al.* [7] the strength of cross-over features is taken to be the average of the two associated resonances.

3. Apparatus and Results

In alkali-metal atoms the D_2 lines have high oscillator strength, and from an experimental perspective rubidium and cesium are ideal species as they have sufficient room-temperature vapour pressure to yield large absorption (10-90%) in cells of modest length (5-10 cm). In addition, for these atoms the transitions occur at wavelengths where inexpensive and reliable diode laser sources exist.

3.1. Experimental Layout

The layout of a typical polarization spectroscopy experiment is shown in Figure 3. For the experimental results presented in this chapter an extended cavity diode laser was used. Light from the laser passes through an optical isolator and subsequently a half-wave plate. The angle of the wave plate dictates what fraction of the light is reflected into the polarization spectrometer, with the transmitted light being available for other applications The beam is spilt by a polarizing beam splitter to form the pump and probe beams. Neutral density filters are used to gain independent control over the powers of these beams. The pump beam passes through a quarter-wave plate in order to make it circularly polarized. The probe beam passes through a half-wave plate to rotate the plane of its linear polarization to be at an angle $\varphi = \pi/4$ with respect to the axis of the polarizing beam splitter in the balanced detector. The beams overlap with a small crossing angle in a vapour cell. The two output beams from the polarizing beam splitter in the balanced detector are focussed onto photodiodes, connected to current-to-voltage circuits which produce an output voltage linearly proportional to the incident power. The polarization spectroscopy signal is obtained by subtracting the voltages from the two photodetectors.

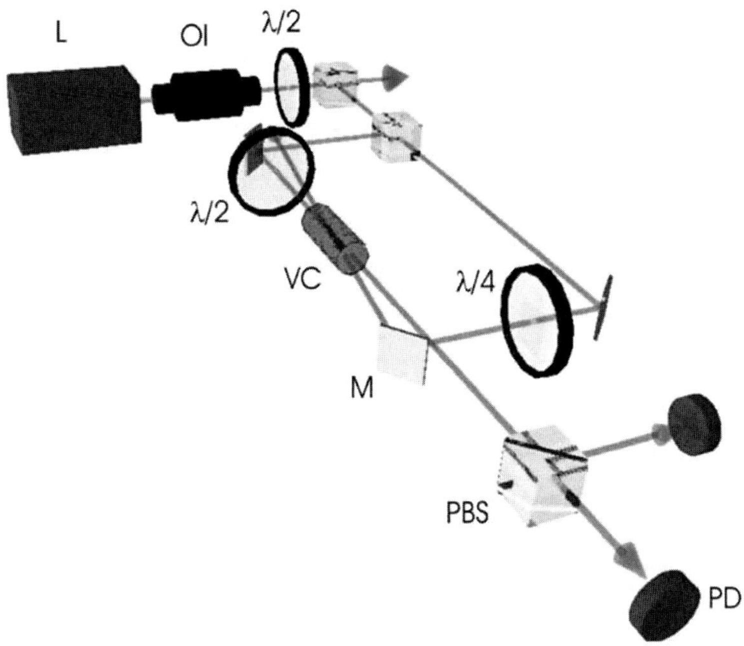

Figure 3. Experimental layout. Light form an extended cavity diode laser (L) passes through an optical isolator (OI). Mirrors (M) and polarizing beam splitters (PBS) are used to steer the pump and probe beams. Quarter and half-wave plate are used to create the appropriate polarization (circular, linear at $\varphi = \pi/4$) for the (pump, probe) beam. The counter propagating pump and probe cross in the vapour cell (VC). Two photodiodes (PD) are used in the balanced detector. Figure courtesy of Paul Griffin.

The circularly polarized light incident on the atoms in the cell may drive σ^+ , σ^- or π transitions depending on the direction of the external magnetic field. To achieve the largest polarization spectroscopy signals it is necessary to establish a preferred magnetic field direction along the vapour cell axis. There are three methods we have used to achieve this: (i) placing the cell inside a solenoid, (ii) placing a cell inside a magnetic shield, and (iii) using 3 pairs of coils along Cartesian axes to control the field. Each method has its own advantages and disadvantages. Empirically we find the best signals are obtained when the transverse ambient field is cancelled leaving a modest field (typically $\simeq 0.1$ mT) along the direction of propagation of the pump beam. Too large a field leads to Zeeman splitting of the levels and complications arising from the Faraday effect.

Although the rubidium spectrum at room temperature produces strong features, occasionally a heated cell is used, especially for weaker transitions (e.g. $F=1 \rightarrow F'$ in ^{87}Rb). By contrast, for cesium the room temperature absorption is too large, and as explained in section 2.1, an optically thick medium is not desirable. Consequently the cesium cell is often used partially submerged in an ice bath.

3.2. Results

3.2.1. Dichroism

To gain insight into the optical anisotropy induced by the pump beam it is possible to measure the dichroism directly. This is done by placing a quarter-wave plate before the polarizing beam splitter in the balanced detector, with the wave plate oriented such that it converts the circular polarization component of the probe which drives σ^+ (σ^-) transitions into vertically (horizontally) polarized light. The output from the arms of the PBS then directly measure the absorption experienced by the components driving σ^+ and σ^- transitions. Figure 4 shows data obtained for ^{85}Rb with the $F=3 \rightarrow F'$ transitions in a 5-cm long cell at room temperature. Parts (a) and (b) both have two traces, corresponding to the pump off and on respectively. In the absence of the pump beam the conventional Doppler-broadened absorption spectrum is obtained. With the pump present a significant modification of the absorption spectrum is observed at some frequencies. This can be mostly attributed to hyperfine pumping into the $F=2$ ground state [22]. However, in the vicinity of the $F=3 \rightarrow F'=4$ transition we expect the absorption to be modified on account of the dichroism induced by the optical pumping from the circularly polarized pump beam. Figure 4 (a) displays the absorption profile for the component of the probe driving σ^+ transitions, showing an enhanced absorption in the region of the closed transition. In contrast the absorption of the component of the probe driving σ^- transitions is reduced for the closed transition, as seen in Figure 4 (b). This demonstrates that at this frequency the medium exhibits circular dichroism, and that anisotropy is established.

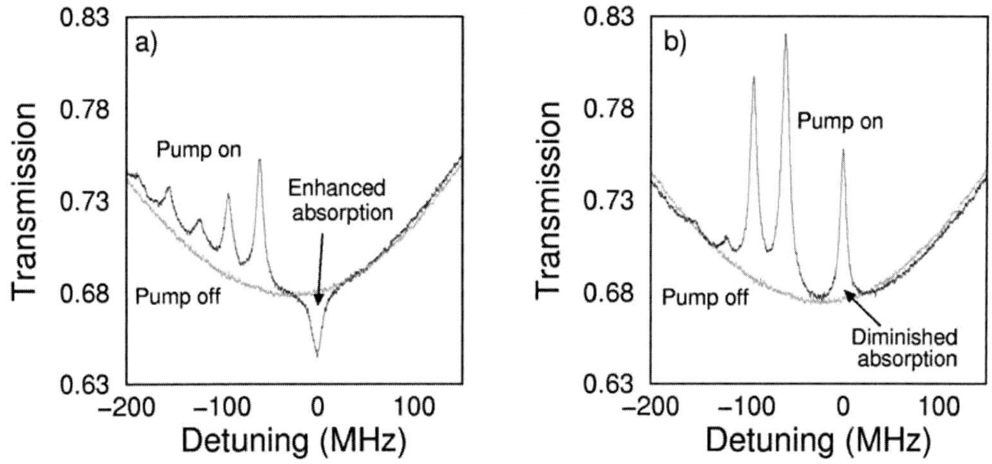

Figure 4. ^{85}Rb spectra obtained with a quarter-wave plate inserted before the beam splitting cube. (a) Absorption profile for the component of the probe driving σ^+ transitions, showing an enhanced absorption in the region of the closed transition. (b) Absorption profile for the component of the probe driving σ^- transitions, showing diminished absorption in the region of the closed transition. The pump-off trace is shown in grey.

3.2.2. The Polarization Spectroscopy Signal

Figure 5 shows data obtained with ^{87}Rb in a room temperature cell. The top two traces correspond to the signals from the separate output arms of the beam splitting cube, proportional to I_y and I_x respectively. Six sub-Doppler features are clearly seen, with asymmetric lineshapes. The trace on the bottom left shows the difference between the upper traces, which gives the polarization spectroscopy signal. The trace on the bottom right is the sum, showing conventional saturated absorption/hyperfine pumping spectroscopy [22]. Zero detuning in this figure is relative to the $F = 2 \rightarrow F' = 3$ transition.

The data of Figure 5 were collected with the cell placed in a magnetic shield. The $F = 2 \rightarrow F' = 3$ transition and neighbouring cross-over are both ideally suited for laser-locking, comprising of large signals with steep gradient through the line centre. Note that by changing the power of pump and probe beams it is possible to enhance the other four lines (at the expense of compromising the quality of the first two) in order to produce adequate locking signals[1]. The spectral resolution of the signals can be degraded deliberately by either (i) power broadening the lines by using a more intense pump and probe beam, or (ii) introducing Doppler broadening by having a larger crossing angle between pump and probe beams. In this way, it is possible to produce error signals which have better lock properties, such as not having zero-crossings within many hundreds of MHz of the lock-point.

[1] All six of the transitions in Figure 5 have been use to lock an external cavity diode laser. Private communication, Ashok Mohapatra.

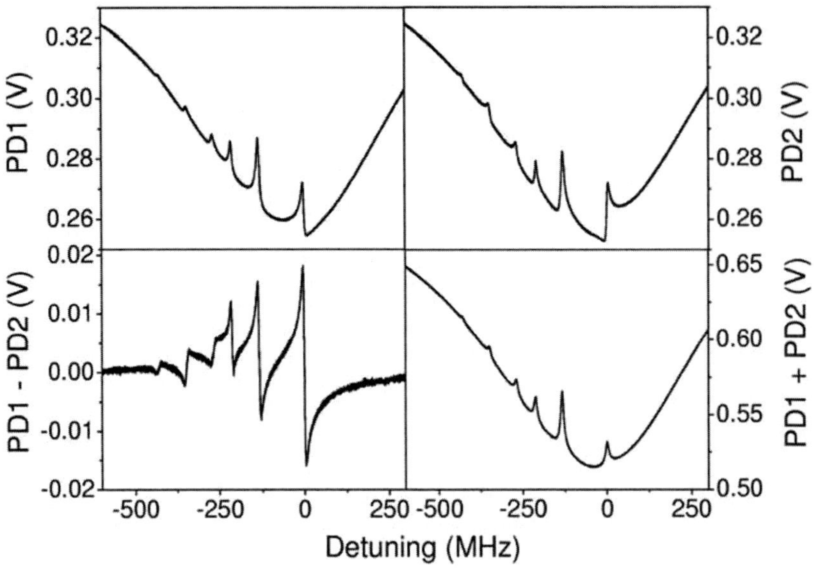

Figure 5. [87]Rb polarization spectroscopy with pump and probe powers $6\,\mu\mathrm{W}$ and $25\,\mu\mathrm{W}$ respectively. The top two traces correspond to the signals obtained from the photodiodes in the separate arms of the beam splitting cube; the trace on the left in the bottom row is the difference between them (the polarization spectroscopy signal); the trace on the right in the bottom row is the sum of the top two (conventional "saturated absorption" and hyperfine pumping spectroscopy). Figure courtesy of Ashok Mohapatra.

3.2.3. Comparison with Theory

Figure 6 shows the comparison between experimental spectra (thick line) and the theoretical spectra (thin line) described in section 2. To reduce saturation effects the experimental spectra were taken with pump and probe beam intensities $\leq 0.1 \times I_{\mathrm{SAT}}$. The spectra shown are of transitions from the upper hyperfine level of the ground state, i.e. the $5\,^2S_{1/2}(F=2) \rightarrow 5\,^2P_{3/2}(F'=1,2,3)$ transitions in ^{87}Rb, and $6\,^2S_{1/2}(F=4) \rightarrow 6\,^2P_{3/2}(F'=3,4,5)$ transitions in ^{133}Cs. The closed transitions $(F \rightarrow F' = F+1)$ are the well-known "cooling" transition used in laser-cooling experiments. Zero detuning is relative to the closed transition, and linear detunings (i.e. $\Delta/2\pi$) are used. For Cs, two sets of data are shown; one taken with the vapour cell at room temperature (dashed line), the other with the cell in an ice bath as described in the previous section (solid line).

Comparing the theoretical and experimental traces shows that the model reproduces experimental features of the upper hyperfine transition spectra. Fine spectral details appear in both experimental and theory curves. The magnitudes of theoretical spectral peaks relative to the large closed transition peaks also agree well with experimental data. Cooling the Cs cell reduces the vapour pressure, bringing closer agreement with theory. At room temperature the

Cs cell is optically thick and, as discussed in section 2.1, this is detrimental to the polarization spectroscopy signal.

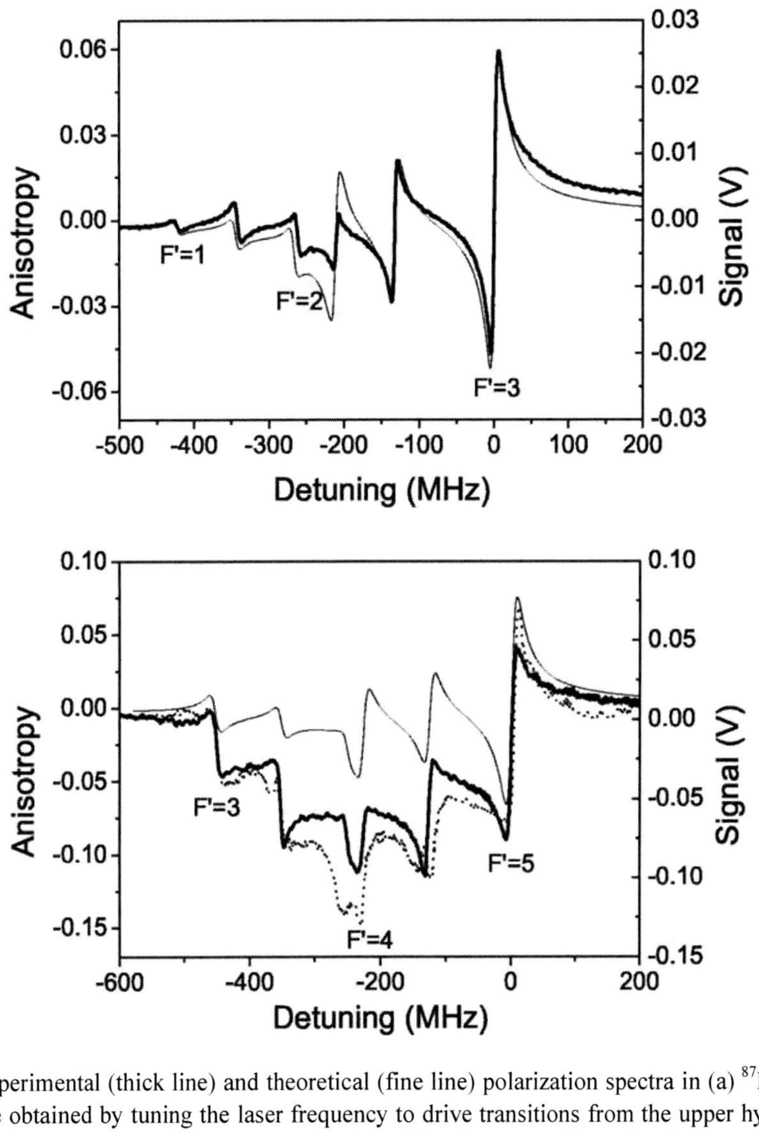

Figure 6. Experimental (thick line) and theoretical (fine line) polarization spectra in (a) ^{87}Rb, (b) ^{133}Cs. Spectra were obtained by tuning the laser frequency to drive transitions from the upper hyperfine level of the atom's ground state. Both spectra are dominated by the strong dispersion feature associated with the closed transition. For Cs, solid (dashed) lines represent spectra taken with the vapour cell at 0°C (23°C).

4. Discussion

Although the spectra taken with transitions from the upper hyperfine level of the ground term show good agreement with theory, spectra taken with transitions from the lower hyperfine level show a more complicated pattern [9]. The (low-frequency) closed transitions in Cs and ^{85}Rb show a characteristic strong dispersion feature, but in ^{87}Rb the largest feature arises from one of the cross-over peaks. In ^{85}Rb closely-spaced transitions generate a polarization spectrum in which individual peaks merge, making exact matching of features and transitions difficult. The most significant approximation in the model that is likely to break down for lower hyperfine transitions is the assumption that only atoms with no axial velocity will contribute to the anisotropy. As was shown in [22], a full description of the spectrum, especially open transitions, must take into account a large velocity class of atoms moving along the beam, not just the "stationary" ones. The contribution of these non-zero velocity classes will be more significant for $F = \mathcal{I} - 1/2$ because the line strength factors for the closed transition are weakest; for $F = \mathcal{I} + 1/2$, line strength factors are strongest for the closed transition, and the stationary-atom approximation is reasonable. Recently Do *et al.* [23] presented calculations for the entire detuning range of the spectra and obtained results which are in good agreement with experimental spectral features, including the resonance and crossover lines. Their model also accounts for the spectra obtained for both rubidium isotopes from both upper and lower hyperfine level of the ground term with great accuracy.

4.1. Comparison with Other Techniques

There exist many other modulation-free techniques for producing an error signal suitable for laser locking. These include the dichroic atomic vapour laser lock (DAVLL) [24]; combination of saturated absorption and DAVLL [25]; Sagnac interferometry [26]; prismatic deflection [27]; using acousto-optic modulators [28]; low-field Faraday polarimetry [29]; and velocity-selective saturated-absorption spectroscopy [30]. The performance and capabilities of the locking techniques differ. There are trade-offs between simplicity, cost, ease of operation, capture range, location of zero crossings with respect to the atomic reference, and sensitivity to external perturbations. DAVLL utilises a single probe beam, therefore the error signal is a Doppler-broadened spectral features with a wide capture range ($\sim 0.5\,\mathrm{GHz}$ for alkali-metal atoms at room temperature). In contrast, the other pump-probe schemes achieve sub-Doppler resolution, with a steeper gradient through the zero-crossings, at the expense of a smaller capture range (typically tens of MHz). The presence of zero crossings associated with six atomic references for the sub-Doppler techniques eliminates a major drawback of DAVLL, namely that the error signal is prone to long term drifts necessitating periodical recalibration.

4.2. Variations on the Polarization Spectroscopy Theme

The method discussed at length in this chapter hinges on producing a circular birefringence (differential refractive index for orthogonal circular polarizations) in a sample with a circularly polarized pump beam. The locking signal is linear over a range approximately equal to the linewidth of the transition. Ratnapala *et al.* presented a method which combines polarization spectroscopy with a measure of the circular dichroism (differential absorption for two orthogonal polarizations) [31]. This technique extends the linear capture range of the error signal as it is linearly proportional to the detuning, even at detunings many times the natural linewidth of the transition.

Tiwari *et al.* have presented a balanced bi-polarimeter [32]. A symmetric, background-free and dispersion-like error signal is obtained by subtracting polarization spectra of two linearly polarized weak probe beams which overlap with two oppositely circularly polarized pump beams. In the polarization spectra presented above the frequency lock point is shifted from the line centre, and is susceptible to temperature dependent birefringence properties of the cell windows and other optical components. In bi-polarization spectroscopy the signal at the line centre is independent of the initial angle of the plane of polarization of the probe beam with respect to the analyzer. Consequently, the error signal is not sensitive to temperature-induced changes in the window birefringence.

4.3. Strong-Field Atomic Polarization Spectroscopy

Typically the pump and probe beam in polarization spectroscopy are chosen to be weak ($\sim 0.1\times I_{\text{SAT}}$ for the experiments in section 3.2.3). For higher pump intensity the optical pumping process is faster, but at the expense of power broadening of the transition. If, however, a very large intensity is used ($\gg I_{\text{SAT}}$) the sign of the polarization spectroscopy signal can be reversed. At these high intensities the population is rapidly pumped into the stretched states. For ^{85}Rb this would correspond to half the population being in $F=3, m_F=3$ and half being in $F'=4, m_{F'}=4$. As a result the component of the probe driving the σ^+ transition $F=3, m_F=3\rightarrow F'=4, m_{F'}=4$ will experience no absorption. In contrast, the much weaker σ^- transition $F=3, m_F=3\rightarrow F'=4, m_{F'}=2$ will occur, and consequently $\Delta\alpha_0$ will have a different sign. In the example discussed above for Rb this is as much a curiosity as a useful result. However, certain atoms have energy levels which do not have conventional polarization spectra, and in these cases the strong-field effect becomes more interesting. For example, the transition $5s^2\,{}^1S_0\rightarrow 5s5p\,{}^1P_1$ in Sr will not exhibit conventional polarization spectroscopy, as a pump beam driving σ^+ transitions can not redistribute population among Zeeman sublevels, as there is only one level ($m_J=0$). However, for a strong enough pump, the population can be excited into the state ${}^1P_1\, m_J=+1$; the component of the linearly polarized probe beam driving the σ^+ transition $J=0, m_J=0\rightarrow J'=1, m_{J'}=1$ will be absorbed less than the component driving the σ^- transition $J=0, m_J=0\rightarrow J'=1, m_{J'}=-1$, hence a finite polarization spectroscopy signal will appear for pump intensities $\geq I_{\text{SAT}}$. This prediction has been verified experimentally.[2]

[2] Private Communication, Matt Jones.

5. Conclusion

In this chapter we have discussed how to realize polarization spectroscopy with alkali-metal atoms. A treatment of the propagation of a linearly polarized probe beam through a circularly dichroic and birefringent medium was presented. The evolution of the anisotropy of the medium as a function of the atom-light interaction time was modelled with rate equations derived from the optical Bloch equations (assuming the coherences have their steady-state values), taking into account the multi-level atomic structure and the different line-strength factors. It was seen that a pump intensity of the order of the saturation intensity is sufficient to generate a significant anisotropy. The presence of this anisotropy was demonstrated by observing the different absorption spectra obtained with the left and right hand circularly polarized components of a probe beam. Using a balanced detector the lineshape of the polarization spectra has the form of a dispersion curve. This modulation-free technique is ideal for frequency stabilization of the laser.

Acknowledgements

The author acknowledges financial support from the UK EPSRC; experimental assistance from P F Griffin, M L Harris, A Mohapatra, W Ng, C P Pearman, D A Smith and M Talbot; theoretical input from N Coppendale, I C McLeod and E Tarleton; stimulating discussions with C S Adams, A S Arnold, S L Cornish, M P A Jones and K J Weatherill.

References

[1] Wieman, C.; Hänsch, T. W. *Phys. Rev. Lett.* 1976, 36,1170–1173.

[2] Demtröder, W *Laser Spectroscopy*; Springer:Berlin, 1998; Chapter 7.

[3] Teets, R. E.; Kowalski, F. V.; Hill, W. T.; Carlson, N.; Hänsch, T. W. *Advances in Laser Spectroscopy* 1977, 113, 80–87.

[4] Hänsch, T. W.; Couillaud, B. *Optics Communications* 1980, 35, 441–444.

[5] Pearman, C. P.; Adams, C. S.; Cox, S. G.; Griffin, P. F.; Smith, D. A.; Hughes, I. G. *J. Phys. B.* 2002, 35, 5141–5151.

[6] Streed, E. W.; Chikkatur, A. P.; Gustavson, T. L.; Boyd, M.; Torii, Y.; Schneble, D.; Campbell, G. K.; Pritchard, D. E.; Ketterle, W. *Rev. Sci. Instrum.* 2006, 77, 023106.

[7] Yoshikawa, Y.; Umeki, T.; Mukae, T.; Torii,Y.; Kuga, T. *Applied Optics* 2003, 42, 6645–6649.

[8] Friedenauer A.; Markert F.; Schmitz H.; Petersen L.; Kahra S.; Herrmann M.; Udem T. Hänsch T W; Schätz T *Appl. Phys. B* 2006, 84, 371–373.

[9] Harris, M. L.; Adams, C. S.; Cornish, S. L.; McLeod, I. C.; Tarleton, E.; Hughes, I. G. *Phys. Rev. A* 2006, 73, 062509.

[10] Thomsen, J. W.; Holme, N. C. R.; Müller, U.; Pedersen, J. O. P. *Meas. Sci. Technol.* 1995, 6, 170–173.

[11] Kim J. B.; Kong, H. J.; Lee, S. S. *Appl. Phys. Lett.* 1988, 52, 417–419.

[12] Nakayama, S. *Jpn. J. Appl. Phys.* 1985, 24 1–7; Nakayama, S. *Phys. Scr.* 1997, *T*70, 64–74.
[13] Jackson, J. D. *Classical Electrodynamics*; John Wiley and Sons: New York, 1999.
[14] Cohen-Tannoudji, C. *Atom-Photon Interactions*; Wiley: New York, 1992; Chapter V.
[15] Gorris-Neveux, M.; Monnot, P.; Fichet, M.; Ducloy, M.; Barbé, R.; Keller, J.C. *Optics Communications*. 1997, 134, 85–90.
[16] Edmonds, A. R. *Angular Momentum in Quantum Mechanics*; Princeton University Press: Princeton, NJ, 1960.
[17] Ye, J.; Swartz, S.; Jungner, P.; Hall, J. L. *Opt. Lett.* 1996, 21, 1280–1282.
[18] Volz, U.; Schmoranzer, H.; *Physica Scripta* 1996, *T*65, 48–56
[19] MacAdam, K. B.; Steinbach, A.; Wieman, C. E. *Am. J. Phys.* 1992, 60, 1098–1111.
[20] Foot, C. J. *Atomic Physics* Oxford Masters Series in Atomic, Optical and Laser Physics; Oxford Univeristy Press: Oxford, 2005; p 142.
[21] Moon, G.; Noh, H-R. *Journal of the Korean Physical Society* 2007, 50, 1037–1043.
[22] Smith D. A.; Hughes I. G. *Am. J. Phys.* 2004, 72, 631–637.
[23] Do H. D.; Moon, G.; Noh, H-R. *Phys. Rev. A*. 2008, 77, 032513.
[24] Chéron, B.; Gilles, H.; Hamel, J.; Moreau, O.; Sorel, H.; *J. de Physique III* 1994, 4, 401–406; Corwin, K. L.; Lu, Z-T.; Hand, C. F.; Epstein, R. J.; Wieman, C. E. *Appl. Opt.* 1998, 37, 3295–3298; Millett-Sikking, A.; Hughes, I. G.; Tierney, P.; Cornish, S. L. *J. Phys. B* 2007, 40, 187–198.
[25] Petelski, T.; Fattori, M.; Lamporesi, G.; Stuhler, J.; Timo, G. M.; *Eur. Phys. J. D* 2002 22, 279–283; Wasik, G.; Gawlik, W.; Zachorowski, J.; Zawadzki, W.; *Appl. Phys. B* 2002, 75, 613–619; Harris, M. L.; Cornish, S. L.; Tripathi, A.; Hughes, I. G. *J. Phys. B* 2008, 41, 085401.
[26] Robins, N. P.; Slagmolen, B. J. J.; Shaddock, D. A.; Close, J. D.; Gray, M. B. *Opt. Lett.* 2002, 27, 1905–1907; Jundt, G.; Purves, G. T.; Adams, C. S.; Hughes, I. G. *Eur. Phys. J. D* 2003, 27, 273–276.
[27] Gilles, H.; Cheron, B.; Hamel, J. *Optics Communications*. 2001, 190, 179–184; Purves, G. T.; Jundt, G.; Adams, C. S.; Hughes, I. G. *Eur. Phys. J. D* 2004, 29, 433–436.
[28] Sukenik, C. I.; Busch, H. C.; Shiddiq, M. *Optics Communications* 2002, 203, 133–137; van Ooijen, E. D.; Katgert, G. van der Straten, P. *Appl. Phys. B* 2004, 79, 57–59.
[29] Kerckhoff, J. A.; Bruzewicz, C. D.; Uhl, R.; Majumder, P. K. *Rev. Sci. Inst.* 2005, 76, 093108.
[30] Park, S. E.; Lee, H. S.; Kwon, T. Y.; Cho, H. *Optics Communications* 2001, 192, 49–55.
[31] Ratnapala, A.; Vale, C. J.; White, A. G.; Harvey, M. D.; Heckenberg, D. R.; Rubinsztein-Dunlop, H. *Opt. Lett.* 2004, 29, 2704–2706.
[32] Tiwari, V. B.; Singh, S.; Mishra, S. R.; Rawat, H. S.; Mehendale, S. C. *Optics Communications* 2006 263 **249**–255; Tiwari, V. B.; Singh, S.; Mishra, S. R.; Rawat, H. S.; Mehendale S. C. *Appl. Phys. B* 2006, 83, 93–96.

In: New Trends in Quantum Coherence...
Editor: Rafael Drampyan, pp. 171-195
ISBN: 978-1-60741-025-6

Chapter 6

COHERENT ACCUMULATION EFFECTS IN ATOMS EXCITED BY A TRAIN OF FEMTOSECOND PULSES

Ticijana Ban[1], Damir Aumiler[2] and Goran Pichler[3]
Institute of Physics, Bijenička cesta 46, HR-10000 Zagreb, Croatia

Abstract

The regular train of pulses with femtosecond duration have a very specific frequency comb structure of extraordinary importance for the further development of atomic, molecular and optical physics. We investigate a few specific consequences of the femtosecond laser frequency comb action on the rubidium atoms in the cell at room temperature. When the resonance lines lifetimes are longer than the time between adjacent pulses of the pulse train accumulation effects appear in populations and coherences. This causes the change of the Doppler profiles into peculiar profiles with peaks and holes separated by repetition frequency of the optical frequency comb. We present the adequate theory which gives a very good comparison with experimental findings. Behaviour in the homogeneous magnetic field is further discussed as well as other interesting applications of the modified direct frequency comb spectrsocopy.

I. Introduction

Mode-locked femtosecond (fs) lasers are a powerful tool for the frequency comb generation [1, 2]. Femtosecond laser frequency comb has a time-domain counterpart in a sequence of pulses with a time separation equal to the inverse of the pulse repetition rate. It consists of a regularly spaced series of sharp lines $f_n = nf_r + f_0$, where n is a large integer, f_r is the laser repetition rate and f_0 is the carrier-envelope offset frequency. In Fig. 1. we show the time and frequency domain representations of a single and multiple fs pulses. The single pulse has a continuum spectrum, while already two adjacent pulses exhibit frequency comb behavior. The comb lines are very broad when only a few pulses are in the train. However, at very large

[1] E-mail address: ticijana@ifs.hr
[2] E-mail address: aumiler@ifs.hr
[3] E-mail address: pichler@ifs.hr

number of pulses in the train each comb line becomes extremely narrow. The half width of the comb lines linearly drops with the number of the subsequent pulses.

Many of the extraordinary features of the optical frequency comb are reviewed in two recent review paper [3, 4]. The most important contribution of the frequency comb useful in precision measurements was introduced by J. Hall and T. Haensch and their coworkers [5, 6, 7]. In addition, the use of the phase-stabilized wide frequency-domain comb output of a femtosecond mode-locked laser has played a key role in recent advances in optical [8, 9, 10], XUV frequency measurements [11, 12, 13], optical frequency synthesizers [14, 15, 16] and all optical atomic clocks [17, 18, 19]. Ultimately, the regularity of the stabilized laser frequency comb will help astronomical spectrograph calibration with broad-spectrum frequency combs, which will enable radial velocity measurements with a precision of 1 cm s^{-1}. This will have a major impact in the search for the extra solar planets in the vicinity of our Sun [20, 21].

In a system with the relaxation time greater than the laser repetition period, the system interacts with the pulse train rather than with the single pulses. Pulse train in time domain corresponds to the frequency comb in the frequency domain. When the time separation between adjacent pulses is smaller than the lifetime of any of the states excited by the frequency comb lines, then these states "detect" the pulse train and undergo changes in population and coherence due to coherent accumulation effects. The interaction of inhomogeneously broadened two-level atoms with a train of fs pulses was theoretically studied in ref. [22]. Obtained analytical results for the case in which the atomic relaxation times are longer than the laser repetition period exhibit the accumulation of population and coherence effects.

The resonant interaction of the fs pulse train with cold rubidium atoms was investigated in literature [23, 24, 25,26]. The group of Jun Ye developed direct frequency comb spectroscopy (DFCS) which uses light from a comb of appropriate structure to directly interrogate atomic energy level structure and to study time-dependent quantum coherence [24]. In ref.[25] DFCS is used for the determination of the absolute atomic transition frequencies anywhere within the comb bandwidth, for one- and two- photon processes in cold ^{87}Rb atoms.

In our recent papers [27, 28] a mode-locked femtosecond laser was used to physically map the laser frequency comb into the velocity comb of the excited Rb atoms at room temperature. Resonant excitation of the rubidium atoms by discrete frequency comb optical spectrum results in the comb-like velocity distribution of the excited state hyperfine level populations and velocity-selective population transfer between the Rb ground state hyperfine levels. We developed a modified direct frequency comb spectroscopy (DFCS) which uses a fixed frequency comb for the ^{85,87}Rb 5 $^2S_{1/2}$→5 $^2P_{1/2,3/2}$ excitation and a weak cw scanning probe laser for ground levels population monitoring. Modulations in the probe laser absorption for both rubidium isotopes were observed as a direct consequence of the velocity-selective optical pumping induced by the frequency comb excitation. In Fig. 2 we show Doppler profiles of ^{85,87}Rb hyperfine lines under the influence of the femtosecond laser pulse train when the frequency comb is imprinted into the velocity distributions. In the present chapter we shall describe some of these results using simple experimental arrangements and available theoretical model based on the semiclassical quantum theory of light-matter interaction already well established in references [29, 30].

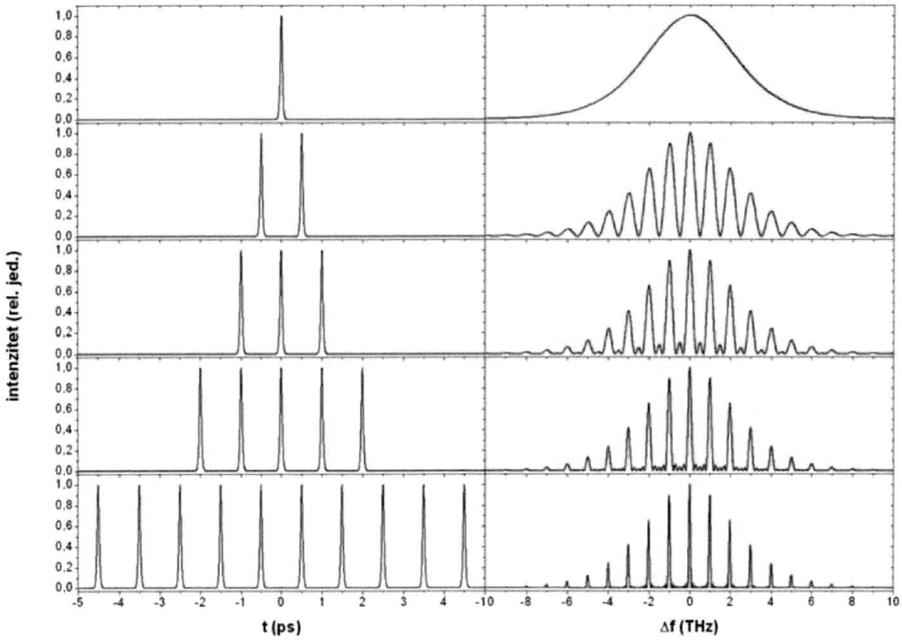

Figure 1. Time and frequency domain of the femtosecond laser pulse train. The calculations have been performed for 1,2,3,5 and 10 pulses of 60 fs duration and repetition rate of 1 THz.

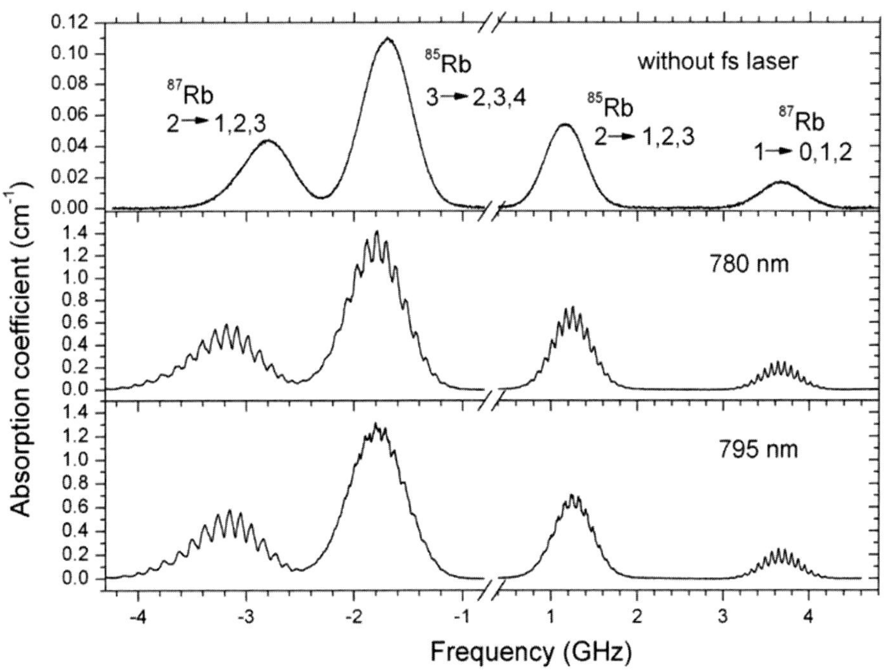

Figure 2. Optical thickness spectra of the rubidium hyperfine lines without and with femtosecond laser pulse train influence. In the lower two panels femtosecond laser was cenetred at D2 (780) nm and D1 (795 nm) Rb resonance lines.

Velocity selective minimum in the velocity distribution made by the c.w. laser is often called a Bennett hole [31]. By virtue of the fact that we have 10^5 simultaneous laser modes in a single femtosecond laser pulse, we may say that within the Doppler profiles of the resonance lines we have multi-Bennett holes at one line from the very ground hyperfine level and corresponding multi-Bennett peaks at the other resonance line stemming from the slightly higher hyperfine level.

One of the ultimate tasks in the emerging field of direct frequency comb spectroscopy would be the method of influencing amplitudes, phases and polarizations of each comb line hidden in the train of the femtosecond pulses.

In what follows we shall describe our recent experimental and theoretical findings in the modified direct frequency comb spectroscopy [32, 33], and will discuss some of potential applications.

II. Experiment

The experimental arrangement is shown in Fig.3 and 4. A Tsunami mode-locked Ti:sapphire laser with pulse duration of about 100 fs generates ~5.8 THz broad optical frequency comb. It uses a radio frequency (rf) synthesizer for active mode-locking by the use of an acusto-optic modulator. Besides this, the laser was not locked to any stable frequency reference, meaning no active stabilization of the laser repetition rate, f_r, and the carrier-envelope offset frequency f_0 was performed.

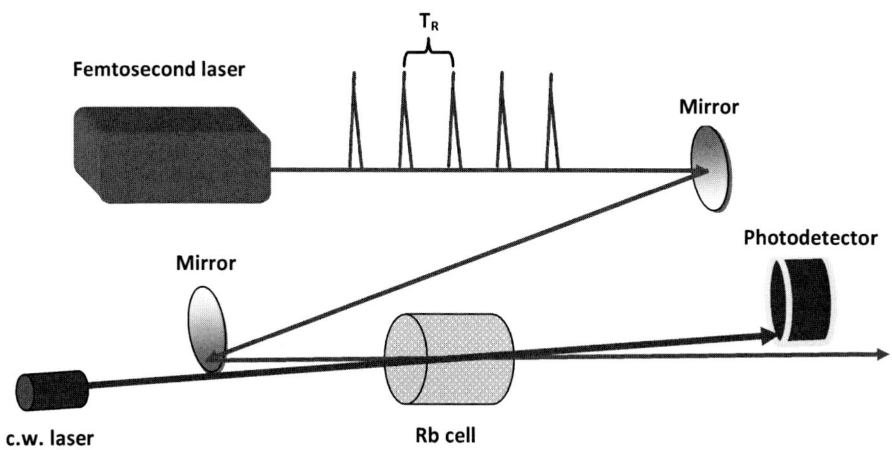

Figure 3. Experimental setup.

However, the experimental spectra are taken in a time scale of about 5 seconds, during which the change in the repetition rate f_r, is negligible. In these conditions, we can say that the comb frequencies were not tunable and the frequency comb was kept constant during the measurements. The laser repetition rate f_r, measured with a fast photodiode amounts 80 MHz. The output of the fs laser was positioned on the Rb $5\,^2S_{1/2} \rightarrow 5\,^2P_{1/2}$ transition at 795 nm (D1 resonance line) or on the Rb $5\,^2S_{1/2} \rightarrow 5\,^2P_{3/2}$ transition at 780 nm (D2 resonance line). The fs laser maximum average power used was up to 1 W with the full width at half maximum around 10 nm. It was chopped (Fig. 4) in order to achieve higher sensitivity with an SR540 mechanical chopper at 3.5 kHz repetition rate, and weakly focused onto the center of the glass cell containing rubidium vapor at room temperature.

The cell was 5 cm long with the outer diameter of 2.5 cm. When needed, the homogeneous external magnetic field was generated with a pair of Helmholtz coils. The external magnetic field was orthogonal to the laser wave polarization and parallel to the laser propagation. The ^{85,87}Rb 5 $^2S_{1/2}$ hyperfine ground state populations were probed with a weak cw diode laser (~2 μW/mm^2), which was propagating nearly collinear with the fs laser, intersecting it under small angle in the center of the cell. The probe laser (TOPTICA DL100, ECDL at 780 nm) provided a continuous single-mode tuning range of up to 15 GHz, with a linewidth of the order of 1 MHz. The probe laser frequency was scanned across the Doppler broadened ^{85,87}Rb 5 $^2S_{1/2} \rightarrow 5\ ^2P_{3/2}$ hyperfine transitions at 3 GHz/ms scanning rate. The probe laser transmission was simultaneously detected with two Si photodiodes (Hamamatsu).

The signal from the photodiode PD2 was directly fed into a digital oscilloscope (Tektronix TDS5140). It measures transmission of the probe laser across all four Doppler broadened $5^2S_{1/2} \rightarrow 5^2P_{3/2}$ absorption lines at 780 nm for the fs laser tuned to $5^2S_{1/2} \rightarrow 5\ ^2P_{1/2}$ transition at 795 nm or $5^2S_{1/2} \rightarrow 5\ ^2P_{3/2}$ transition at 780 nm. The outer absorption profiles result from the ^{87}Rb absorption, whereas the inner two come from ^{85}Rb absorption. The excited state 5 $^2P_{3/2}$ hyperfine levels are not resolved due to the Doppler broadening. Therefore, each of the four absorption profiles consists of three Doppler broadened hyperfine transitions. The modulations of the absorption profiles are the result of the fs pulse train excitation of Rb atoms.

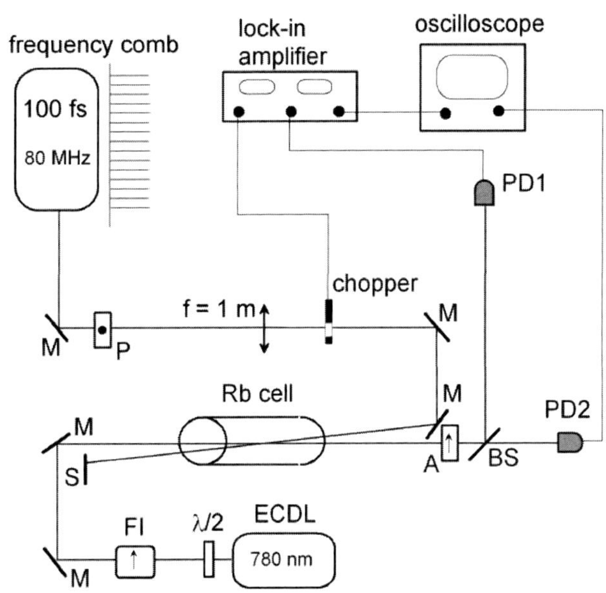

Figure 4. Experimental scheme. M-mirrors, PD-photodiodes, P-polarizer, A-analyzer, FI-Faraday isolator and S-beam stopper.

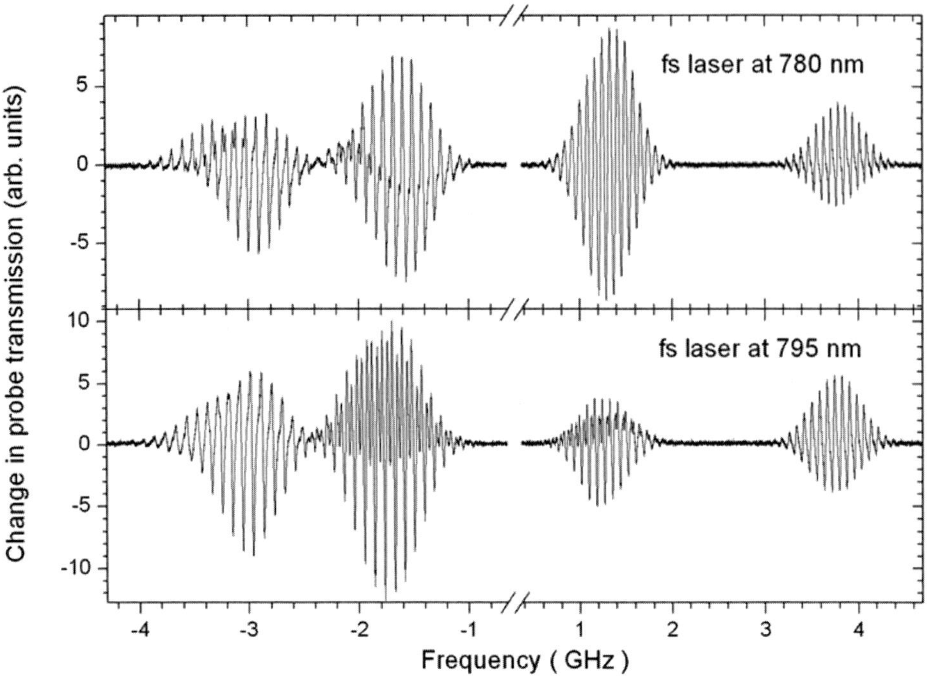

Figure 5. Measured ΔT for fs laser positioned on the Rb 5 $^2S_{1/2} \rightarrow 5\ ^2P_{1/2}$ transition at 795 nm (D1 resonance line) and Rb 5 $^2S_{1/2} \rightarrow 5\ ^2P_{3/2}$ transition at 780 nm (D2 resonance line).

The signal from the photodiode, PD1, was led to the lock-in amplifier (Stanford Research SR510) referenced to the mechanical chopper on the fs laser beam. The lock-in signal represents only the change of the probe laser transmission that was induced by fs laser. The output of the lock-in amplifier was put on the second channel of the digital oscilloscope, and it represent $\Delta T = T_{fs}-T$, where T_{fs} is the probe transmission with and T without fs pulse train excitation. Measured ΔT for fs laser positioned on the Rb 5 $^2S_{1/2} \rightarrow 5\ ^2P_{1/2}$ transition at 795 nm (D1 resonance line) and Rb 5 $^2S_{1/2} \rightarrow 5\ ^2P_{3/2}$ transition at 780 nm (D2 resonance line) is shown in Fig 5. The advantage of the lock-in technique is obvious. The broad Doppler background is eliminated, the signal to noise ratio is greatly enhanced and access to finer details of the modulations is achieved. For the spectra shown in Fig. 5 the fs and probe laser powers were set to 600 mW and 2.6 μW, respectively.

III. Theory

The detailed rubidium atom energy level scheme relevant for the experiment is given in Fig. 6 for both isotopes. Theoretical modeling was carried out utilizing standard density-matrix analysis. The eight-level (two 5 $^2S_{1/2}$ hyperfine ground levels, two $5^2P_{1/2}$ and four $5^2P_{3/2}$ excited hyperfine levels) Rb atoms interacting with the fs and probe laser electric fields were considered.

The density matrix equations of motions are given by the following equations [34]:

$$\dot{\rho}_{nm} = -i\hbar^{-1}\left[\hat{H}, \hat{\rho}\right]_{nm} - \gamma_{nm}\rho_{nm}, \qquad n \neq m, \tag{1}$$

$$\dot{\rho}_{nn} = -i\hbar^{-1}\left[\hat{H}, \hat{\rho}\right]_{nn} + \sum_{E_m \rangle E_n} \Gamma_{nm}\rho_{mm} - \sum_{E_m \langle E_n} \Gamma_{mn}\rho_{nn}. \tag{2}$$

where the subscripts nm refer to the eight hyperfine levels numbered from the lowest to highest energy level. Γ_{nm} gives the rate per atom at which population decays from level m to level n. It includes level to level radiative decay and inelastic collisions. γ_{nm} gives the damping rate of the ρ_{nm} coherence. The damping rates γ_{nm} for the off-diagonal elements of the density matrix are not independent of the damping rates of the diagonal elements. They are represented as:

$$\gamma_{nm} = \frac{1}{2}\left(\Gamma_n + \Gamma_m\right) + \gamma_{nm}^{col}, \tag{3}$$

Γ_n and Γ_m denote the total decay rates of population out of level n and m, respectively. γ_{nm}^{col} is the dephasing rate due to the elastic collisions. The total population decay rate of level n, Γ_n, is calculated as a sum of the radiative decay rates, $\Gamma_{n'n}$.

$$\Gamma_n = \sum_{n'(E_{n'} \langle E_n)} \Gamma_{n'n}. \tag{4}$$

In low density rubidium vapor at room temperature the radiative decay rates dominate among all damping rates. The inelastic collisional rate contributes to the population and coherence decay through the thermalization process. It is given by the product of the collision cross section [35], the average atom velocity, and the atomic number density. In our experimental conditions it is about 8 kHz. The elastic collisions contribute only to the decay of coherence (dephasing). The elastic collision dephasings are deduced from the experimental values for the self broadening of Rb resonance lines [36]. For the rubidium vapor at room temperature they are about 16 kHz. The total radiative decay rates calculated from the lifetimes [37] of $5^2P_{1/2}$ (27.7 ns) and $5^2P_{3/2}$ (26.24 ns) are 36 MHz and 38 MHz (in circular frequencies), respectively. As we can see, both elastic and inelastic decay rates are three orders of magnitude smaller than the radiative decay rates.

The Hamiltonian of the system is $H=H_0+H_{int}$, where H_0 is the Hamiltonian of the free atom and $(H_{int})_{nm} = -\mu_{nm} E_T(t) - \mu_{nm}E_{CW}(t)$ represents the interaction of the atom with the pulse train and cw laser electric field, μ_{nm} being the dipole moment of the electronically allowed ($F_g \rightarrow F_e=F_g$, $F_g\pm1$) transitions, calculated from ref. [38]. Since in our experiment the frequency comb was centered on the $5^2S_{1/2}\rightarrow 5\ ^2P_{1/2}$ transition and the probe laser was scanned across the $5\ ^2S_{1/2}\rightarrow 5\ ^2P_{3/2}$ transitions, every single transition connecting two nm hyperfine levels is affected by just one of this two laser fields. The pulse train electric field is given by:

$$E_T(t) = \left[\sum_{n=0}^{N} \varepsilon(t - nT_R) e^{in\Phi_R} \right] e^{i\omega_L t} = \varepsilon_T(t) e^{i\omega_L t}, \quad (5)$$

where N is a large integer (order of 10^6), $\varepsilon(t - nT_R)$ is the slowly varying envelope of the n-th hyperbolic-secant laser pulse, Φ_R is the round-trip phase acquired by the laser within the cavity, T_R is the laser repetition period and ω_L is the central laser angular frequency. $\varepsilon(t)$ is the slowly varying envelope of the pulse train. The pulse train frequency spectrum consists of a comb of N laser modes separated by $1/T_R$ and centered at $\omega_L + \Phi_R / T_R$. The n-th mode angular frequency is given by $\omega_n = \omega_L + \Phi_R / T_R \pm 2\pi n / T_R$. The probe laser electric field is $E_{CW}(t) = E_0 e^{i\omega_{CW} t}$, where E_0 is amplitude of the monochromatic electric field and ω_{CW} is angular frequency.

From equation (1) and (2) a system of 36 coupled differential equations for the slowly varying density-matrix elements was obtained. The population of the n-th atomic level is given by the diagonal density-matrix element ρ_{nn}, whereas off diagonal elements σ_{nm}, where $\sigma_{nm} = \rho_{nm} e^{-i\omega_L t}$, represent the slowly varying envelope of the coherences. The system of coupled differential equations was integrated using a standard fourth-order Runge-Kutta method.

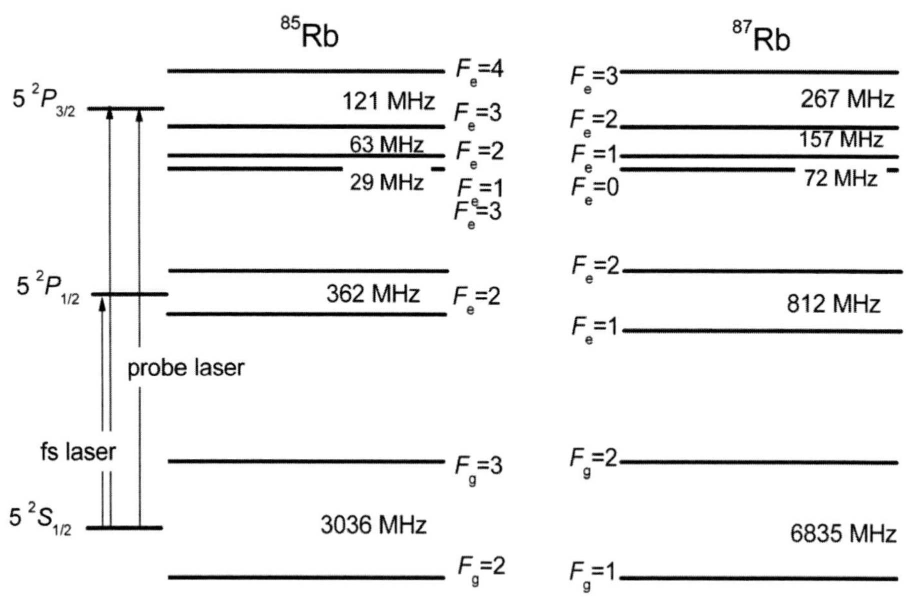

Figure 6. Hyperfine energy level scheme of the ^{85,87}Rb 5 $^2S_{1/2}$, 5 $^2P_{1/2,3/2}$ ten-level system

In the case of weak field approximation and linear absorption regime, the populations of the two ground levels are not noticeably altered by the interaction with the radiation field. Therefore, for the probe laser intensities which are far below the saturation intensity for Rb

$5^2S_{1/2} \rightarrow 5^2P_{3/2}$ transition (2.5 mW/cm^2) the influence of the probe laser on the ground levels population is negligible. Measured optical thickness is then directly proportional to the ground hyperfine level populations ρ_{11} and ρ_{22}, which are modified due to the $5\ ^2S_{1/2} \rightarrow 5\ ^2P_{1/2}$ fs pulse train excitation. In these conditions, the above described eight level system reduces to a four level system (two $5\ ^2S_{1/2}$ ground and two $5^2P_{1/2}$ excited levels) interacting only with pulse train electric field.

Typical time evolution of ^{87}Rb($5\ ^2P_{1/2}$) hyperfine excited level populations (ρ_{33}, ρ_{44}) and ^{87}Rb($5\ ^2S_{1/2}$) hyperfine ground state populations (ρ_{11}, ρ_{22}) are shown in Figs. 7 and 8. The calculations were performed for the electrical field peak amplitude $0.3{\cdot}10^6$ V/m, T_R=12.5 ns, Φ_R=0 and fs laser ω_L equal to ^{87}Rb F_g=1→F_e=2 transition frequency. The pulse train effects are clearly seen. Due to the pulse repetition period T_R which is smaller than the relaxation times of the system, the system can never completely relax between two consecutive laser pulses. Therefore, the atoms accumulate excitation in the form of coherence and excited state population, as observed in Fig. 7 for F_e=2 hyperfine excited level. The stationary state is reached after about 15 pulses with the coherent enhancement factor around 3 times larger than the incoherent enhancement factor (one pulse excitation). The stationary state ^{87}Rb($5\ ^2P_{1/2}$) population is less than 1% due to the four-level system without optically closed transition and weak pulse electrical field. The significant difference in the fractional hyperfine ground state populations is clearly seen in Fig. 8. This is due to the effective optical pumping process for which a stationary state solution is obtained on a time scale of the order of 10 μs (final hyperfine state population). In the case of ^{85}Rb the calculations show qualitatively similar behavior.

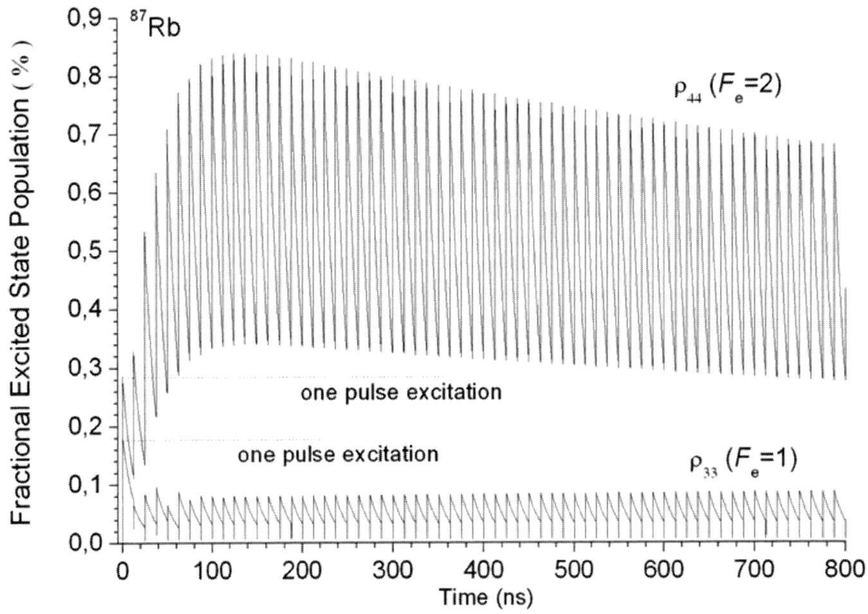

Figure 7. Typical time evolution of ^{87}Rb($5\ ^2P_{1/2}$) hyperfine excited level populations calculated for fs laser positioned at ^{87}Rb F_g=1→F_e=2 transition frequency.

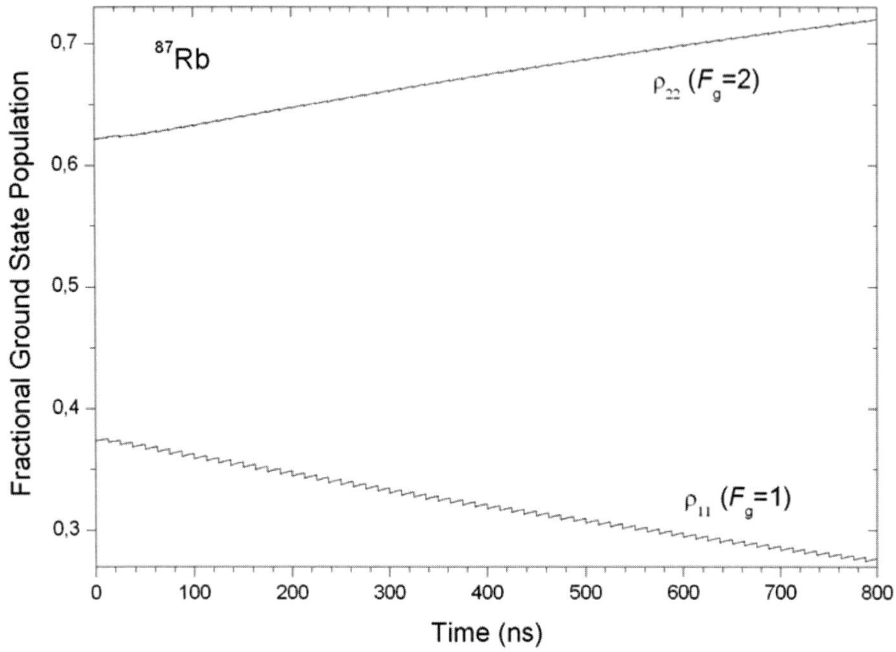

Figure 8. Typical time evolution of ^{87}Rb(5 $^2S_{1/2}$) hyperfine ground level populations calculated for fs laser positioned at ^{87}Rb F_g=1→F_e=2 transition frequency.

For the rubidium vapor at room temperature, the inhomogeneous Doppler broadening (of about 500 MHz) is significantly larger than the homogeneous broadening. Therefore, the atomic transition frequency ω_{ge} must be replaced with $\omega'_{ge} = \omega_{ge} + \vec{k} \cdot \vec{v}$, where $\vec{k}$ is the laser wavevector and $\vec{v}$ is the atomic velocity. Different velocity groups correspond to different frequency detuning, $\delta = \vec{k} \cdot \vec{v}$, so for a given ω_n and a given 5 $^2S_{1/2}(F_g)$→5 $^2P_{1/2}(F_e)$ hyperfine transition there is a velocity group (δ_n detuning), which fulfills $\omega_n = \omega'_{ge}$ resonance condition. Since the pulse train frequency spectrum consists of a comb of laser modes separated by $1/T_R$ (80 MHz), the resonance condition is also satisfied for velocity groups with detuning $\delta = \delta_n \pm 2\pi k / T_R$, where k is a positive integer, corresponding to resonance with different comb modes. Therefore, different velocity groups are in different situation with respect to the excitation (accumulation) process, which leads to the velocity selective optical pumping of ground hyperfine levels and velocity selective comb-like population in the excited hyperfine levels. Calculated ^{87}Rb hyperfine ground and excited state levels populations for different atomic velocity groups (different detuning δ) in the case of 5 $^2S_{1/2}$→5 $^2P_{1/2}$ fs excitations are shown in Fig. 9. The fractional level populations are weighted by the Doppler profile (only a central part of the Doppler is presented). The calculations were performed for the electric field amplitude $1.5 \cdot 10^6$ V/m, T_R=12.5 ns, Φ_R=0 and ω_L equal to ^{87}Rb F_g=1→F_e=2. The level populations exhibit unique oscillatory structure which varies with 80 MHz period. This is a direct consequence of the pulse train excitation having the frequency comb spectrum. However, the level populations themselves are determined by the 5 $^2S_{1/2}$, 5 $^2P_{1/2}$ and 5 $^2P_{3/2}$ hyperfine energy splitting and relative transition probabilities. The optical pumping of the hyperfine ground levels and velocity selective excitation, i.e. mapping of the frequency comb to the excited levels velocity comb are clearly seen.

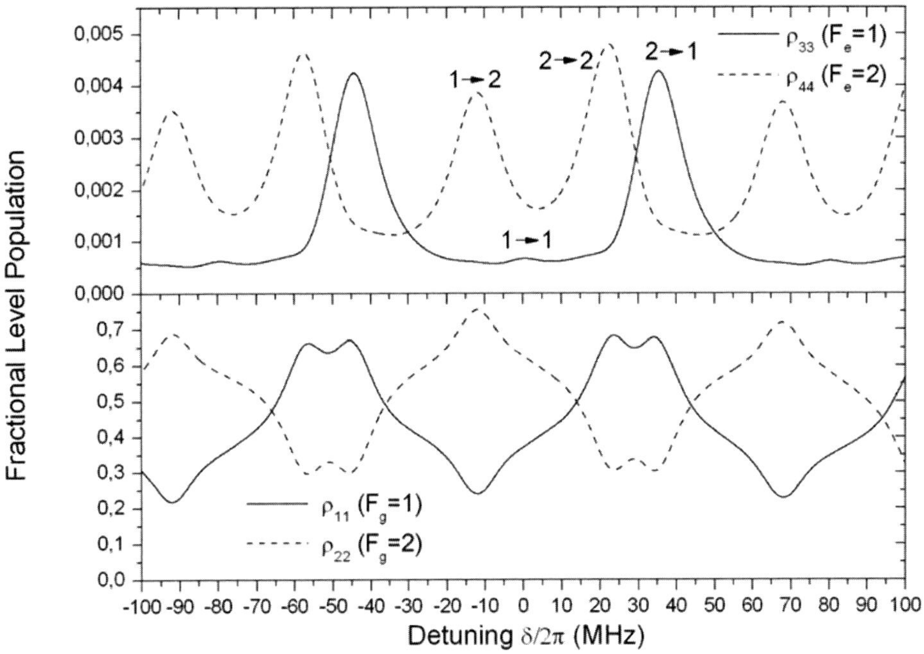

Figure 9. ^{87}Rb calculated ground (ρ_{11}, ρ_{22}) and excited (ρ_{33}, ρ_{44}) states hyperfine level populations for different atomic velocity groups in the case of $5\,^2S_{1/2}\rightarrow 5\,^2P_{1/2}$ fs laser excitation.

Calculated absorption coefficient for the rubidium vapor at room temperature interacting with fs pulse train for ^{87}Rb isotope is shown in Figs. 10. Each hyperfine absorption profile (shown with thin line in Fig. 10) was calculated as a convolution of the velocity distribution of the ground state population and the Lorentzian of natural width. The modulation in each hyperfine transition line is determined by the $5^2S_{1/2}$ and $5^2P_{1/2}$ hyperfine energy splitting and $5\,^2S_{1/2}(F_g)\rightarrow 5\,^2P_{1/2}(F_e)$ transition probabilities. The periodic behavior is given by the 80 MHz mode separation in the frequency comb. The difference in modulation structure for the case of ^{87}Rb and ^{85}Rb isotopes is explained in detail in ref. [28]. Each $5^2S_{1/2}(F_g)\rightarrow 5^2P_{3/2}(F_g, F_g\pm 1)$ absorption line is calculated by adding the contributions of three hyperfine components.

From calculated optical thickness and Beer Lambert law the probe laser transmission with (T_f) and without (T) the fs laser are calculated. By subtracting these two spectra, we simulate the change in the probe laser transmission due to the fs pulse train excitation, $\Delta T = T_{fs} - T$. Simulated and measured ΔT in the case of weak probe laser scanning across all four $5^2S_{1/2}\rightarrow 5^2P_{3/2}$ absorption lines for ^{87}Rb atoms excited by the fs laser positioned at 795 nm are shown in Fig. 11. The agreement between theory and experiment confirms this new method as a high sensitivity technique for the frequency comb excitation processes monitoring. By monitoring ΔT, this technique enables us to study the influence of different experimental parameters (such as central fs laser wavelength and output powers of the fs and probe lasers) on the velocity selective optical pumping.

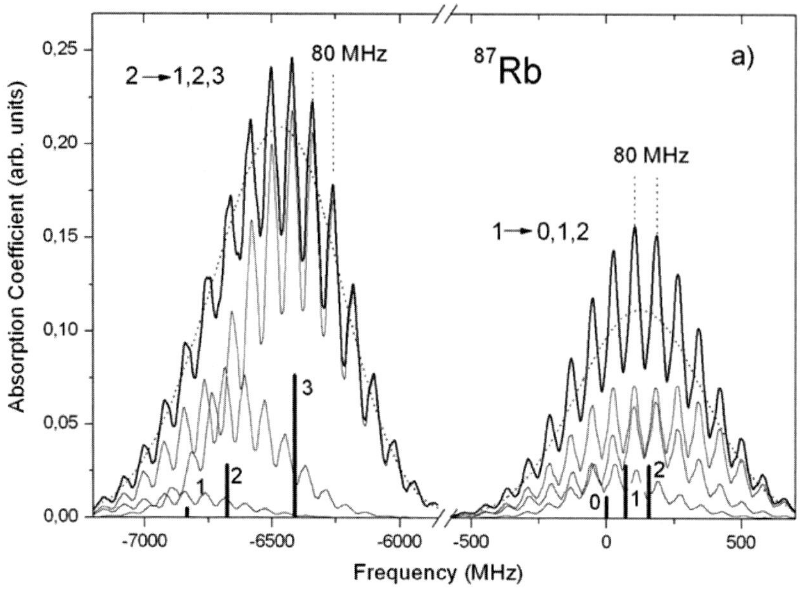

Figure 10. Calculated optical thickness for ^{87}Rb isotope in the case of $5^2S_{1/2} \rightarrow 5\ ^2P_{1/2}$ fs laser excitation. An absorption line (thick line) is calculated by adding the contributions of three hyperfine components (thin lines). Doppler profiles are shown with dotted lines. Thick vertical lines denote the frequencies and relative transition dipole moments of hyperfine lines.

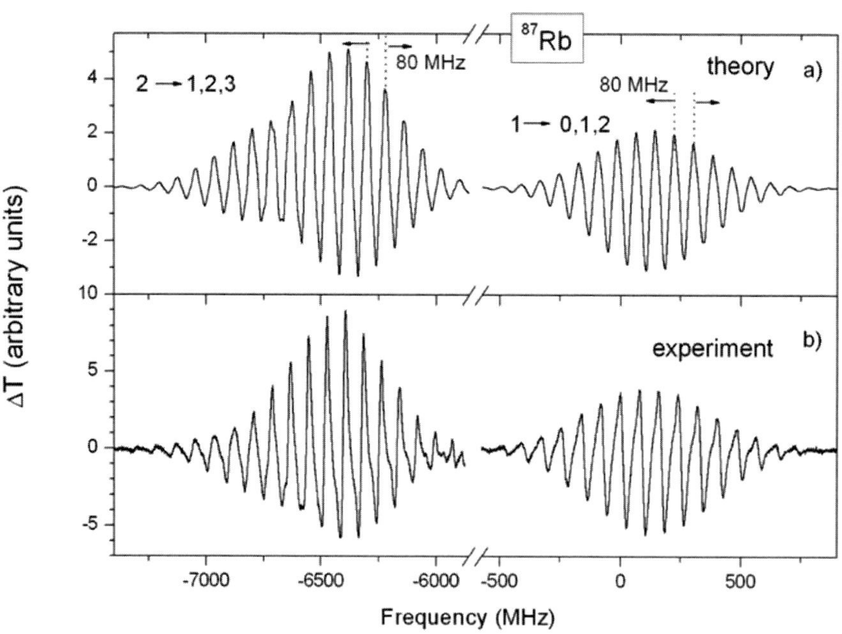

Figure 11. Comparison between calculated and measured ΔT for ^{87}Rb isotope in the weak probe case. $\Delta T = T_{fs} - T$, where T_{fs} and T are the probe transmission with and without fs laser excitation. In the experiment the fs laser was adjusted at 795 nm with 600 mW of output power. The probe laser of 2.6 μW output power was scanned across $5^2S_{1/2} \rightarrow 5^2P_{3/2}$ absorption.

Utilizing presented experimental methods and theoretical approach we investigated different coherence accumulation effects in rubidium vapor due to the fs pulse train excitation.

IV. Influence of Magnetic Field

In an external magnetic field the Rb ground and excited state levels are not characterized any more by total angular momentum quantum numbers F_g and F_e, but they are a mixture of these states created by the external field. The magnetic sublevel energies are dependent on the external magnetic field strength. For very low field strengths each hyperfine level experiences linear Zeeman effect. As the field strength increases, Zeeman energy becomes close to the hyperfine splitting energy and the energy level pattern becomes even more complex.

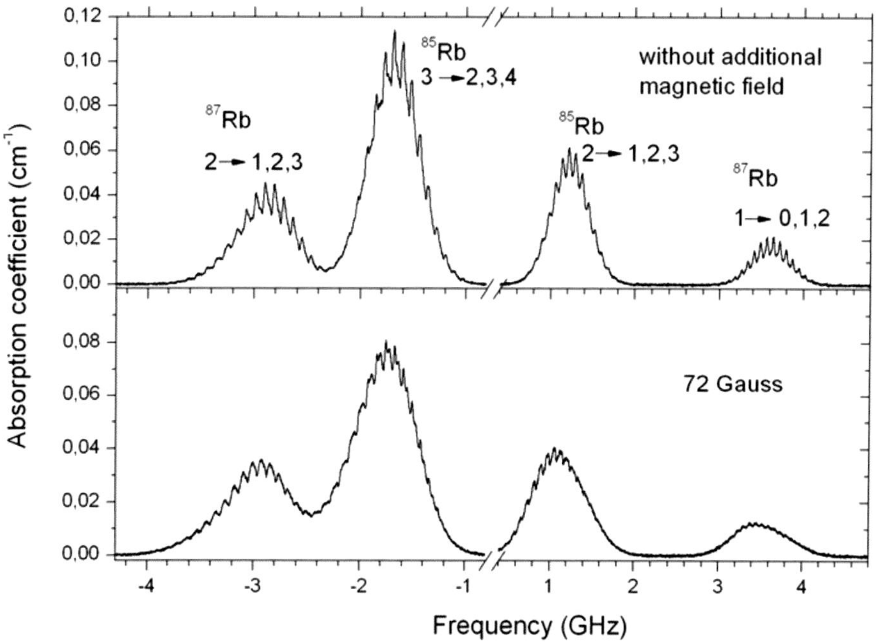

Figure 12. Measured ^{85,87}Rb $5\ ^2S_{1/2} \rightarrow 5\ ^2P_{3/2}$ absorption spectra for $5^2S_{1/2} \rightarrow 5\ ^2P_{3/2}$ fs laser excitation in the conditions without (upper panel) and with external magnetic field (lower panel).

Measured absorption spectra for $5^2S_{1/2} \rightarrow 5\ ^2P_{3/2}$ fs laser excitation with and without external magnetic field are shown in Fig. 12. In the external magnetic field the shape of the modulations changes, due to the change in the splitting structure. The relative modulation intensity generally decreases in an undulatory fashion. Relative modulation intensity dependence on external magnetic field for the ^{87}Rb F_g=1→F_e=0,1,2 absorption line shows damped oscillatory behavior, as shown in Fig. 13. Subsequent maxima are separated by approximately 30 Gauss. As previously discussed, the observed absorption line modulations strongly depend upon energy level splitting, transition probabilities and fs frequency comb spectrum and its intensity. For the field strength which leads to a favorable energy level splitting (the frequency separation of different transitions is a multiple of 80 MHz) one would expect stronger modulations.

Elaborate theoretical calculations reveal the main features observed experimentally. By increasing the magnetic field strength the relative modulation of the peaks in the line profile diminishes up to about 20 Gauss, and then it rises again to about 30 Gauss, however, not attaining large values of relative modulation as for zero magnetic field strength. After the second maximum, there is a minimum and the third maximum at about 60 Gauss. This undulating character in relative modulation of the peaks within the single line profile indicates some additional resonances in complex energy level structure of the relevant energy levels in magnetic field of different strengths. Careful simulations reveal similar relative modulations of the multi-Bennett holes and peaks, pointing out that some important details have to be investigated further in the near future. Our preliminary theoretical description of the relative undulations dependence upon the magnetic field took into account two level atom picture. Thus it cannot reproduce experimental findings in every detail, although the rough comparison is rather satisfactory. An analytical model approach would be more appropriate and we intend to do that in the near future.

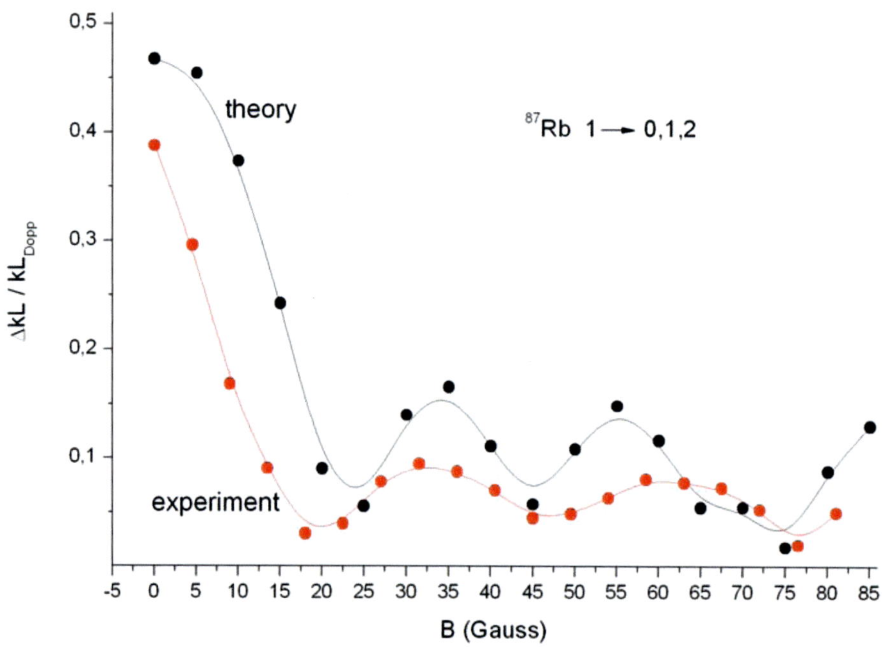

Figure 13. Relative modulation intensity of ^{87}Rb F_g=1→F_e=0,1,2 absorption line as a function of external magnetic field strength, dots. Solid line is an eye-guiding fit to the experimental data.

The behaviour of the relative undulations in a homogeneous electric field would be of considerable importance, because the Stark splitting is degenerate in magnetic quantum numbers $+m_F$ and $-m_F$. We hope to perform appropriate measurements, using metal plates to generate homogeneous electric filed, in the near future.

V. Influence of the Strong Probe Laser

In the strong probe case the population time evolutions change significantly. Strong optical pumping dominates the atom dynamics, but it is in this case governed predominately by the strong probe laser. No accumulation of population and coherence induced by the fs laser excitation was observed, see Fig. 14. Since accumulation of population and coherence are typical pulse train effects, we can conclude that by increasing probe laser power the interaction of the system with fs laser changes from the pulse train interaction to the single pulse after single pulse interaction. In other words, the effective atomic coherence relaxation time decreases when the probe laser power is increased. The experimental investigation of the change in the probe transmission due to the fs laser excitation, $\Delta T = T_{fs} - T$, for different probe laser powers is shown in Fig. 15. The fs laser was tuned to $5^2S_{1/2} \rightarrow 5^2P_{1/2}$ transition at 795 nm and the average output power was 600 mW. The probe laser was scanned across all four $5^2S_{1/2} \rightarrow 5^2P_{3/2}$ absorption lines. The probe laser output power was increased gradually from 2.6 μW (weak probe case) up to 3.9 mW. By increasing the probe laser power the structure and depth of the modulations in ΔT change significantly. In the strong probe case, the atoms interact with the broad continuous spectrum of the single pulse and not with the frequency comb. Thus by sudden increase of the probe laser power we can change the experimental condition from the train of femtosecond pulses to a single femtosecond pulse arrangement. Such a switch could possibly find certain application in the field of quantum computing.

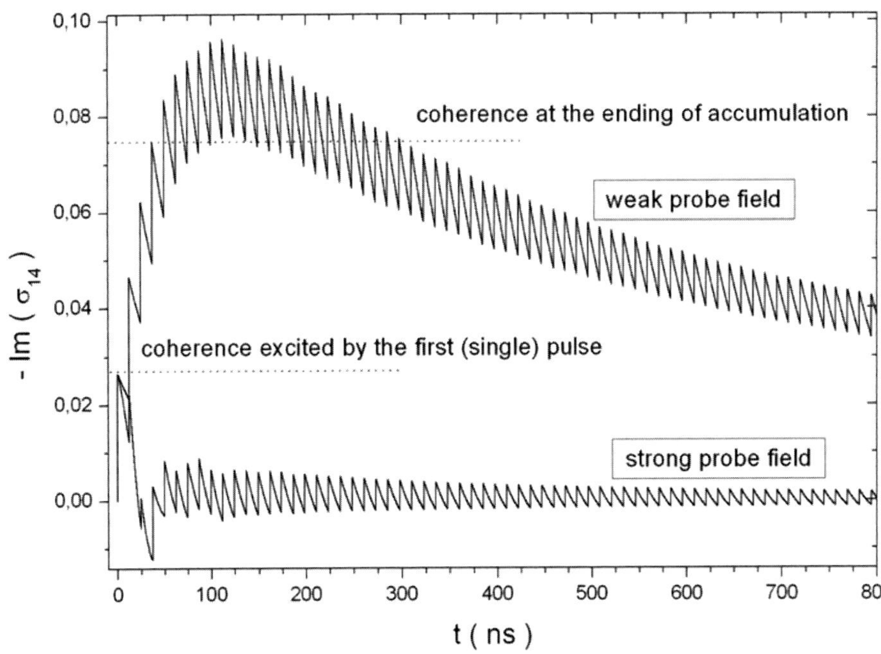

Figure 14. Calculated time evolution of the slowly varying envelope of the coherence σ_{14} for the transition resonantly excited by the fs laser, in the case of weak (18 V/m) and strong (300 V/m) probe fields. fs laser central frequency was set to $5\ ^2S_{1/2}(F_g=1) \rightarrow 5\ ^2P_{1/2}(F_e=2)$ hyperfine transition, while the probe laser frequency was set to $5\ ^2S_{1/2}(F_g=1) \rightarrow 5\ ^2P_{3/2}(F_e=1)$ hyperfine transition frequency.

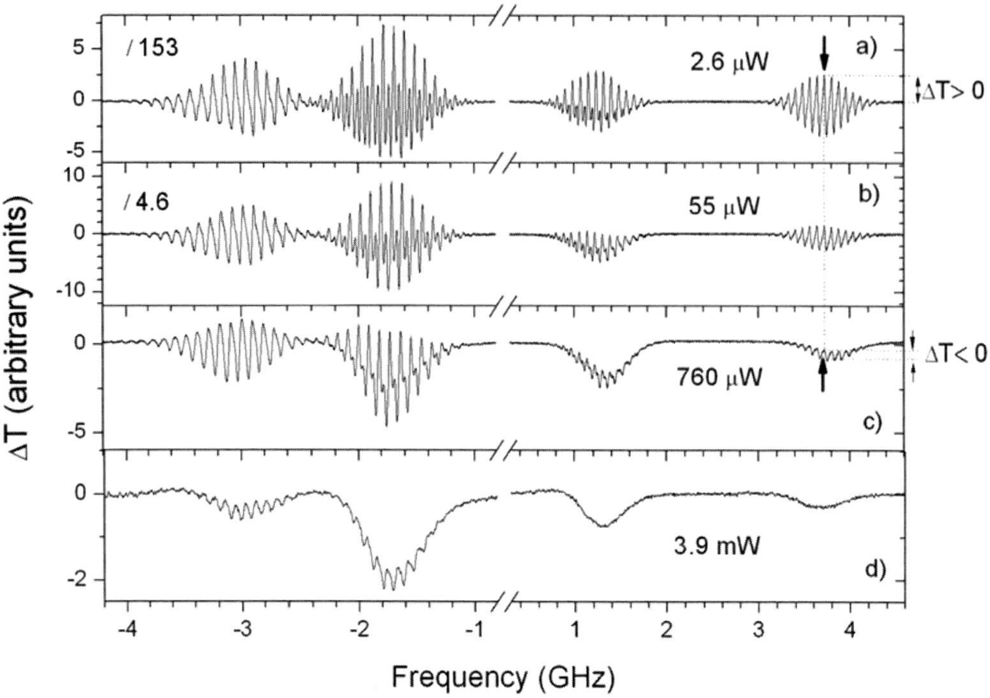

Figure 15. The dependence of the probe laser transmission modulations ΔT on the probe laser output power. The fs laser is tuned to 795 nm, with 600 mW of average output power. The probe laser is scanned across $5^2S_{1/2}\rightarrow5^2P_{3/2}$ absorption lines at 780 nm.

VI. Atomic Filter

We have seen that a femtosecond (fs) mode-locked laser pulse train excitation makes a direct mapping of the optical frequency comb onto the Rb atom velocity comb. The induced comb-like structure in the Rb ground state hyperfine level populations at different velocity groups is monitored by a weak cw scanning probe laser. However, by introducing an additional Rb cell, as shown in Fig.16, a hyperfine atomic filter is realized, which modifies the amplitude of the selected frequency comb lines underneath the Doppler profiles. Depending on the particle density in the hyperfine atomic filter cell we may obtain partial or full attenuation of those comb lines which are closer to the center of the Doppler profile. The comb lines on both sides of the Doppler profile sense a different index of refraction, which leads to the change in the shape of the pulses in their time domain.

Our results indicate the possibility of manipulating the amplitude and phase of the frequency comb in the specific interval of frequencies under the profile of the hyperfine components of the Rb resonance lines. Natural extension of this technique towards alkali diatomic molecules is possible. In such experiments the molecular filtering and modified frequency comb may be employed as a way to selectively excite the heteronuclear molecules, AB, by placing a cell with A and a cell with B on the path of the femtosecond laser beam. In doing so, the frequency comb would come to the main cell filled with A+B, cleaned from the components that would excite A_2 and B_2 molecules. This would be particularly interesting in the case of alkali heteronuclear dimers or alkali hydrides where the absorption bands of hetero- and homonuclear molecules overlap with each other. Currently we are working with isotopically pure rubidium absorption cells as atomic hyperfine filters so that only one of the ^{87}Rb or ^{85}Rb isotopes would be attenuated.

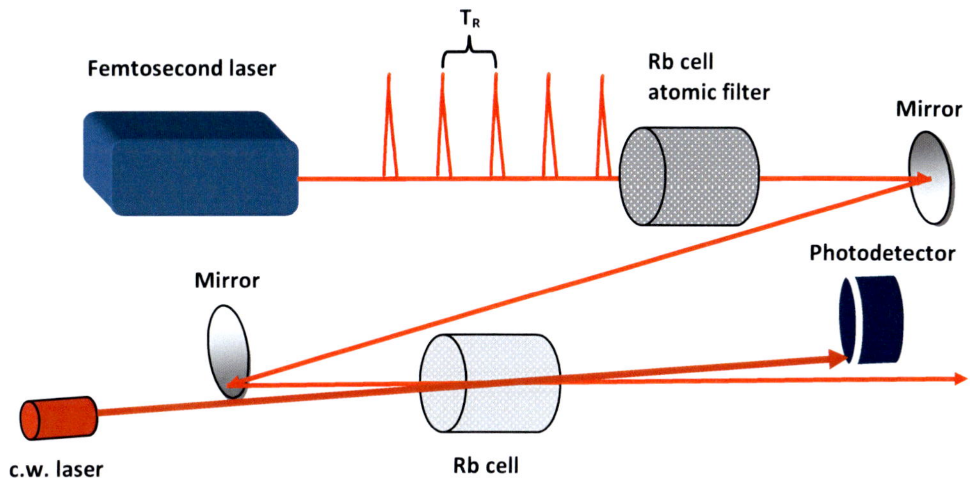

Figure 16. Experimental setup for the atomic hyperfine filter measurement.

The application of the atomic hyperfine filter using fs laser frequency comb may be in the photoassociation of ultracold atoms. At certain interatomic distances some of the fs comb lines within the Doppler profile at room temperature can be extracted by using an atomic hyperfine filter, as presented in the current work. The rest of the frequency comb may strike one or more resonances that lead to the photoassociation of ultracold atoms.

VII. Different Repetition Rates of the Frequency Comb

It is interesting to note that higher repetition rate of the femtosecond laser will lead to even more peculiar pattern of the final line profile. Higher repetition rates mean that the necessary condition for the onset of the accumulation effects is is even better fulfilled. The consequence of the optical pumping is well visible in Fig. 17. We performed the same type of calculations as in previous cases, except that we changed the repetition rate from 100 MHz to 500 MHz. We may readily observe that at locations of the peak absorption in one hyperfine component there is always a dip in the absorption of another hyperfine component at least up to about 150 MHz repetition rates. At even higher repetition frequencies the line profiles appear very peculiar. Instead of obtaining rather regular separated Bennett holes and corresponding peaks at two hyperfine resonance lines from lowest two hyperfine levels, the Doppler profiles exhibit rather irregular forms. Further investigations certainly require more elaborate experiments followed by adequate simulations.

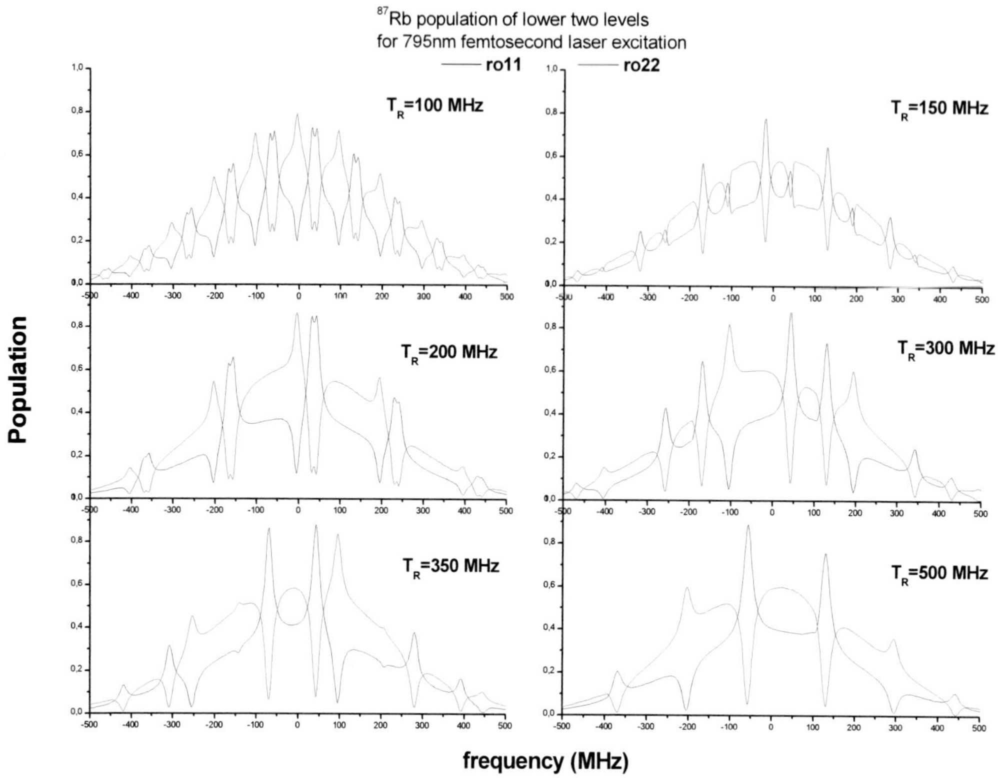

Figure 17. Distrubution of populations (matrix elements of ro11 and ro22) in the lower two hyperfine levels excited by the fs laser comb of six different repetition rates. The center of the fs comb was centered at 795 nm.

Another experimental vision would be the application of two frequency combs with slightly different repetition rates as it was recently discussed by Newbury et al. [39] The manipulation of coherent accumulation of populations and coherences could be, in that case, studied in the microwave spectral region by beating of the neighboring frequency comb line pairs.

VIII. Precision Measurements

Simultaneous recording of the multi-Bennett peaks and holes in the line profiles of rubidium resonance lines together with the saturation resonances may be used in precise measurements of the frequency comb lines. A new method for determination of the frequency comb offset frequency f_0 would help in easy and simple absolute frequency determination. We exploit the effect of velocity selective optical pumping induced by frequency comb excitation to effectively map the frequency comb spectrum onto the velocity distribution of hyperfine level populations in room-temperature Rb atoms. Since the optical pumping process is dependent upon hyperfine energy level splitting and relative transition dipole moments, this physical mapping is not straightforward and a full time-dependent theoretical modeling of atom-laser field interaction is needed for interpretation of the observed phenomena. On the other hand, the theoretical modeling provides not only qualitative but also quantitative description of the interacting atom-laser system, as has been illustrated by the excellent agreement with the experimental results. Based on this agreement, using experimentally well-known system of room-temperature Rb atoms as 'detectors' for the frequency comb, we suggest a method for determination and monitoring of the frequency comb offset frequency f_0. It is based on the fitting of the theoretical model results (taking frequency comb parameters as free parameters) to the experimentally observed modulations in the transmission profiles of the probe laser [40].

In order to calculate the hyperfine level distributions, one has to solve the set of equations separately for every velocity group. The time evolution of the hyperfine level populations is obtained and the population values at time T_0 are taken as the final populations. The time T_0 is given by the average time of interaction of atoms with the fs laser and is deduced from the fs laser beam cross-section and the average atom velocity (in our experimental conditions T_0 corresponds to a sequence of about 100 pulses). This is where the advantage of the analytical solution of equations (1) and (2) is clearly evident since the full numerical calculation of the results shown in Fig. 8 and 9 takes several hours, while the application of the analytical solution gives the same result in less than a second.

IX. The Case of Cesium

Femtosecond laser pulse train effect on Doppler profile of cesium resonance lines was carried out in a direct observation of the velocity-selective optical pumping of the Cs ground state hyperfine levels induced by the femtosecond (fs) laser oscillator centered at either D2 (6s $^2S_{1/2} \rightarrow$ 6p $^2P_{3/2}$, 852 nm) or D1 (6s $^2S_{1/2} \rightarrow$ 6p $^2P_{1/2}$, 894 nm) cesium line. We utilized previously developed modified direct frequency comb spectroscopy (DFCS) which uses a fixed frequency comb for the excitation and a weak cw scanning probe laser centered at the ^{133}Cs 6s $^2S_{1/2} \rightarrow$ 6p $^2P_{3/2}$ transition (D2 line) for ground levels population monitoring. The frequency comb excitation changes the usual Doppler absorption profile into a specific periodic, comb like structure. The mechanism of the velocity selective population transfer between the Cs ground state hyperfine levels induced by fs pulse train excitation is verified in a theoretical treatment of the multilevel atomic system subjected to a pulse train resonant field interaction [32]. We used all the data concerning cesium resonance transitions from ref. [41] and our experimental parameters as in all previous cases of rubidium atom excitations.

The cesium atom serves as the primary standard of time and it plays an important role in many experiments on laser cooling, trapping, atom interferometry and the precise measurements. It would be interesting to investigate the same effect on cesium with cw laser scanning over the D1 line (894 nm), due to the relatively simple transition hyperfine 4-level structure. The large hyperfine structure of the ground level 9.2 GHz and excite-state hyperfine splitting of over 1 GHz makes cesium case especially interesting for many future applications. Beside this the hyperfine spectrum is additionally simplified due to only one natural isotope.

X. The Case of Potassium

Potassium has the smallest hyperfine structure and the D1 resonance line is an interesting overlap of two hyperfine components stemming from the ground levels of ^{39}K atoms, Fig. 18. Other isotopes, and especially the fermion ^{40}K exhibits similar structure. All different isotopes of potassium should be carefully investigated. The vicinity of D1 and D2 resonance lines makes experiments easier to perform due to the fact that a single laser diode may cover both resonance lines. However, the small hyperfine structure complicates the simulation calculations. In an actual experiment, which we intend to perform, it will be necessary to heat the glass cell filled with potassium in order to attain sufficient density for the accumulation effect to become visible, at least with the present equipment described in the experimental part of this chapter. From the Table 1 it may be readily seen that the ground level hyperfine separation of all isotopes is very small, and therefore the Doppler broadened components will be heavily overlapped. For the case of fermionic isotope ^{40}K, where the separation is larger, 1286 MHz, it would be possible to observe resolved Doppler broadened hyperfine components. The excited state 4p $^2P_{3/2}$ of potassium ^{40}K and ^{41}K atoms have hyperfine level separations from a few MHz to several tens of MHz and will be not resolved [42].

Table 1. Data for the different potassium isotopes.

Isotope	%Natural abundance	Hyperfine splitting (MHz)
^{39}K	93.3	462
^{40}K	0.012	1286
^{41}K	6.7	254

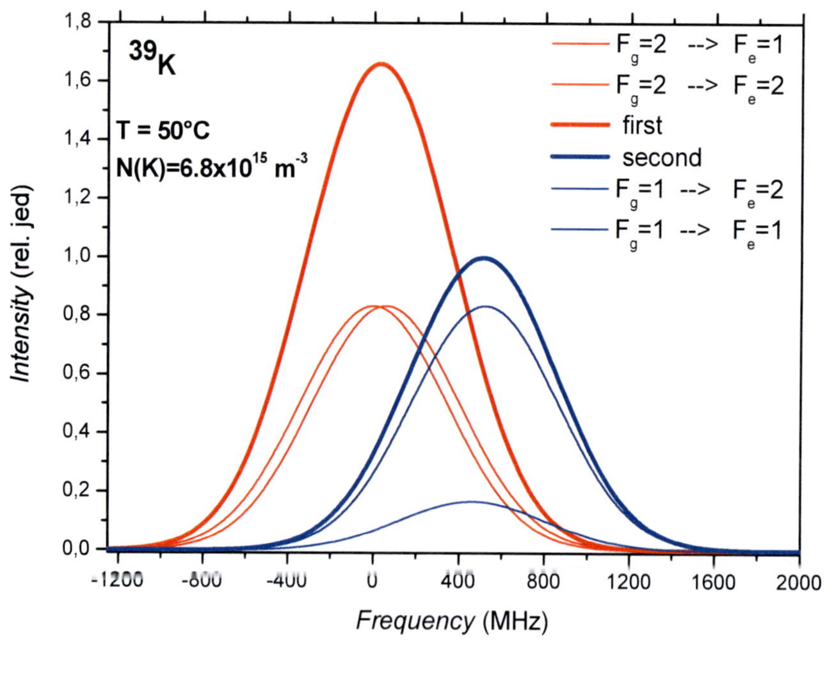

Figure 18. Doppler profiles of different hyperfine components of 39K isotope.

XI. Discussion and Outlook

We have presented enhanced sensitivity measurements of the velocity selective optical pumping of the Rb ground state hyperfine levels population induced by the pulse train excitation. Using this approach we were able to directly measure the change of the probe laser transmission as a result of the resonant frequency comb excitation of the Rb atoms. The experimental method is verified by the theoretical treatment developed in our previous paper and applied for the investigation of the resonant frequency comb excitation in the strong probe case. The effects of the fs laser excitation on the strong probe laser absorption are calculated by applying an extended theoretical model. The calculated results confirm the measured increase of the probe laser absorption. Additionally, we show that the strong probe laser field decreases the relaxation time of the Rb system leading to the single pulse by single pulse excitation experimental condition. Therefore, by changing the probe laser power, gradual change from the frequency comb excitation to the pulse by pulse excitation of Rb atoms is observed.

The applied laser scheme (strong probe case) could be correlated to the V-type electromagnetically induced transparency (EIT) scheme [43], where strong probe laser acts as a pump laser and the frequency comb modes represent an array of the weak probe lasers. However, to observe the EIT resonance in such a scheme the absorption of an individual frequency comb mode has to be measured. This is an additional experimental challenge which will be treated in the future experiments.

Velocity selection within the ground state alkali atoms is certainly a very attractive technique to manipulate ensemble by pulse shaping of the femtosecond laser frequency comb. Numbers of new experiments are needed in which we intend to show possibilities to transfer atoms from one velocity group to another velocity group.

It would be interesting to subject the atoms to simultaneous action of two stabilized frequency combs that differ slightly in repetition rates. If the rates could be tunable in a certain interval we might influence the overall picture of the velocity distributions.

Femtosecond laser frequency comb generated by the fiber laser should have several amplification stages in order to achieve sufficient power to efficiently affect the velocity distribution of atoms.

D1 line of the potassium, rubidium or cesium atoms would be much more amenable for the further experimental and theoretical studies, because one could immediately compare calculated and observed velocity profiles, practically online. In another words any changes obtained by the special laser pulse shaping would be visible almost in the real time. In Fig. 19 we present preliminary measurements of the Doppler profile of the pure isotope ^{87}Rb D1 resonance line with almost entirely separated components stemming from the upper level hyperfine splitting of 812 MHz. Small undulations on four Doppler profiles stem from the influence of the femtosecond laser frequency comb centered at 795 nm. The separation between adjacent hyperfine components is 812 MHz, which is about factor of two larger than the Doppler profile width. Therefore, this experimental profile can be directly compared with a theoretical profile. We plan detailed measurements with scanning c.w. monochromatic laser over the Doppler profiles of ^{87}Rb and ^{85}Rb 795 nm resonance lines in the cells with isotopic pure rubidium vapor.

It would be of considerable interest to apply the present technique for the case of the laser cooling procedure in the magneto-optical trap (MOT). We would need all six femtosecond laser beams to be pulse shaped by the spatial light modulator (or similar device) in order to extract from the frequency comb all the necessary comb lines and to supress those which are not essentially needed. Such an experiment has not been performed yet, although several announcements in different arrangements have been presented at different conferences and workshops.

In addition to above, we see the application of the method presented in this paper in the field of frequency metrology. The rubidium velocity comb measurements shown in this paper can serve as simple frequency markers for other experiments. By applying the saturation spectroscopy simultaneous to the modulation measurements, the absolute determination of the measured frequency scale in the vicinity of the hyperfine transitions could be obtained. Also, the parallel implementation of the presented experimental arrangement and theoretical model could be used as a simple tool for the absolute frequency determination of the frequency comb lines position.

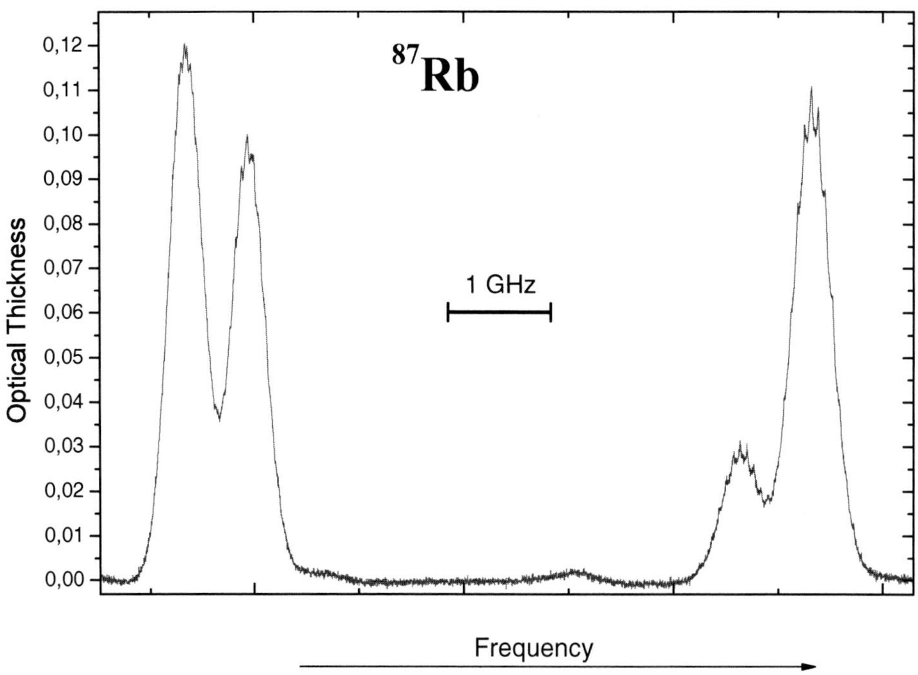

Figure 19. Doppler profile of the pure isotope ^{87}Rb D1 resonance line with almost separated components from the upper level hyperfine splitting. Small undulations on four Doppler profiles stem from the influence of the femtosecond laser frequency comb centered at 795 nm. Separation between adjacent hyperfine components is 812 MHz.

The most recent experiments at LAC showed how the femtosecond frequency comb can be used in the vibrational cooling of translationally ultracold molecules. The group of P. Pillet [44] succeeded in accumulation of population in the ground vibrational state of the ultracold cesium dimer. They performed effective blocking of the excitation from the v=0, using the frequency comb for X to B excitation at 760 nm. In a short amount of time the cycling of optical pumping and spontaneous emission accumulated vast population into the v=0 level.

In a special ring cavity with appropriate chirped mirrors the femtosecond laser frequency comb generated in the laser oscillator of relatively small energy could be amplified to such a level when the high harmonic generation in xenon beam becomes feasible [45]. This certainly widens the spectrum of the frequency comb applications of the accumulation effects into the shorter wavelength spectrum.

Maintaining the highest form of alertness for the technological results in neighboring fields one may hope to find the way for influencing amplitudes, phases and polarizations of each comb line hidden in the train of the femtosecond pulses.

XII. Conclusion

Jumping to conclusion, at this early stage of the frequency comb spectroscopy development, is certainly an unthankful job, because the regular behavior of the frequency comb still surprises and excites increasing number of researchers. This means that the near future will show a great number of new findings and appropriate theoretical models for the action of the optical frequency comb on atoms, molecules and clusters at different experimental conditions. Even the very interesting and important medical applications in the form of the human breath analysis [46] by using cavity enhanced frequency comb is only the beginning of the wide new world of femtosecond laser frequency comb spectroscopy.

Acknowledgments

We acknowledge the support from the Ministry of Science and Technology of Republic of Croatia (Projects 035-0352851-2857), European Commission Research Training Network (FW-5) and Alexander von Humboldt Stiftung, Germany. This work would not be possible without a considerable help in extensive theoretical and experimental work by H. Skenderović, S. Vdović and N. Vujičić. We are grateful to Jun Ye who sent us an important literature at the early phase of our work on femtosecond laser frequency comb which greatly facilitated our studies of the accumulation effects of the optical frequency comb.

References

[1] Cundiff, S. T. ; Ye, J. *Rev. Mod. Phys.* 2003, 75, 325.

[2] Ye, J.; Cundiff, S. T. *Femtosecond Optical Frequency Comb Technology*, Springer, Boston, 2005.

[3] Stowe, M. C.; Thorpe, M. J.; Pe'er, A.; Ye, J; Stalnaker, J. E.; Gerginov, V.; Diddams, S. A. *Adv. At. Mol. Opt. Phys.* 2008, 55, 1.

[4] Thorpe, M. J.; Ye, *Appl. Phys. B* 2008, 91, 397.

[5] Udem, Th.; Reichert, J.; Holzwarth, R.; Hänsch, T. W. *Phys. Rev. Lett.* 1999, 82, 3568.

[6] Ye, J.; Yoon, T. H.; Hall, J. L.; Madei, A. A.; Bernard, J. E.; Siemsen, K. J.; Marmet, L.; Chartier, J.-M.; Chartier, A. *Phys. Rev. Lett.* 2000, 85, 3797.

[7] Niering, M.; Holzwarth, R.; Reichert, J.; Pokasov, P.; Udem, Th.; Weitz, M.; Hänsch, T. W. *Phys. Rev. Lett.* 2000, 84, 5496.
[8] Udem, Th.; Reichert, J.; Holzwarth, R.; Hänsch, T.W. *Phys. Rev. Lett.* 1999, 82, 3568.
[9] Ye, J.; Yoon, T.H.; Hall, J.L.; Madei, A.A.; Bernard, J.E.; Siemsen, K.J.; Marmet, L.; Chartier, J.-M.; Chartier A. *Phys. Rev. Lett.* 2000, 85, 3797.
[10] Udem, Th.; Diddams, S. A.; Vogel, K. R.; Oates, C. W.; Curtis, E. A.; Lee, W. D.; Itano, W. M.; Drullinger, R. E.; Bergquist, J. C.; Hollberg, L. *Phys. Rev. Lett.* 2001, 86, 004996
[11] Witte, S.; Zinkstok, R.Th.; Ubachs, W.; Hogervorst, W.; Eikema, K. S. E. *Science* 2005, 307, 400.
[12] Jones, R. J.; Moll, K. D.; Thorpe M. J.; Ye, J. *Phys. Rev. Lett.* 2005, 94, 193201.
[13] Hannemann, S.; Salumbides, E.J.; Witte, S.; Zinkstok, R.Th.; Van Duijin, E.-J.;Eikema K.S.E.; Ubachs, W. *Phys. Rev. A* 2006, 74, 062514.
[14] Cundiff, S.T.; Ye J.; Hall, J. L. *Rev. Sci. Instrum.* 72, 3749 (2001).
[15] Holzwarth, R.; Udem, Th.; Hänsch, T. W.; Knight, J. C.; Wadsworth W. J.; Russell, P. St. J. *Phys. Rev. Lett.* 2000, 85, 2264.
[16] Jost, J.D.; Hall, J.L.; Ye, J. *Optics Express* 2001, 87, 270801 .
[17] Diddams, S. A. *et al.*, *Science* 2001, 293, 825.
[18] Ye, J.; Ma, L.S.; Hall, J.L. *Phys. Rev. Lett.* 2001, 87, 270801.
[19] Ludlow A.D. et al., *Phys. Rev. Lett.* 2006, 96, 033003.
[20] Braje, D. A.; Kirchner, M. S.; Osterman, S.; Fortier T.; Diddams, S. A. The European Physical Journal D - Atomic, Molecular, Optical and Plasma Physics 2008, 48, 57.
[21] Li, Chih-Hao; Benedick, A. J.; Fendel, P.; Glenday, A. G.; Kaertner F. X.; Phillips, D. F.; Sasselov, D.; Szentgyorgyi A.; Walsworth, R. L. *Nature*, 2008.
[22] Felinto, D.; Bosco, C. A. C.; Acioli, L. H.; Vianna, S. S. *Opt. Commun.* 2003, 215, 69.
[23] Snadden, M. J.; Bell, A. S.; Riis, E.; Ferguson, A. I. *Opt. Commun.* 1996, 125, 70.
[24] Marian, A.; Stowe, M. C.; Lawall, J. R.; Felinto, D.; Ye, J. *Science* 2004, 306, 2063.
[25] Marian, A.; Stowe, M. C.; Felinto, D.; Ye, J. *Phys Rev. Lett.* 2005, 95, 023001.
[26] Stowe, M.C.; Cruz, F.C.; Marian, A.; Ye, J. *Phys. Rev. Lett.* 2006, 96, 153001.
[27] Aumiler, D.; Ban, T.; Skenderović, H.; Pichler, G. *Phys. Rev. Lett.* 2005, 95, 233001.
[28]]Ban, T.; Aumiler, D.; Skenderović, H.; Pichler, G. *Phys. Rev. A* 2006, 73, 043407.
[29] Felinto, D.; Acioli, L. H.; Vianna, S. S. *Phys. Rev. A* 2004, 70, 043403.
[30] Felinto, D.; Bosco, C. A. C.; Acioli, L. H.; Vianna, S. S. *Phys. Rev. A* 2001, 64, 063413.
[31] Bennett, W. R. *Phys. Rev.* 1962, 126, 580.
[32] Vujičić, N.; Vdović, S.; Aumiler, D.; Ban, T.; Skenderović, H.; Pichler, G. *Eur. Phys. J. D* 2007, 41, 447.
[33] Ban, T.; Aumiler, D; Skenderović, H.; Vdović, S.; Vujičić, N.; Pichler, G. *Phys. Rev. A* 2007, 76, 043410.
[34] Boyd, R. W. *Nonlinear optics* (Academic Press, San Diego, 2003).
[35] Cohen-Tannondji C.; Kastler, A. *Prog. Opt.* 1966, 5, 3.
[36] Demtröder, W. *Laser Spectroscopy* (Springer-Verlag, Berlin, 2003).
[37] Steck, D. A. *Rubidium 87 D Line Data*, http://steck.us/alkalidata.
[38] O. Axner, O; Gustafsson, J; Omenetto N; Winefordner, J. D. *Spectrochimica Acta B* 2004, 59, 1.
[39] Coddington, I; Swann, W. C.; Nebury, N. R. *Phys. Rev. Lett.* 2008, 100, 013902.

[40] Aumiler D.; Ban T.; Vujičić N.; Vdović S.; Skenderović H.; Pichler G.; submitted to *Phys. Rev. A* 2008.

[41] Steck, D.A. *Cesium D line data*, http://steck.us/alkalidata.

[42] Catani, J ; Maioli, P ; De Sarlo, L ; Minardi, F ; and Inguscio, M ; *Phys. Rev. A* 2006, 73, 033415.

[43] Lukin M. D. *Rev. Mod. Phys.* 2003, 75, 457.

[44] Viteau M.; Chotia A.; Allegrini M.; Bouloufa N.; Dulieu O.; Comparat D.; Pillet P.; *Science* 2008, 321, 232.

[45] Ozawa, A.; Rauschenberger, J.; Gohle, C.; Herrmann, M.; Walker, D. R.; Pervak, V.; Fernandez, A.; Graf, R.; Apolonski, A.; Holzwarth, R.; Krausz, F.; Hänsch, T.W. and Udem, Th.; Phys. *Rev. Lett.* 2008, 100, 253901.

[46] Michael J. T.; Balslev-Clausen, D.; Kirchner, M. S.; and Ye, J.; *OPTICS EXPRESS* 2008, 16, 2387.

In: New Trends in Quantum Coherence... ISBN: 978-1-60741-025-6
Editor: Rafael Drampyan, pp. 197-219

Chapter 7

Quantum Coherence in the Rubidium 5S-5P-5D Ladder System: Applications in Frequency Conversion, Imaging, and Squeezing

R.E. Scholten**, *S.C. Bell, D.V. Sheludko, J.D. White, L.D. Turner, T. Meijer, B. Smeets, M. Jeppesen, C.S. Hofmann and M. Jasperse
Centre of Excellence for Coherent X-Ray Science,
School of Physics, The University of Melbourne, Victoria 3010, Australia

Abstract

The rubidium three-level ladder system consisting of the 5S, 5P and 5D states provides a readily accessible and powerful tool for extreme non-linear optics. We consider some phenomena and applications based on this system, including efficient frequency upconversion, state-selective cold atom imaging, and quantum squeezing.

Introduction

The alkali atoms provide a rich resource for investigating the many manifestations of light-atom interactions, with their readily accessible optical transitions and high atomic densities for moderate source temperatures. Even considering them as simple two-level atoms with a single laser field is sufficient for the fundamentals of laser cooling and the mechanical effects of light on atomic trajectories. But the alkalis also provide multiply coupled transitions, allowing controllable optical nonlinearity through application of two optical fields to three or more levels, and coherent superposition of the atomic eigenstates [1]. Atomic coherence has led to an explosion of developments in nonlinear optics, such as electromagnetically induced transparency (EIT) and lasing without inversion (LWI) [2], slow and fast light pulse propagation [3], four-wave mixing [4] and high-harmonic generation [5], and applications in spectroscopy [6], atomic frequency standards [7, 8], quantum information processing and

* E-mail address: scholten@unimelb.edu.au

memory [9], magnetometry and other sensing [1]. For three-level systems, the states can be arranged in a lambda (Λ) or vee (V) pattern, or a linear "ladder" (also called "cascade") configuration [10]. Here we consider two applications of using atomic coherence for controlling optical nonlinearity, based on the rubidium 5S-5P-5D two-colour ladder system. In the first, we have produced continuous coherent blue laser light from a hot vapour cell, with very high efficiency relative to conventional nonlinear crystal techniques. In the second application, rubidium atoms are cooled and trapped in a magneto-optical trap (MOT), which also inherently excites the atoms to the 5P state. The excited-state atoms are selectively imaged using the 5P-5D transition.

The motivation for our work stems from an inherent interest in understanding the many manifestations of light-atom interactions, but also from an applied perspective. By controlling the atomic system we can enhance or suppress particular light-atom interactions, for example to improve the efficiency of the frequency upconversion process, or to control the refractive index of a cold atom cloud for imaging contrast enhancement. We hope to use these tools in the production of frozen Rydberg gas (FRG) and ultracold plasma (UCP) [11], in which rubidium will be cooled and trapped on the 780 nm 5S to 5P transition, and then excited from the 5P to Rydberg states or photoionised to produce a cold plasma. FRG and UCP are new and exotic states of matter which promise exciting new avenues for atomic physics research, for example to study long-range dipole-dipole interactions, quantum dynamics, and strongly coupled plasmas [11].

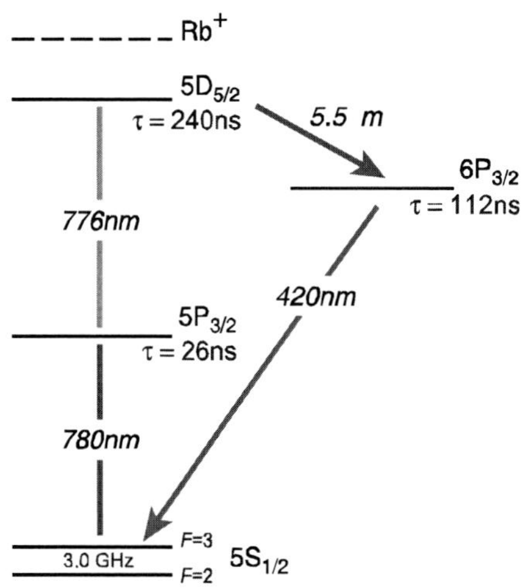

Figure 1. Simplified ^{85}Rb level scheme showing used for frequency upconversion and state-selective imaging. Two lasers (780 and 776 nm) excite ^{85}Rb on the 5S-5P-5D ladder transitions. Decay via the 6D state leads to blue fluorescence at 420 nm.

Frequency Upconversion by Four-Wave Mixing in Rubidium Vapour

A simplified energy level scheme for rubidium is shown in Figure 1. In our frequency upconversion experiments (described briefly elsewhere [12]), two low-power (20 mW) lasers at 780 nm and 776 nm couple the 5S-5P and 5P-5D ladder states. Atoms can decay from the 5D5/2 to the 6P3/2 state via infrared emission at 5.5 μm. From the $6P_{3/2}$ level they can then decay to the $5S_{1/2}$ ground state, emitting at 420 nm. The blue fluorescence was readily observed over a broad range of conditions. With careful laser tuning and within a relatively narrow temperature range, we observed emission of a spatially and temporally coherent 420 nm light beam with output power of up to $P_{420} = 40\ \mu$W.

Related work has been reported for relatively high-power pulsed laser excitation, for example in potassium [13] and lead [14]. Our work is an extension of the scheme of Ref. [15] in which continuous frequency upconversion in Rb was demonstrated at low laser power. In repeating those experiments we were surprised to find much greater upconversion efficiency when the lasers were detuned far from the two-step resonance. After a number of experiments to probe the upconversion system, and with the aid of a semiclassical model, we found the improvement to be due to a combination of effects. With appropriate detunings, we couple both ground-state hyperfine levels (F =2, 3) to the 5P and 5D states, inducing coherences between four levels. Since both hyperfine ground states are involved, we have stronger coupling to a larger fraction of the atomic population, and optical pumping into a dark hyperfine level is minimised. The 780 nm pump laser is far from resonance with either ground-state hyperfine level, and thus single-photon absorption is small. Finally, cancellation of the optical phase shifts for the two ground states optimises phase-matching for the up-conversion process at the appropriate detuning.

Experiment

A simplified version of the experimental apparatus is shown in Figure 2. Two low-power infrared laser beams from external cavity diode lasers (ECDLs) [16, 17] were co-propagated through the primary rubidium vapour cell, which was heated to a temperature of $T = 100$ to 200°C. The cell was 8 cm in length, with natural isotopic abundance of Rb and no buffer gas. The power in each laser beam was around 20 mW, focused to a beam waist of diameter 0.3 mm using a lens of focal length $f = 50$ cm.

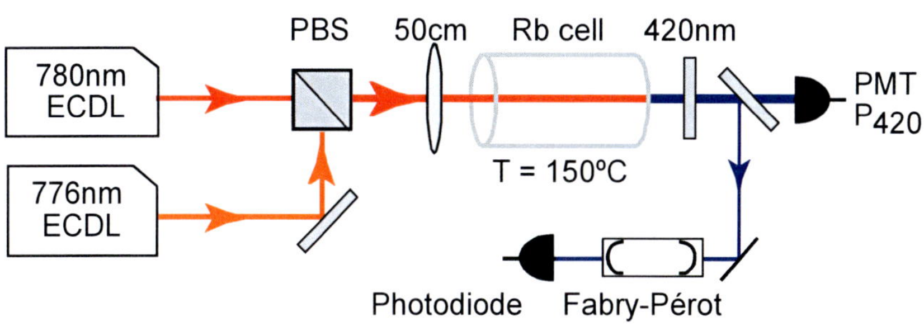

Figure 2. Simplified experimental arrangement. Two lasers excite ^{85}Rb on the 5S-5P-5D transitions, and coherent 420 nm radiation is emitted. PBS polarizing beam splitter, PMT photomultiplier tube.

Intense blue fluorescence was observed in the vapour cell for a broad range of laser detunings, consistent with the Doppler broadening of the gas (Figure 3). The fluorescence indicates significant two-step 780 nm + 776 nm excitation to the upper 5D level. Atoms in the

5D state can decay via the 6P intermediate state, and then to the ground state, emitting blue fluorescence.

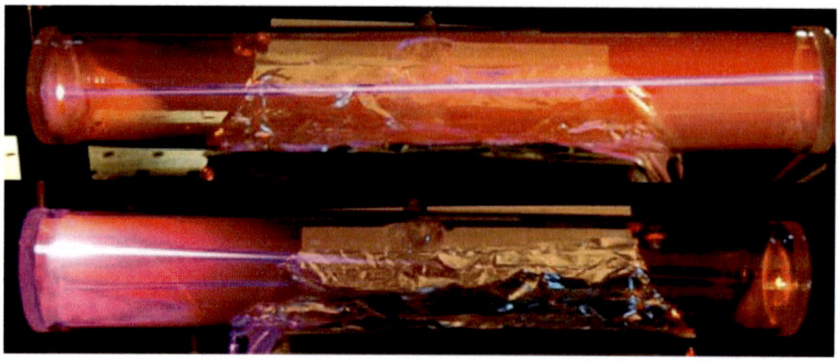

Figure 3. Photos of fluorescence (unfiltered; include 420 and 780 nm fluorescence) in a 20 cm vapour cell. The 780 and 776 nm laser beams are propagating from left to right. The upper image shows the cell heated to 150°C. As temperature increased, so did absorption. In the lower image the temperature had been increased to 300°C, with total absorption within a few centimetres into the cell.

We measured the frequency upconversion process as a function of the 776 nm laser detuning Δ_{776}, for different specific 780 nm laser detunings Δ_{780}. The 780 nm laser detuning was determined using saturated absorption spectroscopy in a separate vapour cell (not shown), relative to the 5S $F = 3 \rightarrow$ 5P $F = 4$ resonance ($\tau_{5P} = 26$ ns). The 776 nm laser frequency was determined from the blue fluorescence spectrum in a third vapour cell, heated to ~ 70°C, with counter-propagating 780 nm and 776 nm light beams (Figure 4). The net detuning of both lasers $\Delta_{780} + \Delta_{776} \equiv 0$ when the two-step excitation is exactly on the 5S $F = 3$ $\rightarrow$ 5D resonance, indicated by the blue fluorescence intensity.

Note that for two counter-propagating lasers of similar wavelength, the frequency sum (for two-step excitation) is almost Doppler-free. The sum $\nu_{780} + \nu_{776}$ is Doppler shifted by $v_z(\nu_{780} - \nu_{776})/c$ where v_z is the atomic velocity along the laser beams. For 776 nm and 780 nm the Doppler-broadened absorption width is then below 3 MHz, two orders of magnitude smaller than the single-beam Doppler width. Thus all atoms in the thermal velocity distribution are resonant, though with decreasing coupling strength for higher v_z.

Measurements

Blue fluorescence was readily observed for a broad range of vapour cell temperatures, and at high temperatures, for a wide range of laser detunings. For specific combinations of cell temperature and laser detunings, a bright blue beam was emitted from the cell in the direction of propagation of the two input pump beams. The output was linearly polarised, and both spatially and temporally coherent. The coherence was demonstrated superficially with two-slit diffraction, which showed high-contrast interference fringes (Figure 5), and more precisely from the linewidth measured with a confocal Fabry-Pérot (FP) etalon, discussed further below.

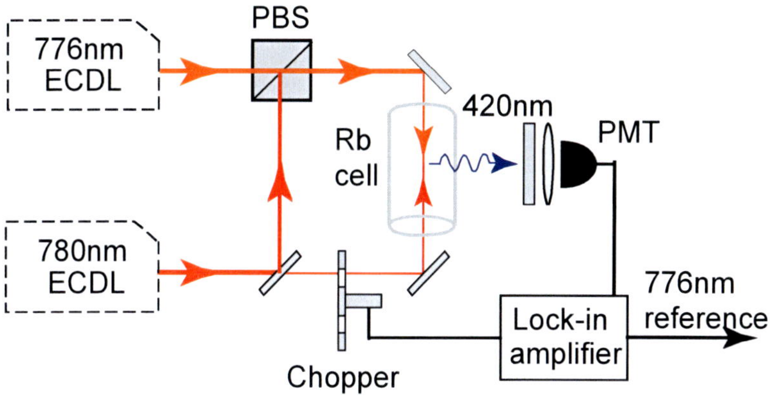

Figure 4. Frequency reference measurement system for 776 nm laser. Light from the two lasers shown in Figure 2 counter-propagated through a separate heated vapour cell (70°C). Blue fluorescence at 420 nm was detected via a filter and lens using a photomultiplier tube. The 780 nm beam was modulated with a mechanical chopper, and a lock-in amplifier extracted the fluorescence signal from the background, to provide a signal dependent on the 776 nm laser frequency.

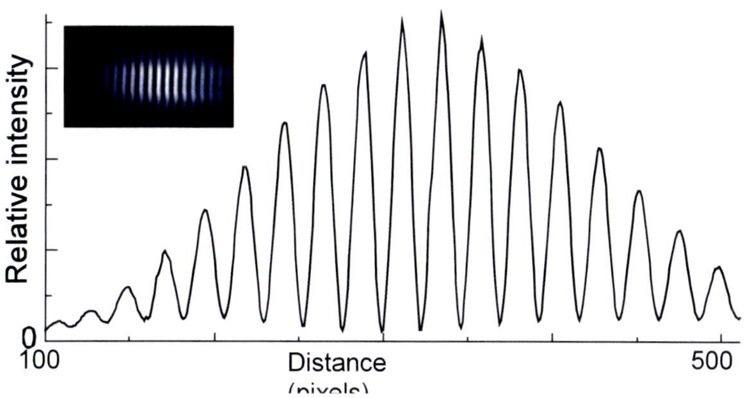

Figure 5. Double-slit interference pattern in false colour (inset), generated from the frequency upconversion output beam at 420 nm. Slit separation 0.25mm, width 0.04mm. A single pixel row profile across the diffraction pattern is shown.

Figure 6 shows the relative output power P_{420} of the coherent beam as a function of the 780 nm laser detuning, for a few specific 776 nm laser detunings. As expected, when the 776 nm laser was near resonance with the 5P-5D transition ($\Delta_{776} \sim 0$), a blue beam was found for 780 nm detuning across the Doppler spread for the $F = 3$ ground state (red curve, $\Delta_{776} = 0.1$ GHz), as found in Ref. [15]. Blue output was also found for $\Delta_{780} = +3$ GHz, $\Delta_{776} = 0$, corresponding to the $F = 2$ ground state, though only for higher temperature (200°C; not shown).

Surprisingly, greater coherent blue output was observed with both lasers detuned far from their respective resonances, specifically when the 780 nm laser was detuned between the two ground state hyperfine levels, provided the net two-colour detuning $\Delta_{780} + \Delta_{776}$ was close to zero. At $\Delta_{780} \approx 2$ GHz and $T \approx 200$°C we found a maximum of approximately 40 μW, three times greater than observed previously achieved [15] for similar input power.

We investigated the spectral properties of the blue output for several specific detunings, using a Fabry-Pérot (FP) etalon with a free spectral range (FSR) of 1.5 GHz (Figure 7). Spectra A,B were measured with net detunings $\Delta_{780} + \Delta_{776} = 0$, for which we expect blue output to the $F = 3$ ground state, while C,D were measured with $\Delta_{780} + \Delta_{776} = 3$ GHz, corresponding to coupling with the $F = 2$ ground state. Structure due to the 6P hyperfine splittings ($F = 4, 3, 2, 1$; separations 40, 20, 10 MHz [18]) is evident from the double peaks in spectra A,B and the broad peaks of C,D which suggest overlap of several lines of similar strength.

The apparent frequency shifts of the spectra are modulo the 1.5 GHz FSR of the etalon. No correction was made for drifts due to temperature and pressure variations in the etalon, which were up to a few hundred MHz. The frequency of the peaks in curve B are offset from the on-resonance peaks of curve A by an amount $|\Delta \bmod 1.5\ \text{GHz}| = 1$ GHz, which is also consistent with an offset of, for example, −0.5 or −2 GHz. The large offsets of the non-resonant cases B, C scale approximately with the square root of the 780 nm pump power and are attributed to ac Stark shift of the dressed ground-state level.

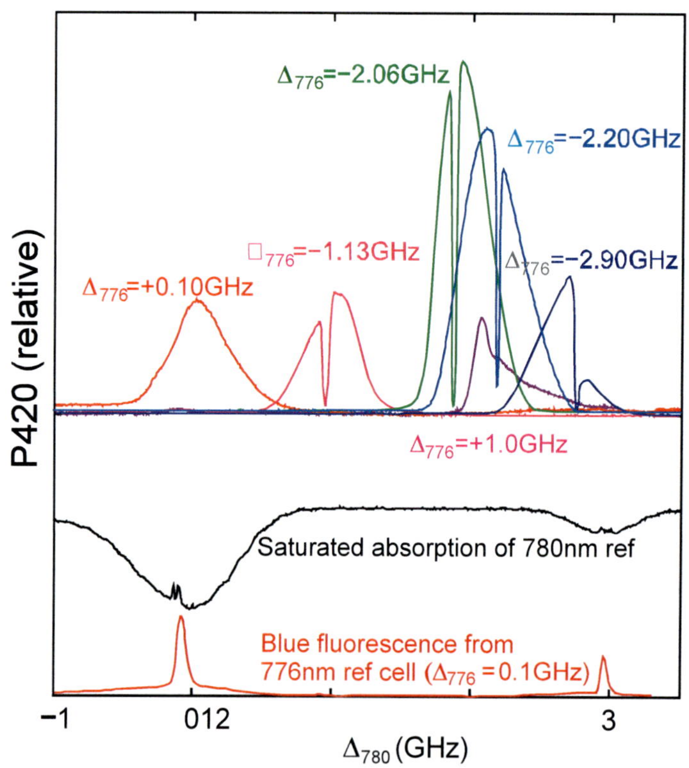

Figure 6. Blue output power for different pump laser detunings, at $T \approx 150$°C. The maximum output was observed when the 780 and 776nm input fields were orthogonally polarized, with $\Delta_{780} \approx 2$GHz. The lowest two curves show frequency reference spectra.

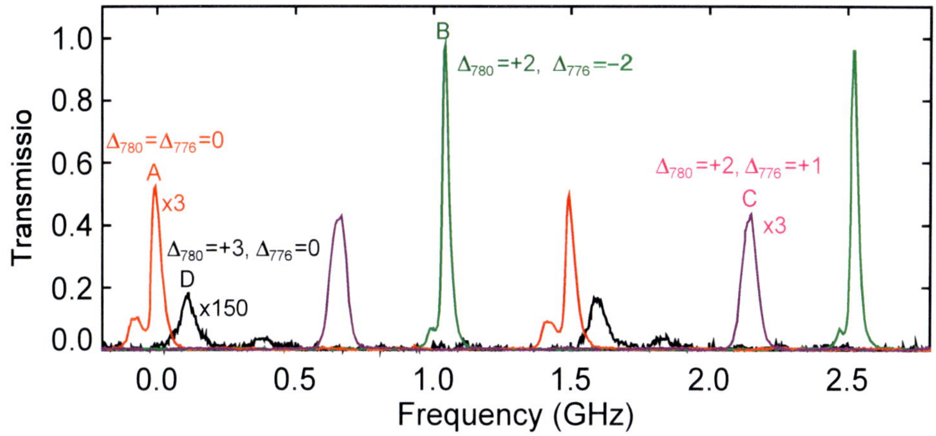

Figure 7. Spectra of the coherent blue output, measured with a Fabry-Pérot of 1.5 GHz free spectral range, normalised to spectrum B.

Model

We have modeled the system using a semiclassical Maxwell-Bloch approach [2, 19]. Four classical fields (780, 776, 5500, 420 nm) were described by Maxwell's wave equation linking the electric field **E** to the induced polarisation of the atoms **P**:

$$\left(\nabla^2 - \mu_0 \varepsilon_0 \frac{\partial^2}{\partial t^2}\right)\mathbf{E} = \mu_0 \frac{\partial^2 \mathbf{P}}{\partial t^2} \tag{1}$$

P is given by the appropriate off-diagonal element of the atomic density matrix ρ, where ρ was determined by solution of the Bloch equations [27]. We have calculated the steady-state atomic density matrix with four, five and sixteen levels. In the simplest case we included only the four primary levels (5S, 5P, 5D, 6P). The effect of both hyperfine ground states (5S F =2, 3) was included in a five-level model, and the 16-level calculation included all hyperfine levels of the four primary states. Magnetic sub-states were neglected in all cases. The calculations did not include the Maxwellian atomic velocity distribution and associated Doppler effects.

The time evolution of the atomic density matrix ρ is given by

$$\frac{\partial \rho}{\partial t} = -\frac{i}{\hbar}\left[\rho, \hat{H}\right] + \hat{L}\rho \tag{2}$$

where H is the Hamiltonian including atomic and interaction terms, and we specifically include relaxation $\hat{L}$. In the dipole approximation the interaction is $\hat{H}_{\text{int}} = -\mathbf{d}.E$ which we quantify in terms of the Rabi frequency for a given transition between levels i, j; that is, $\left(\hat{H}_{\text{int}}\right)_{ij} \equiv \hbar\Omega_{ij}$. We transform to the interaction basis, make the rotating wave approximation, and find steady-state solutions via standard numerical matrix methods.

The amplitude of a particular field (e.g. the 420 nm beam) increases or decreases along the propagation direction, depending on the gain which is proportional to the appropriate off-diagonal density matrix element. Thus the respective Rabi frequency of the propagating field depends on the associated off-diagonal element:

$$\frac{\partial}{\partial z}\Omega_{ij} \propto \mathrm{Im}\left(\rho_{ij}\right) \tag{3}$$

We are particularly interested in gain on the 420 nm lines, thus $\rho_{5S(F=2,3),6P}$.

Figure 8 shows the relative 420 nm gain dependence on the 780 nm detuning, for the two $6P_{3/2}$ to $5S_{1/2}$ ($F = 2, 3$) transitions, in the five-level model, with Rabi frequencies applicable to our experimental conditions. The gain is greatest where the 780 nm laser is above the single-laser 5S-5P resonance; that is, tuned between the two ground states, rather than when resonant with either. The maximum gain to the upper ($F = 3$) ground state is predicted for $\Delta_{780} = +2.1$ GHz which is consistent with measurements (Figure 6). For the $F = 2$ ground state, maximum gain is at $\Delta_{780} = +0.8$ GHz, but the gain is also strong for Δ_{780}, $\Delta_{776} = +2, +1$ GHz corresponding to Fabry-Pérot spectrum C (Figure 7). Sixteen-level calculations including the 5P and 5D hyperfine levels also predict two gain regions at similar detunings, while in the four-level model there can only be a single gain peak, which is found at $\Delta_{780} = 0$. Note that in the calculations the sum detuning $\Delta_{780} + \Delta_{776}$ is fixed, whereas in the experiments Δ_{780} is varied for a given Δ_{776}. Thus the measurements show peaks which are much narrower than the gain curves, but consistent with Doppler-broadened widths.

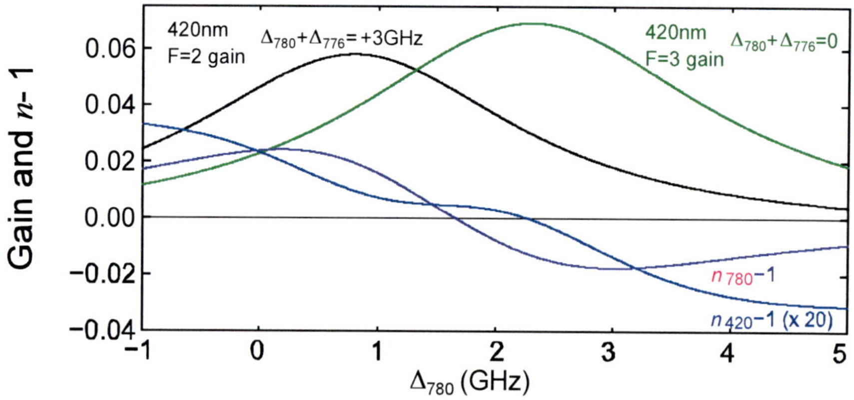

Figure 8. Relative gain and refractive index of 6P-5S blue transition for varying Δ_{780}, in the five-level model, calculated from $\mathrm{Im}(\rho_{ij})$ and $\sum_F \mathrm{Re}\left(\rho_{5S_F,j}\right) \equiv n-1$ respectively, where $F = 2, 3$ and j is 6P (420 nm) or 5P (780 nm).

In a four-wave mixing process, the efficiency is strongly dependent on phase matching the four beams; that is, we expect $k_{420} + k_{5.5} = k_{780} + k_{776}$ where the wavevectors $k\lambda$ depend on the refractive index of the gas. The refractive index is proportional to the real part of the off-diagonal density matrix elements. In the 5-level model, these components are small and slowly varying around 2 GHz, reducing phase mis-match for both the 780 nm pump and 420 nm output fields (Figure 8). The zero-crossing and maximum gain frequencies are sensitive to the intensities of the pump laser fields. Since the latter are strongly varying radially, and also axially due to absorption, these simple models can provide only qualitative guidance.

Comments on Frequency Upconversion

Both the experiments and calculations show that the frequency upconversion process is enhanced for pump laser frequencies that are off-resonance with the individual 5S-5P and 5P-5D transitions. With the 780 nm pump laser detuned between the hyperfine ground states, we couple equally both $F = 2$ and $F = 3$ population fractions of 5/12 and 7/12, which also inherently minimises losses due to optical pumping between hyperfine states. Coupling both ground-state levels induces coherence between four levels, indicated by strong off-diagonal elements of the calculated density matrix.

The vapour is also nearly transparent to the 780 nm laser alone (without 776 nm laser) at these detunings, even at elevated temperatures where the atomic density is high and the vapour is optically very dense on-resonance. At 200°C, the length of the blue fluorescence along the cell was notably shorter for $\Delta_{780}\approx 0$ than off-resonance ($\Delta_{780} = +2$ GHz). Off resonance, two-photon 780 and 776 nm absorption occurs for a greater length along the cell, at greater density, thus increasing the active population.

The FP spectra (Figure 7) show that the blue output beam was also frequency-shifted relative to the on-resonant case, and hence not resonant with unperturbed atoms. While we were unable to measure the absolute frequency of the blue beam, we did find that it was not absorbed by an independent vapour cell, demonstrating that it was not resonant with the normal 5S-5P transition.

Finally, calculations show that phase mis-match is reduced for detuning between the ground-state hyperfine levels. From a simple viewpoint, the refractive index dispersion curves for the 780 nm laser relative to the $F = 2$ and $F = 3$ ground states will sum to zero at some detuning between the two resonances (see Figure 9), approximately where optimum blue output was observed.

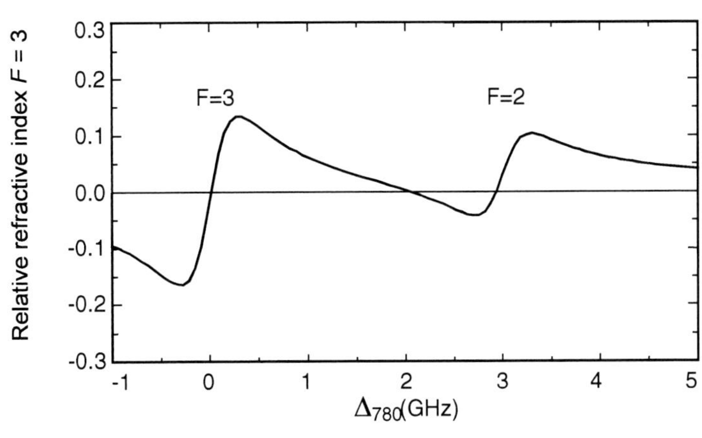

Figure 9. Relative refractive index dependence at 780 nm. The contributions from the two ground-state hyperfine levels will cancel for a detuning between the two resonances.

There is considerable scope for further work on this system. Our model is overly simplistic, and a quantitative model would provide useful insight and guidance on improving the frequency upconversion efficiency. The model could be enhanced by including all the atomic magnetic substates and the laser polarisation, the atomic velocity distribution, and spatial modes of the laser fields. While challenging, there are many interesting phenomena to explore, such as the apparent resonance in the blue light near optimum conditions, which suggests an EIT-like coherence effect.

Even without an improved model, many questions could be answered experimentally. For example, the blue output is not resonant with unperturbed atoms, but we have not verified that it matches the four-wave mixing resonance condition. We have seen indications of possible enhancement of the blue output with the addition of a second 780 nm pump laser, such that one is close to resonance with each of the two hyperfine levels, but have not been able to explore this further. We note that the blue output power increases with increased pump power, but have not had high-power lasers available to find the limits. The dependence on vapour cell length, or rather on the vapour length-density product, could be explored by using two cells sequentially, each with independent temperature control. Nonlinear optical processes are often enhanced using an external cavity, but this has not been attempted for an atomic four-wave mixing process such as that described here.

There is also the question of whether the technique can be applied to other atoms. For example, even ^{87}Rb rather than ^{85}Rb, or the D1 transition in Rb at 794 nm. It has been suggested [10] that the pump fields should have well-matched wavelengths, to reduce the effect of residual Doppler linewidths, so the success with Rb at lower powers may be fortuitous.

Excited State Imaging

There has been a surge of interest recently in applying laser cooling techniques to produce ultracold samples of atoms in Rydberg states, or to produce ultracold plasma in which the electron and ion energies are comparable to or smaller than their Coulomb binding [11, 20, 21]. Excitation to the Rydberg states, or photoionisation to form plasma, is normally a two-step process in which the laser cooling pre-excites the atoms (see Figure 10). In such experiments it can be beneficial to determine the excited-state distribution of the cooled atom cloud; for example in the case of Rb, to image the atoms in the 5P state specifically.

In principle it might seem straightforward to acquire such images, for example by taking the difference between two conventional ground-state absorption images with and without excitation. However, the imaging laser and excitation laser couple the same states (see Figure 1), and it is not generally possible to separate the imaging and excitation laser-atom interactions. We have developed an approach based on imaging with a 776 nm laser tuned to the transition from the excited 5P state to the upper 5D level. Because the excitation fraction is limited to 50%, and the linewidth of the 776 nm transition is narrow (600 kHz), the absorption is very weak. In combination, the signal-to-noise ratio (SNR) for 776 nm absorption imaging is at least 50 times poorer than for ground-state 780 nm imaging.

To further complicate the problem, the excitation field (780 nm) must remain on during the imaging exposure, because the 5P lifetime is only 26 ns, much shorter than typical imaging exposure durations (50 to 100 μs). The atomic state is perturbed by this field, causing significant ac Stark shift of the 5P energy level, influencing the effective detuning and consequently the absorption and refractive index for the 776 nm imaging transition.

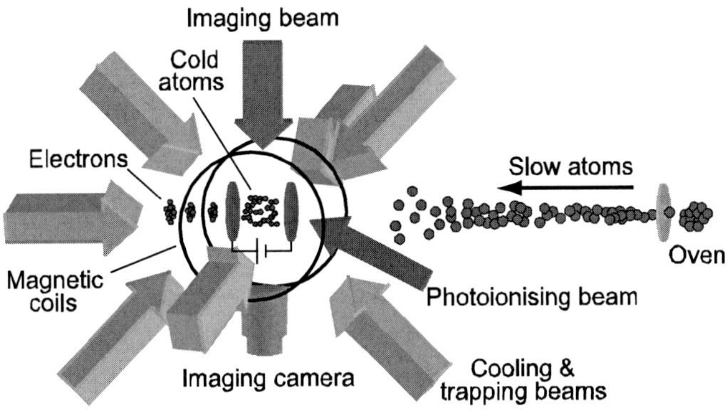

Figure 10. Frozen Rydberg gas and ultracold plasma can be created by photoexcitation or photoionisation of an ultracold neutral gas in a magneto-optic trap. Electron (or ion) pulses can be extracted with an electric field as shown.

Imaging Theory

We use an imaging approach based on diffraction of a propagating probe beam to render phase gradients into intensity variations [22–24]; see Figure 11. An off-resonance imaging beam is incident on a cloud of cold atoms, which introduces spatially varying absorption and phase shift, and the light then propagates to an imaging detector such as a CCD camera. The Fresnel diffraction pattern is inverted to generate a quantitative measurement of the column density of the sample [22]. The relation between the Fourier transform of the normalised contrast of the diffraction pattern, $F\,[(I - I_0)/I_0]$, and the Fourier transform of the column density at the object, F the diffraction pattern, $F\,[(I - I_0)/I_0]$, and the Fourier transform of the column density at the object, $F\;[\rho(x)]$, is given by [22]:

$$\mathrm{F}\left\{\frac{I-I_0}{I_0}\right\} = 2k\left[\delta\sin\left(\pi\lambda z u^2\right) - \beta\cos\left(\pi\lambda z u^2\right)\right]\mathrm{F}\left\{\rho(x)\right\} \tag{4}$$

where $k = 2\pi/\lambda$, λ is the wavelength of the probe laser, z is the propagation distance and u is the spatial frequency conjugate to $\mathbf{x}$. δ and β are the phase and absorption coefficients, where $\varphi = k\delta\rho$ and $\mu = k\beta\rho$ are the phase shift and absorption of the atomic cloud. The imaging can be lensless, and is insensitive to imprecision in the defocus distance and detuning, making it a robust quantitative technique compared to standard absorption imaging.

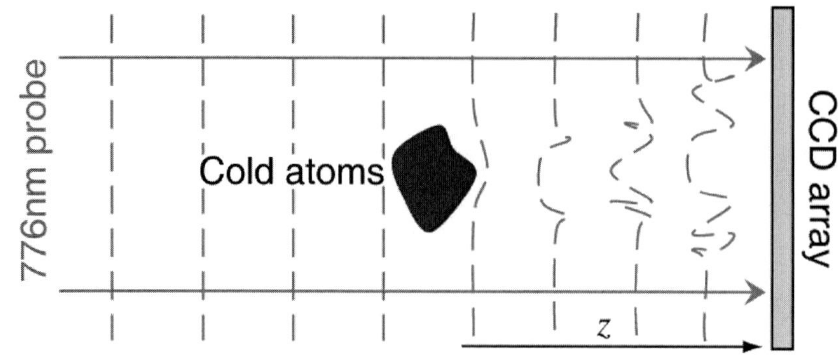

Figure 11. Diffraction contrast imaging records the diffraction pattern of an off-resonant plane wave incident on a cold atom sample.

The diffraction contrast factor relating the two Fourier terms has nulls at spatial frequencies which depend on the relative absorption and phase. These nulls are removed by Tikhonov regularisation [25], which has minimal effect at low spatial frequencies dominant for imaging cold atom clouds, and leaves only minor artefacts at higher frequencies.

Refractive Index

To retrieve the column density of the atomic sample, we require the *ratio* of the phase and absorption, δ/β, for all transverse position coordinates. In general, of course we do not know either absorption or phase (the real part of the refractive index) separately, otherwise we would not need to image the sample. However, for a two-level atom the ratio depends only on the detuning of the imaging laser [22]:

$$\frac{\delta}{\beta} = -2\Delta. \tag{5}$$

For excited-state imaging, a two-level approximation is not appropriate due to the perturbing effect of the excitation laser field. Instead we find the ratio by solution of the optical Bloch equations (OBEs, eq. 2), for a three-level 5S-5P-5D system. The ratio depends only on known atomic and laser parameters: the spontaneous decay rates, the laser detunings Δ_{780}, Δ_{776} and the intensities described in terms of the Rabi frequencies, Ω_{780}, Ω_{776}.

The refractive index of a medium is defined as $n = \sqrt{\varepsilon}$, where $1+\chi_e$ is the electric permittivity and χ_e the electric susceptibility. For small χ_e, $n \approx 1+\chi_e/2$ [26]. Treating the atom-field interaction semiclassically, the polarization **P** of a monatomic gas is related to the field **E** by **P** $=\chi_e$ **E**. The polarization of the atomic gas can be determined using a density matrix formalism, where P is given by the off-diagonal elements of the density matrix for the system, ρ [2]. The density matrix elements were calculated using optical Bloch equations as above (see eq. 2) for three-level atoms (5S, 5P, 5D; see Figure 1) with two coupling laser fields. The system is described by six coupled differential equations, which can be solved by numerical integration for steady-state conditions, but an analytical expression for the density matrix elements was achieved by recasting the optical Bloch equations into vector form.

The OBEs are derived from the Liouville equation (eq. 2) which can be recast into vector form:

$$\frac{d\vec{\rho}}{dt} = \mathbf{M} \cdot \vec{\rho} \tag{6}$$

where $\mathbf{M}$ is a 9×9 matrix for a three level system. Since we want the steady-state solution, the left hand side of Eq. (6) is $\vec{0}$. Finding the steady-state solution amounts to finding the nullspace of the matrix $\mathbf{M}$, simplifying the problem considerably. The density matrix elements, and hence the refractive index, are given by the eigenvectors of the system.

The exact solutions for both the real and imaginary parts of the susceptibility were found using this method, each of which contained approximately 100 terms. In the weak coupling limit on the 776 beam, a first-order Taylor expansion gives a more manageable result for the absorption ratio, β/δ:

$$\frac{\beta}{\delta} = \frac{\Gamma_{776}\left(\Gamma^2 + 4\Delta_{776}^2\right) + \Gamma\Omega_{780}^2}{2\Delta\left(\Gamma^2 + 4\Delta_{776}^2\right) - 2\Delta_{776}\Omega_{780}^2} \tag{7}$$

where Γ_λ and Δ_λ are the linewidth and detuning of the transitions at wavelength λ, $\Gamma \equiv \Gamma_{780} + \Gamma_{776}$, $\Delta \equiv \Delta_{780} + \Delta_{776}$, and $\mathbf{\Omega}_{780}$ is the Rabi frequency of the 780 nm laser [29]. The absorption ratio depends entirely on known parameters: beam detunings, natural linewidths, and the cooling beam Rabi frequency, which can be calculated from the measured excitation beam intensity. The calculated absorption and phase components of the 776 nm susceptibility are shown in Figure 12, for $\Delta_{780} = -3\Gamma$. Autler-Townes energy level splitting [30, 31] is evident in the predicted spectrum of the 776 nm laser, caused by the strong perturbing influence of the applied 780 nm field. The energy level splitting (ΔE) increases with 780 nm detuning and with 780 nm power [32]:

$$\Delta E = \hbar\sqrt{\Omega_{780}^2 + \Delta_{780}^2}\,. \tag{8}$$

To maximise the phase shift and thus imaging contrast, the 776 nm probe must be detuned, e.g. to $+40\Gamma_{776}$. Although not investigated here, imaging as a function of Δ_{776} provides a detailed measurement of the Autler-Townes effect, with spatial resolution. Note that near resonance, the 6 MHz linewidth of the 780 nm transition broadens the effective linewidth of the 776 nm transition. The broadening is reduced by any detuning of the 780 nm excitation beams, so that the expected SNR for excited-state imaging remains low.

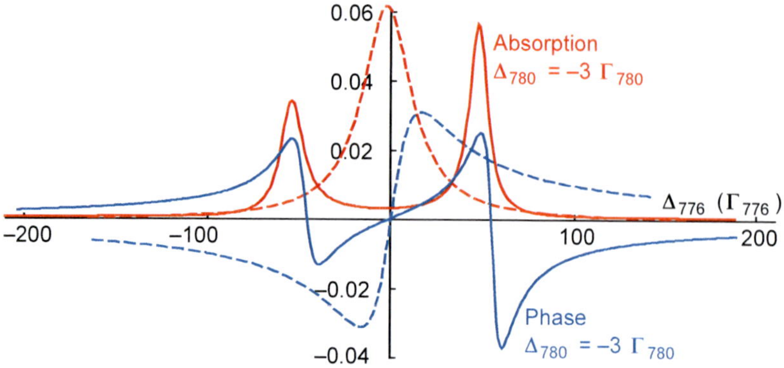

Figure 12. Analytical result for the phase and absorption on the 776nm transition for a three-level model of Rb, in the weak coupling limit on the 776nm beam. The two-level results (dashed) are clearly not valid for excited state imaging.

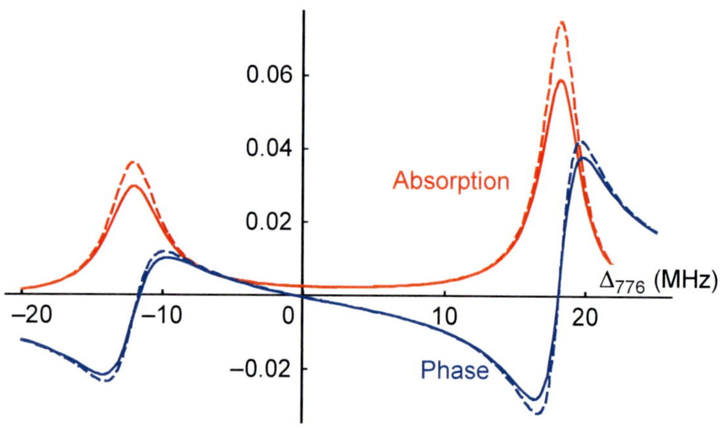

Figure 13. Calculated absorption and phase for the 776nm transition, showing exact analytic (solid) and first-order approximation (Eq. 7; dashed), for typical experimental parameters ($\Delta_{780} = -3\Gamma$, $\Omega_{780} =$ 30 MHz).

A comparison of the first-order approximation (Eq. 7) with the exact result is shown in Figure 13, for typical experimental parameters. The energy level splitting shows good agreement. Note that in our image retrieval results we use the exact result for the ratio.

Autler-Townes

Before attempting excited state imaging, we verified our Bloch equation model by measuring the Autler-Townes splitting of the 776 nm absorption line. Figure 14 shows the 420 nm blue fluorescence from the MOT as the 776 nm was tuned through the double 5P-5D resonance. The Δ_{780} uncertainty is around $\pm\Gamma/2$ and so the agreement in splitting is very good, and the relative peak heights are also consistent.

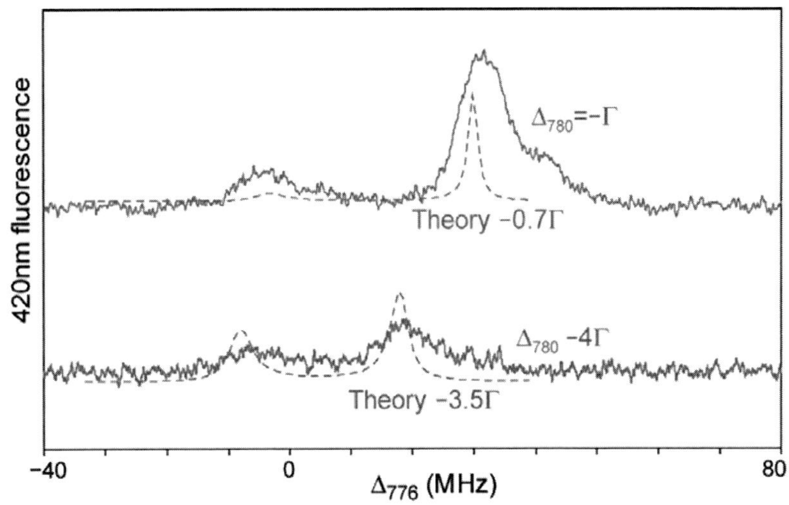

Figure 14. Theoretical (dashed) and experimental (solid) 420nm fluorescence from the MOT for 780nm cooling beam detunings of $-\Gamma$ and -4Γ, for varying 776nm detuning. The two fluorescence peaks show that the 5P-5D transition is split by the Autler-Townes (ac Stark) effect.

Apparatus

The atoms were cooled and trapped using the $5S_{1/2}$ F=3 → $5P_{3/2}$ F = 4 cycling transition of ^{85}Rb, with a repump laser on the F = 2 → F = 3 transition. The ECDLs were frequency stabilised using saturated absorption spectroscopy [17]. A tapered amplifier [33] on the cooling laser provided a total cooling beam power of 150 mW at the MOT, after fibre coupling. The cooling beam diameter at the MOT was approximately 20mm ($1/e^2$). Anti-Helmholtz coils were used to generate a quadrupole field with a gradient of 1 mT/A cm. Rubidium atoms were produced by a dispenser source with resistive heating.

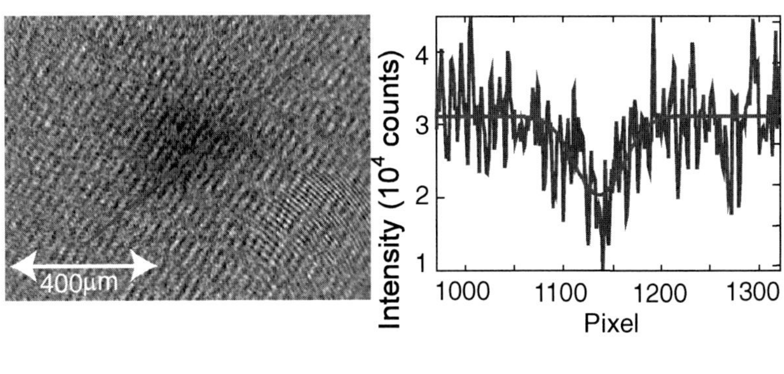

Figure 15. In-focus absorption image of atoms in the 5P state using 776 nm probe beam and atomic absorption filter to block 780 nm fluorescence. A line profile with Gaussian fit is also shown.

The imaging laser was frequency stabilised as described for the frequency upconversion experiments, with the addition of a double-pass acousto-optic modulator (AOM) to allow precise control of the detuning, and another AOM to allow rapid switching to produce 100 μs imaging pulses. A 780 nm laser beam was also frequency-shifted and switched by an AOM. The two imaging beams were then combined on a beam-splitter cube and coupled into a single-mode fibre to transport the beam to the vacuum chamber, where the light was collimated to provide a stable Gaussian beam profile with 10 mm diameter (FWHM). After passing through the atom cloud, the light was collected with a standard camera lens (200 mm/f4 micro-Nikkor) and imaged with a cooled CCD array (Apogee U2000).

780 nm Filter

Typical imaging beam power was 150 μW and duration 100 μs. The minimum camera exposure time was 1 ms. The signal from background 780 nm fluorescence due to the cooling lasers during the 1 ms exposure was comparable to that from the short 776 nm pulse, and thus dramatically reduced the image contrast. To separate the 776 nm imaging probe light from the 780 nm fluorescence, a very narrow bandbass filter was needed. We used a simple atomic absorption filter based on a heated Rb vapour cell of 10 cm length, which reduced the 780 nm intensity by more than 40 dB at 85°C.

776 nm Images

In the first instance, on-resonant in-focus 776 nm absorption images were obtained (Figure 15). The contrast is poor, as predicted given the small excitation fraction (≈30%) and small transition cross-section.

To increase the imaging contrast, the density of the cold atom cloud was increased using a compressed MOT (CMOT) technique [34]. Atoms were cooled and trapped at low magnetic field gradient (10 G/cm), and the cloud was then compressed by increasing the field up to 60 G/cm over in 4 ms.

The 776 nm imaging results for a compressed cloud are shown in Figure 16, together with a ground-state image for comparison. The excited state population is 45 ± 6%. Using a two level approximation for our experimental beam powers and detunings, we expect a theoretical excited state fraction of 46%, in good agreement with our results.

The apparent size of the excited-state cloud is much smaller than that of the ground state cloud (0.42 mm versus 1.18 mm). We attribute the difference to the strong magnetic field gradient used to compress the cloud, and the corresponding Zeeman shift of the transition, which we estimate at ±3 MHz at the edges of the cloud where the atom density is half the central peak density. By switching the magnetic field coils off following the compression stage, and then imaging, we expect that it will be possible to directly measure the Zeeman-shift artefact.

DCI results were consistent using defocus distances between 40 and 150 mm and detunings up to 5Γ from the main absorption peak. Due to the narrow linewidth of the 776 nm transition, the reduction in sensitivity to imaging beam detuning compared to absorption imaging makes DCI a robust measurement technique for excited-state imaging.

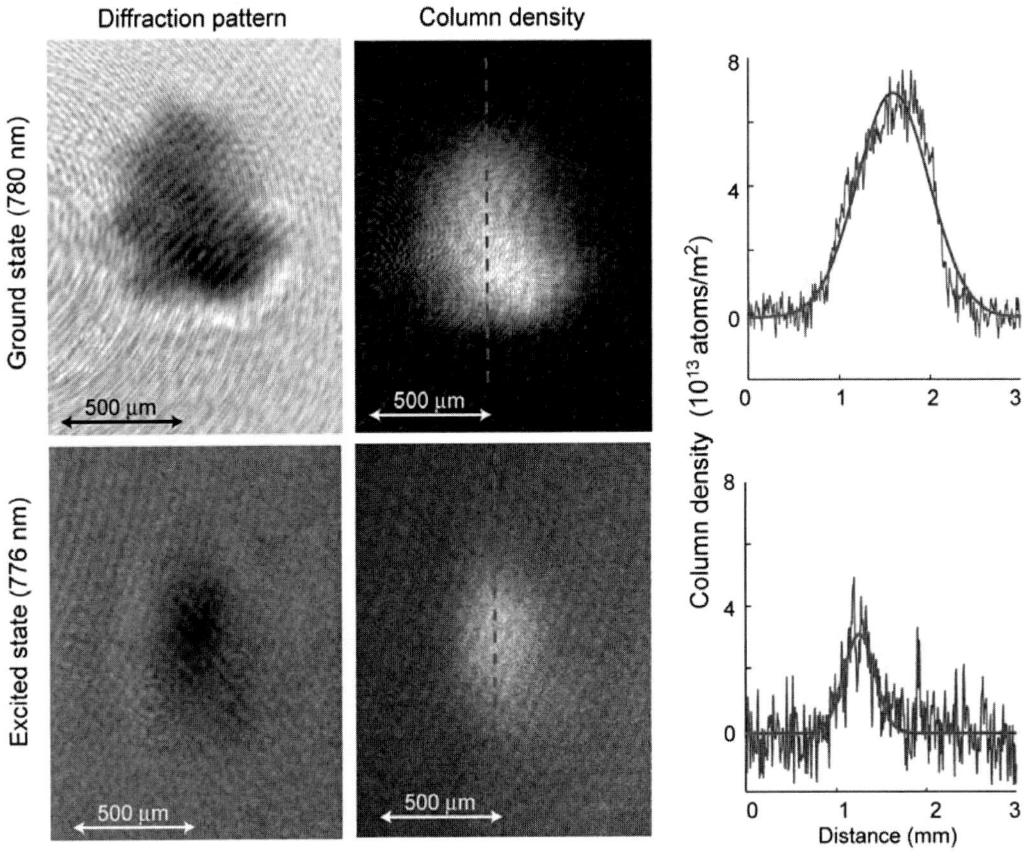

Figure 16. Ground and excited state diffraction patterns, column density reconstructions and column density line profiles. Experimental parameters for ground state image: 60 G/cm peak magnetic field gradient, 4 ms field ramp, −6 MHz maximum cooling beam detuning, 850 μW probe beam power, 30 μs pulse duration, 1 ms exposure, +2.5 Γ_{780} probe detuning, 120 mm effective defocus. For excited state image: 150 μW probe beam power, 100 μs pulse duration, +12.5 MHz probe detuning (+2 Γ_{776} from absorption peak), 130 mm effective defocus.

Quantum Squeezing

The first demonstrations of quantum squeezed states of light were based on four-wave mixing (4WM), with free alkali atoms in an optical cavity [38]. After two decades of development, primarily using optical parametric oscillators rather than 4WM, has led to squeezed beams of light with noise more than 10dB below the standard quantum limit (SQL) [39]. Recently, efforts have returned to 4WM, and a relatively simple system in rubidium vapour has been demonstrated [40], achieving quantum noise reduction of more than 8.8dB [41].

The recent 4WM results use a double-lambda energy level configuration in the D1 transition of ^{85}Rb, coupling the two $5S_{1/2}$ hyperfine ground states (separation $\Delta_G \sim$ 3GHz for ^{85}Rb) to the $5P_{1/2}$ levels (see Figure 17). The $F = 2$ ground state |1> is coupled to the $5P_{1/2}$ state |3> via a strong "pump" beam. The $F = 3$ ground state |2> is coupled to |3> with a weak "probe" beam. When the detunings of the two beams, Δ_{pump} and Δ_{probe}, are equal and blue of resonance, Raman transitions are excited between |1> and |2> through state |3> which is far (~ 1GHz) from resonance with the $5P_{1/2}$ state. Parametric gain occurs when two pump photons are converted into one probe photon and one conjugate photon, with quantum correlation between the two output photons. The experiment (Figure 18) is intrinsically simple, and only one laser is required, with a modulator to generate the probe with appropriate frequency offset Δ_G.

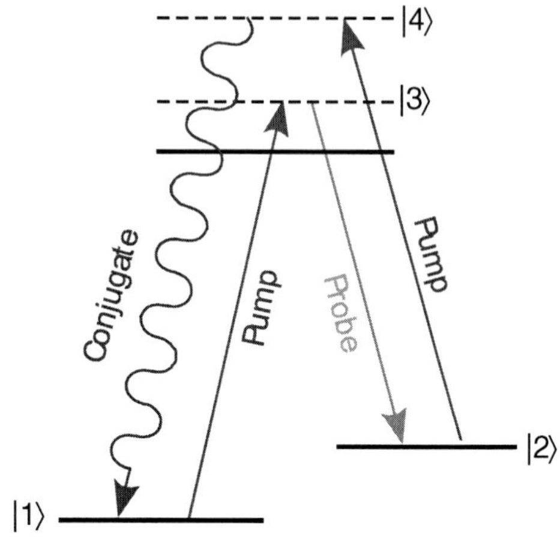

Figure 17. Double-lambda scheme showing degenerate pump beams, off-resonant probe and generated conjugate.

The experiments have demonstrated relative intensity squeezing comparable to that of conventional optical parametric oscillator approaches, at much lower Fourier frequencies (of order kHz), and in higher-order spatial modes such that sub-quantum spatial imaging has been possible [42].

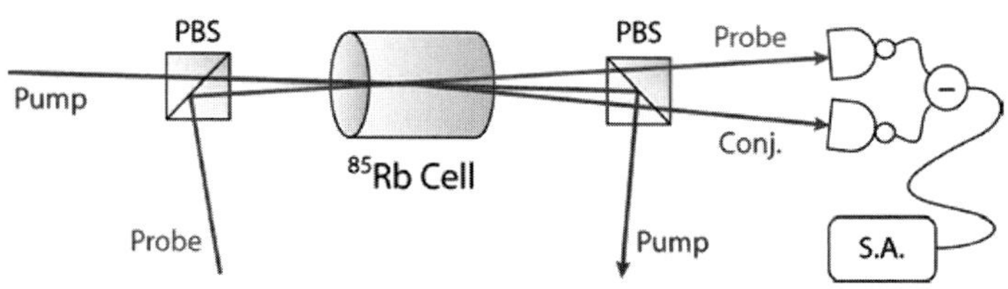

Figure 18. Experimental configuration for producing relative intensity squeezed light by four-wave mixing in an atomic vapour (from Ref. [41]). Two orthogonally polarised aser beams are incident on a heated Rb vapour cell. The amplified probe and generated conjugate beams are detected and their difference analysed on a spectrum analyser.

Based on our experience with the frequency upconversion experiments described above, we propose an alternative scheme for producing intensity squeezed light, using the Rb 5S-5P-5D ladder system (Figure 19). The scheme will produce 776nm light relative intensity squeezed with respect to a 780nm beam. Two pump lasers at 780nm and 776nm will establish a coherent superposition of the 5S and 5D states (|1> and |4>) via state |2> which is off-resonance with the 5P state. A far-detuned 776nm probe beam $\vec{k}_3$ will then trigger stimulated

Raman transition via state |3>, producing gain on the probe and the emission of a conjugate 780nm beam $\vec{k}_c$.

It is important to note that the 5D state has narrow linewidth and long spontaneous decay lifetime. This long decoherence time, analogous to the (much longer) $F = 3$ to $F = 2$ decoherence of the double-lambda system, is important in allowing strong non-linear interactions dependent on the atomic coherence, and allows the high efficiency observed for frequency upconversion and promises similarly high efficiency for our proposed squeezing system.

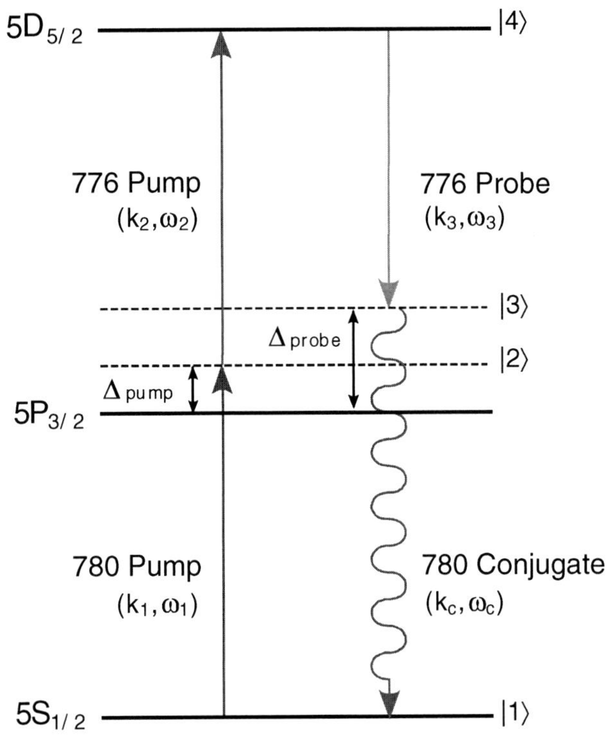

Figure 19. Proposed scheme based on the Rb 5S-5P-5D ladder system. Non-degenerate pump beams at 780nm and 776nm establish a strong coherent superposition of the 5S and 5D states. A 776nm probe triggers production of a 780nm conjugate beam. The 776nm probe and 780nm conjugate are expected to be relative intensity squeezed.

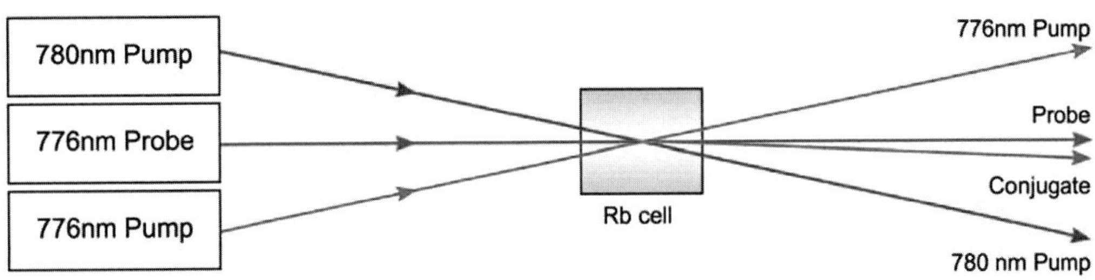

Figure 20. Experimental configuration for the proposed ladder system. Note that the phase matching conditions allow all beams to be spatially separated.

The experimental configuration is shown in Figure 20. The 776nm pump and probe beams can be generated from one laser using a modulator. The 780nm and 776nm pump frequencies are not critical, but the combination must be locked to the 5S to 5D two-step excitation. The frequency locking can be achieved as for the frequency upconversion experiments described earlier.

Phase Matching

For strong non-linear interactions, energy and momentum conservation phase-matching conditions must be met. In vacuum, the refractive index is not frequency dependent so conservation of energy is only consistent with conservation of momentum when the pump and probe beams are perfectly overlapped (spatially degenerate). In the atomic media, particularly with strong atomic coherence, there is strong dispersion near the transitions. As the detunings are different for each beam in the 4WM process, so is the refractive index deviation and it is possible to satisfy the phase-matching with a small nonzero angle between the beams. The conjugate then also propagates in a distinct non-overlapping output beam direction, allowing all the beams to be individually isolated. The spatial non-degeneracy has been observed experimentally in the double-lambda system [40]. In principle, the dispersion profile can be calculated using the optical Bloch equations and the phase-matching condition solved to determine the optimum angle between the beams.

Semiclassical Model

We have modeled the system using optical Bloch equations (OBEs) to estimate the degree of atomic coherence, and thus the nonlinear gain in the coherent atomic medium. From the OBEs we can then calculate the external fields using the standard Maxwell-Bloch formalism as developed for our frequency upconversion system.

Figure 21 shows the results of our model for the conventional double-lambda system with 300mW of pump laser power, and for the proposed cascade system, assuming only 30mW of pump laser power. For the double-lambda system, strong gain is predicted for a small relative detuning between probe and pump, corresponding to AC Stark shift induced by the high power pump field. The light shift is seen to increase as the pump frequency approaches resonance with the 5S-5P transition, as expected.

For the cascade system, much higher gain is predicted at only 30mW of pump power, but the 3-photon detuning – and hence the conjugate detuning – is relatively small and absorption will therefore have greater influence on the effective output power. At increased pump detuning, and conjugate detuning, comparable gain is predicted but at the expense of greater sensitivity to the 2- and 3-photon detunings.

At present we are working towards experimental verification of the scheme, while also further developing the model to incorporate phase-matching conditions. We look forward to demonstrating yet another remarkable application of the versatile Rb three-level cascade system.

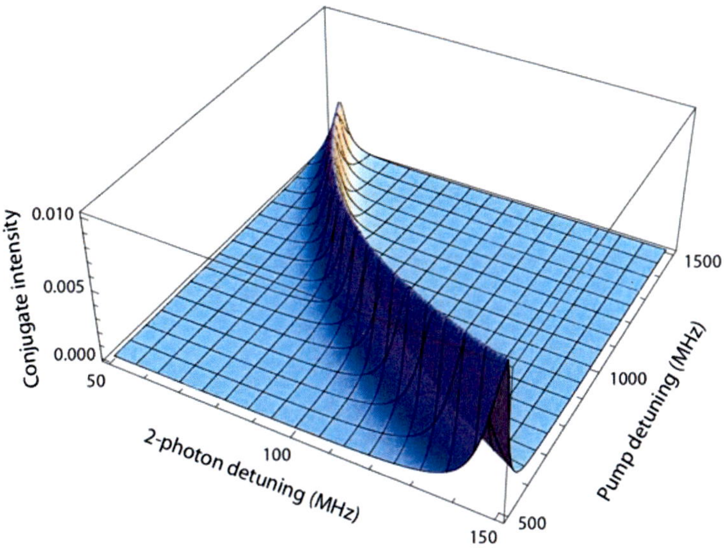

Figure 21. Relative conjugate gain calculated using the optical Bloch equation model, for the conventional double-lambda system, at 300mW pump power. Note that the maximum gain is predicted for non-zero probe-pump (2-photon) detuning, due to the non-trivial AC Stark shift for such high pump power.

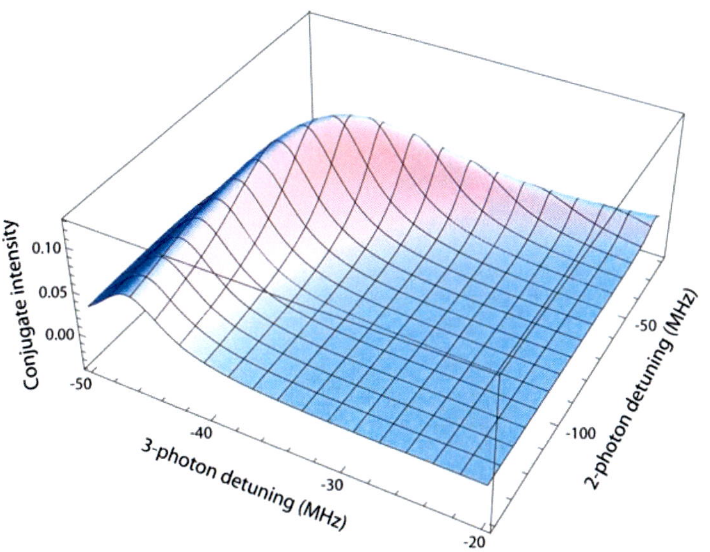

Figure 22. Relative conjugate gain calculated using the optical Bloch equation model, for the proposed double-ladder system, at just 30mW pump power. The calculations are for a pump detuning of 150MHz.

Conclusion

The seemingly simple three-level ladder system of rubidium offers a remarkably diverse array of interesting phenomena in nonlinear and quantum-atom optics. A range of future research areas remain open, for example in efficient low-power quantum squeezing [35], parametric downconversion [36], or electromagnetically induced transparency [37], in addition to Rydberg excitation and formation of ultracold plasma.

This research has been supported by the Australian Research Council Centre of Excellence for Coherent X-ray Science and the Australian International Science Linkages programme.

References

[1] Fleischhauer, M.; Imamoglu, A.; Marangos, J. P. *Rev. Mod. Phys.* 2005, 77, 633.
[2] Scully, M. O.; Zubairy, M. S. *Quantum Optics,* Cambridge University Press, 1997.
[3] Hau, L.V.; Harris, S.E.; Dutton, Z.; Behroozi, C.H. *Nature* 1999, 397, 594.
[4] Harris, S. E. *Phys. Rev. Lett.* 1990, 85, 4032.
[5] Hakuta, K.; Marmet, L.; Stoicheff, B. P. *Phys. Rev. Lett.* 1991, 66, 596.
[6] Arimondo, E. *Prog. Opt.* 1996, XXXV, 259.
[7] Hemmer, P.R.; Ontai, G.P.; Ezekiel, S. *J. Opt. Soc. Am. B* 1986, 3, 219.
[8] Bell, S.C.; Heywood, D.M.; White, J.D.; Close, J.D.; Scholten, R.E. *Appl. Phys. Lett.* 2007, 171120.
[9] van der Wal, C.H.; Eisaman, M.D.; André, A.; Walsworth, R.L.; Phillips, D.F.; Zibrov, A.S.; Lukin, M.D. *Science* 2003, 301, 196.
[10] Fulton, D.; Shepherd, S.; Moseley, R.; Sinclair, B.; Dunn, M. *Phys. Rev. A* 1995, 52, 2302.
[11] Bergeson, S.; Killian, T.C. *Physics world* 2003, 16, 37.
[12] Meijer, T.; White, J.D.; Smeets, B.; Jeppesen, M.; Scholten, R.E. *Opt. Lett.* 2006, 31, 1002 – 1004.
[13] Efthimiopoulos, T.; Movsessian, M.; Katharakis, M.; Merlemis, N. *J. Appl. Phys.* 1996, 80, 639.
[14] Jain, M.; Xia, H.; Yin, G.; Merriam, A.; Harris, S. *Phys. Rev. Lett.* 1996, 77, 4326.
[15] Zibrov, A.S.; Lukin, M.D.; Hollberg, L.; Scully, M.O.; *Phys. Rev. A* 2002, 65, 051801R.
[16] Hawthorn, C.J.; Weber, K.P.; Scholten, R.E. *Rev. Sci. Instrum.* 2001, 72, 4477.
[17] MOGlabs laser diode controller, www.moglabs.com.
[18] Arimondo, E.; Inguscio, M.; Violino, P. *Rev. Mod. Phys.* 1977, 49, 31.
[19] Scholten, R. E.; O'Kane, T. J.; Mackin, T. R.; Hunt, T. A.; Farrell, P. M. *Aust. J. Phys.* 1999, 52, 493.
[20] Gallagher, T.F. *Rydberg Atoms*, Cambridge University Press, 1994.
[21] Killian, T.C.; Kulin, S.; Bergeson, S.D.; Orozco, L.A.; Orzel, C.; Rolston, S. *Phys. Rev. Lett.* 1999, 83, 4776.
[22] Turner, L.D.; Domen, K.F. E.M.; Scholten, R.E. *Phys. Rev. A* 2005, 72, 1403.
[23] Turner, L.D.; Dhal, B.; Hayes, J.; Mancuso, A.; Nugent, K.A.; Paterson, D.; Scholten, R.E.; Tran, C.; Peele, A. *Optics Express* 2004, 12, 2960.
[24] Sheludko, D.V.; Bell, S.C.; Vredenbregt, E.J.D.; Scholten, R.E. *J. Phys.: Conf. Ser.* 2007, 80, 012040.
[25] Tikhonov, A.; Arsenin, V. *Solutions of Ill-posed Problems*, V. H. Winston & Sons, 1977.

[26] Loudon, R. *The Quantum Theory of Light*, Clarendon Press, 1973.
[27] Maguire, L.P.; van Bijnen, R.M.W.; Mese, E.; Scholten, R.E. *J. Phys. B: At., Mol. and Opt. Phys.* 2006, 39, 2709.
[28] Whitley, R.M.;Stroud, Jr, C.R. *Phys. Rev. A* 1976, 14, 1498.
[29] Steck, D.A. Th. Div.; Los Alamos Nat. Lab. 2001, available online at http://steck.us/alkalidata.
[30] Autler, S.H.; Townes, C.H. *Phys. Rev.* 1955, 100, 703.
[31] Streed, E.W.; Chikkatur, A.P.; Gustavson, T.L.; Boyd, M.; Torii, Y.; Schneble, D.; Campbell, G.K.; Pritchard, D.E.; W. Ketterle, Rev. Sci. Instrum. 2006, 77, 023106.
[32] Cohen-Tannoudji, C.; Dupont-Roc, J.; Grynberg, G. *Atom-Photon Interactions: Basic processes and applications*, John Wiley and Sons, Inc., 2004.
[33] Wilson, A.C.; Sharpe, J.C.; McKenzie, C.R.; Manson, P.J.; Warrington, D.M. *Appl. Opt.* 1998, 37, 4871.
[34] Petrich, W.; Anderson, M.H.; Ensher, J.R.; Cornell, E.A. *J. Opt. Soc. Am. B* 1994, 11, 1332.
[35] McCormick, C.; Boyer, V.; Arimondo, E.; Lett, P. *Optics Letters* 2007, 32, 178.
[36] Paul, B.; Cooper, J.; Gallagher, A.; Raymer, M. *Physical Review A* 2002, 66, 63816.
[37] Badger, S.; Hughes, I.; Adams, C. *J. Phys. B: At., Mol. and Opt. Phys.* 2001, 34, L749.
[38] Slusher, R.; Hollberg, L.; Yurke, B.; Mertz, J.; Valley. J. *Phys. Rev. Lett.* 1985, 55, 2409–2412.
[39] Vahlbruch, H.; Mehmet, M.; Chelkowski, S.; Hage, B.; Franzen, A.; Lastzka, N.; Goßler, S.; Danzmann, K.; Schnabel, R. *Phys. Rev. Lett.* 2008, 100, 033602.
[40] McCormick, C.; Boyer, V.; Arimondo, E.; Lett, P. *Opt. Lett.* 2007, 32, 178–180.
[41] McCormick, C.; Marino, A.M.; Boyer, V.; Lett, P. *arXiv:quant-ph* 2008, 0703111.
[42] Boyer, V.; Marino, A.M.; Pooser, R.C.; Lett, P.D. *Science* 2008, 321, 544–547.

In: New Trends in Quantum Coherence…
Editor: Rafael Drampyan, pp. 221-241
ISBN: 978-1-60741-025-6

Chapter 8

DOPPLER-FREE NONLINEAR FARADAY ROTATION SPECTROSCOPY OF RB ATOMS

***Rafael Kh. Drampyan*[1,a], *Andrew D. Greentree*[2,3,b] *and Alan V. Durrant*[2,c]**

[1]Institute for Physical Research, National Academy of Sciences of Armenia, 0203, Ashtarak-2, Armenia

[2]Department of Physics and Astronomy, The Open University, Walton Hall, Milton Keynes, MK7 6AA, UK

[3]School of Physics, University of Melbourne, Victoria 3010, Australia

Abstract

The nonlinear Faraday rotation in the scheme with counter-propagating light beams was investigated in atomic Rb vapor. The Doppler-free Faraday rotation signals were observed for each hyperfine and crossover resonance. Additionally, rotation spectra as a function of pump and probe intensity, pump polarization and external magnetic field strength were studied. The investigations were performed in a regime where the pump beam power was sufficiently low relative to the probe power to avoid polarization rotation due to pump-induced anisotropy. Our results are analyzed in the context of Bennett structures due to the velocity selective optical pumping of atoms. The possible applications of the effect are discussed.

I. Introduction

Despite the resonant Faraday or Macaluso-Corbino effect [1] (polarization rotation of light propagating through an atomic vapor in a longitudinal magnetic field *B*) having been discussed over a century ago, the nonlinear version of this effect is finding a renewed interest due to the discovery of coherent effects, such as coherent population trapping (CPT) [2], electromagnetically induced transparency (EIT) [3] and electromagnetically induced

---

[a] E-mail address: rdramp@ipr.sci.am
[b] E-mail address: andrew.greentree@ph.unimelb.edu.au
[c] E-mail address: A.V.Durrant@open.ac.uk

absorption (EIA) [4]. In typical vapor cell experiments with *weak* light fields Faraday rotation is to first order independent of light power (linear Faraday rotation). The linear resonant Faraday effect is proven as a powerful linear spectroscopy method [5].

For nearly monochromatic light and zero frequency detuning from resonance the Faraday rotation angle φ is determined by [6,7]:

$$\varphi \sim \{2g\mu B/h\,\Gamma/[(1+(2g\mu B/h\Gamma)^2]\}\; l\,/\,l_0 \qquad (1)$$

where the Zeeman splitting of levels is $\nu_Z = g\mu B/h$, g is the Lande factor, μ is Bohr magneton, B is magnetic field strength, h is Plank's constant, Γ is the resonant transition linewidth, l is the medium length, l_0 is the absorption length. From eq.1 we see that the rotation angle φ is linearly proportional to the applied magnetic filed B for Zeeman shifts ν_Z smaller than resonant linewidth Γ, has a peak at $2\nu_Z \sim \Gamma$ and decreases as $1/B$ in the limit of large fields. In the linear Faraday rotation regime, the spectral width of Faraday rotation features is dominated by Doppler broadening, which is of order of 1 GHz for optical transitions at room temperature.

In nonlinear spectroscopy it is possible to achieve much narrower linewidth for optical transitions, than in linear optics [8]. Because the small magnetic field Faraday rotation is inversely proportional to Γ, the rotation angle can be enhanced by many orders of magnitude for nonlinear Faraday rotation (NFR), when $2\nu_Z \sim \Gamma$ [6].

The main physical phenomena responsible for NFR are related to two physical processes [9], the light-induced coherences between Zeeman components of atomic states [2] and hole burning in the atomic velocity distribution (Bennett structures) [10].

Bennett structures [10] are "peaks" and "holes" in the velocity distributions of atomic state populations caused by optical pumping. In saturation spectroscopy (see, for example, [8,11,12]) these narrow structures are analyzed with a probe laser beam, with the ultimate spectral resolution determined by the *natural linewidth* of the transition. In single beam NFR experiments these narrow features are revealed by the magnetic field dependence of the rotation signal. However, there are only limited studies of NFR related to Bennett structure [9, 13, 14].

Much narrower spectral width can be obtained due to the coherence effects in nonlinear spectroscopy. CPT is one effect responsible for *subnatural* width resonances. In the simplest case, CPT occurs in three-level atomic systems interacting with two coherent laser fields [2]. In most alkali vapor realizations, two ground-state hyperfine levels are optically coupled via a common excited state (Λ -system). When the frequency difference of the two coupling fields equals the frequency separation between two ground states (i.e. the ground states are coupled via a two-photon resonance) the atoms are prepared in a non-absorbing (or dark) superposition state.

The effect of optically-induced coherence in NFR has been the subject of extensive investigations (see, for example, [6, 7, 9, 15-23]). In early experiments in atomic Na vapors the strong narrowing of the forward scattering signal dependence on magnetic field observed by Gawlik *et al* was attributed to the Zeeman coherence effects [15]. In [9, 16, 17] NFR was observed in samarium vapor. Most NFR experiments have been performed in alkali metal vapors. The narrowest feature, yet observed in such systems, with a width 2π Hz was observed by Budker *et al* in a paraffin coated Rb cell in [24]. In these experiments the magnetic field dependence of rotation angle exhibits dispersion-like features determined by the Bennett structure (with a width of a few Gauss) and atomic ground state Zeeman coherence (with width much less than 1 G). The nonlinear magneto-optical rotation may appear at light powers as low as a few μW, which is the result of specific lower state structure and long lifetime of atoms in the ground state. A review of resonant nonlinear magneto-optical effects in atoms is given in [25].

Despite the presence of narrow features in the magnetic field dependence of magneto-optical rotation in single beam experiments, the rotation spectra remain Doppler-broadened (see, for example, [6, 13]). From a spectroscopic point of view and for applications in laser-frequency stabilization, it is more interesting to explore natural linewidth NFR signals without Doppler-broadening.

The Zeeman effect in polarization spectroscopy, first introduced by Wieman and Hänsch [26], was studied by Nakayama *et al* [27] on the Na D_1 line in a magnetic field of 50 G ($2\nu_Z > \Gamma$) with pump and probe intensities 25 μW/mm^2 and 0.5 μW/mm^2 (where the saturation intensity was ~0.5 μW/mm^2). The polarization rotation of the probe was identified as being partly due to Faraday rotation, partly to dichroism. At such high fields all Zeeman components of the pertinent energy levels were resolved and the very large number of crossover resonances between split transitions was observed. Recently, Doppler-free magnetically induced nonlinear circular dichroism (when ~10 G longitudinal magnetic field was applied) was studied by Wasik *et al* [28] for Na and ^{85}Rb atoms in the counter-propagating pump-probe configuration. The dispersive signals were obtained by subtracting the σ^+ and σ^- probe beam transmission and were proposed for frequency stabilization of lasers to atomic transitions. Doppler-free Faraday resonances in a counter-propagating pump-probe scheme have been employed for laser stabilization using both Rb and Cs vapors [29, 30]. The systematic study of Faraday rotation spectra in schemes with both single and counter-propagating pump and probe beam geometries was reported in [31]. The use of counter-propagating beam geometry enabled the first observation of Faraday rotation for each Doppler-free hyperfine resonance, as well as at the crossover resonances. The experiment was performed for both *linear orthogonal and* $\sigma\pm$ *circular polarizations* of the pump beam and in magnetic fields B = 0-10 G. The dispersive shape signals obtained at crossover resonances were suggested for frequency stabilization of diode lasers used in laser-cooling experiments, where the laser frequency is red detuned by a few linewidth from a resonant transition.

Nonlinear magneto-optical effects are important tools in laser spectroscopy and have applications to high precision magnetometry [24, 32, 33], searching for parity violation in atoms [34] and have more recently been investigated in the context of quantum information processing [35]. Exploiting ultranarrow linewidth zero-field nonlinear magneto-optical rotation resonances for magnetometry allows achieving sensitivity ~10^{-11} G/Hz$^{1/2}$. However, the disadvantage of this method is its limitation to near zero fields. Since many applications (particularly, measurements of geomagnetic fields) require high sensitivity at magnetic fields of ~1 G, it is necessary to extend the range of ultra-precise magnetometers to the stronger magnetic fields. For this purpose in [36-38] the frequency [36,37] and amplitude [38] modulated light was used. In both methods the ultra-narrow resonances at zero magnetic field are replicated to higher fields, thereby extending the ultra-sensitive magnetometry technique to geomagnetic field ranges.

The sensitivity of δB to magnetic fields along the direction of light propagation is determined by [32,33]

$$\delta B = (\partial \varphi / \partial B)^{-1} \delta\varphi \qquad (2)$$

where $\delta\varphi$ is sensitivity to light polarization rotation, measured in rad $Hz^{-1/2}$, and $\partial \varphi / \partial B$ is the rate of change of optical rotation with respect to the longitudinal magnetic field. The shot noise limit of $\delta\varphi$ is inversely proportional to the square root of the transmitted light power and large light power is required to minimize the shot noise. On the other hand high light power leads to the light induced frequency shifts (ac Stark shift) and power broadening of the atomic resonances [33, 39] which reduces the steepness $\partial \varphi / \partial B$. There is, therefore an optimum light intensity to obtain maximum precision. The sensitivity δB as a function of atomic density, light intensity and frequency was studied for D_1 and D_2 lines of ^{85}Rb in [32]. Single beam NFR at high light powers, where the rotation is significantly modified by ac Stark shifts, was investigated for Rb vapor in [20]. These shifts are shown to change the overall sign of rotation for closed $F - F+1$ transitions compared to the low light power limit. The quantum limit of optical magnetometry in the presence of ac Stark shift was studied theoretically in [33]. An experimental study of ac-Stark shift in the single beam nonlinear Faraday effect for ^{87}Rb D_1 transition and estimation of maximum sensitivity of an NFR based optical magnetometer ($B_{\min} \approx 6 \times 10^{-12}$ G $Hz^{-1/2}$) were performed in [39]. In [21,22] it was shown experimentally for ^{87}Rb atoms, that simultaneous increase of laser intensity and atomic density enhances NFR by many orders of magnitude. At the same time atomic coherences giving rise to EIT leads to a linear decrease of light transmission through the atomic medium, instead of usual exponential Beer's law. Estimates show [21,22], that the sensitivity of magnetic field measurements for realized case is comparable with that of low-power methods [33].

In this chapter we present a detailed description of Doppler-free nonlinear Faraday rotation spectroscopy for Rb atoms in a scheme of counter-propagating pump and probe beams. For comparison we also give the results of spectral studies for a single beam experiment. Investigations were performed for the full experimental parameter space, including pump polarization, pump and probe relative intensities and magnetic field strength. The counter-propagating beam (CPB) experiment was performed for both *linear orthogonal and $\sigma^{\pm}$ circular polarizations* of the pump [31]. Here we present the case of $\sigma^{\pm}$ circular polarizations of backward pump beam, which corresponds to the polarization spectroscopy in an external magnetic field. However, the novelty of our experiment is that the backward pump beam was sufficiently low in power compared to the forward probe beam, thereby avoiding polarization rotation due to the anisotropy induced by backward circularly polarized "pump" beam at B=0. The CPB geometry enabled the first observations of Doppler-free Faraday rotation signals for each hyperfine resonance, as well as at the crossover resonances.

Some of the results, presented in this chapter, were published in [31] and our experimental motivation was briefly reported in Ref. [40]. At the same time we present new experimental results for both single beam and CPB geometries.

II. Experimental Arrangement

The experimental set-up is shown in Fig.1. An external cavity cw diode laser with wavelength 780 nm and linewidth <1 MHz was used in the experiment. Out-coupling of the laser was achieved via an intra-cavity beamsplitter, which produced two output beams: the signal beam, used in the experiment, and a reference beam used for frequency determination. The reference beam is monitored using conventional saturated absorption (SA) spectroscopy in an 8 cm long glass Rb cell.

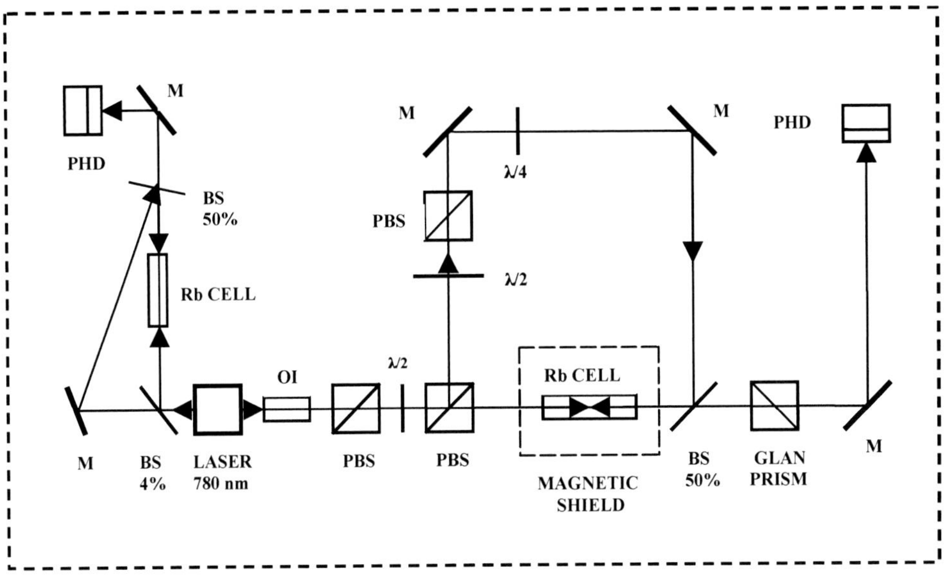

Figure 1. Scheme of experimental arrangement. LASER 780 nm – is an external-cavity grating controlled diode laser, which produced two beams, one monitored with a standard saturation absorption setup for wavelength determination, the other used for the main experiment. OI is an optical isolator, BS - beam splitter, PBS - polarizing beam splitter, λ/2 and λ/4 - half-wave and quarter-wave plates, Rb - rubidium cell, PHD - photodiode, M – mirror.

The laser wavelength was scanned across the ^{87}Rb $S_{1/2}$ - $P_{3/2}$, $F_g = 2 - F_e = 1,2,3$ transitions. The relevant energy levels are shown in Fig.2. The Doppler width of this transition at room temperature is 530 MHz, the natural linewidth of the $5P_{3/2}$ state is $\Gamma \sim 6$ MHz and the ground state Zeeman splitting is 0.71 MHz/G [4]. A 6 cm long sealed off sapphire cell with garnet windows, containing Rb vapor, was placed between crossed polarizers P_1 and P_2. The reason of using the sapphire cell with garnet windows is the absence of birefringence for these windows, which makes such cell more suitable for the study of polarization effects, particularly Faraday rotation.

The cell was placed centrally in the magnetic field produced by three pairs of square, mutually orthogonal Helmholtz coils. These coils were used to cancel the laboratory magnetic field to an accuracy better than 100 mG and also to apply an additional bias field as required. The mentioned value 100 mG is determined by the accuracy of magnetometer used. In fact, the compensation effect is checked by Faraday and Voigt rotation itself and is determined by the absence of rotation signal within the accuracy of the extinction ratio 10^{-5} of the polarizers used in the experimental arrangement.

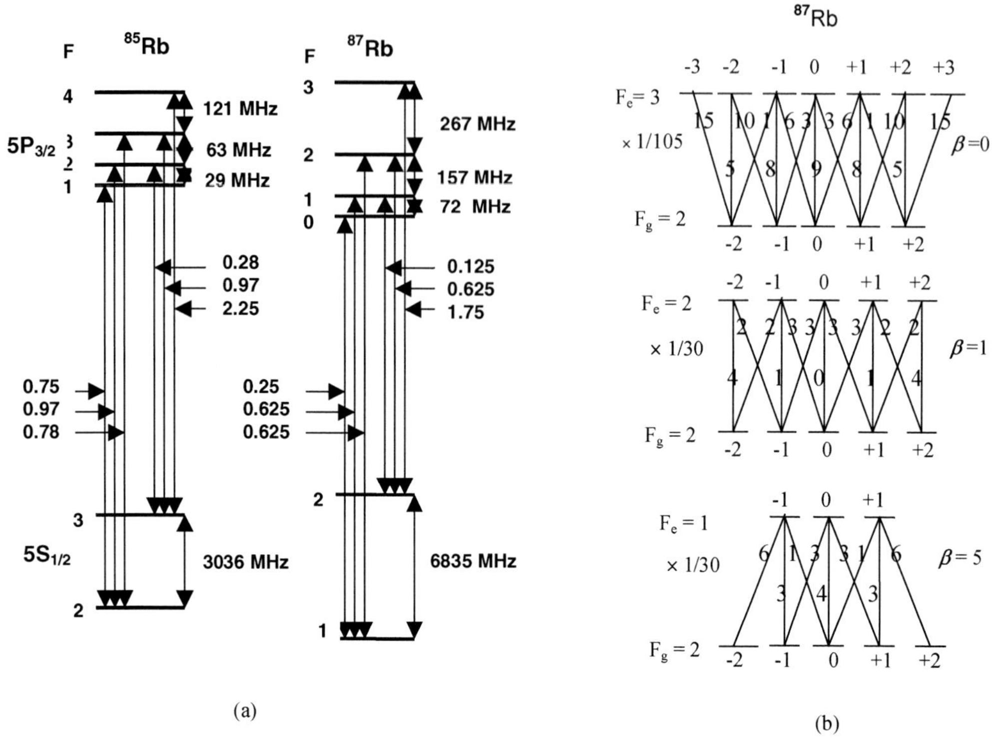

Figure 2. (a) Energy level schemes for the ^{85}Rb and ^{87}Rb D_2 transition. The numbers near the horizontal arrows show the relative transition probabilities between hyperfine levels. (b) Normalized transition probabilities between magnetic sublevels of ^{87}Rb D_2 $F_g=2 - F_e = 1,2,3$ transitions. The numbers near the lines, showing each transition, are the relative transition probabilities. These relative probabilities must be multiplied by the factor on the left side of the diagram to obtain the branching rations for individual transition within each $F_g - F_e$ transition. Normalization is such that the total probability of transitions within each $F_g - F_e$ transition is 1. The factor on the right side of the diagram is the decay rate of the excited state F_e to the other ground state $F_{g'}$, external to the F_g - F_e transition, determined by the ratio $\beta = \Gamma_{Fe-Fg'}/\Gamma_{Fe-Fg}$ (see Ref. [43]).

The longitudinal magnetic field strength was varied in the range 0 - 10 G. The relative powers of the forward (FW) propagating probe and backward (BW) propagating pump beams were varied in the range 20-150 μW using a polarizing beam splitter (PBS) and $\lambda/2$ plate. The probe beam was linearly polarized. A quarter-wave plate generated the $\sigma^{\pm}$ circularly polarized pump beams as required. The beam diameter was 1 mm. The forward traveling probe was detected through the crossed polarizers. The simple blocking of BW beam provided the single beam scheme for NFR. The forward scattering (FS) signal due to the Faraday rotation and reference saturated absorption spectra were detected by photodiodes connected to a Tektronix TDS 520 B oscilloscope. The detected signals were averaged 200 times. The measurements were performed at room temperature 23-24 °C, corresponding to a Rb atomic density ~10^{10} cm^{-3}.

III. Nonlinear Faraday Rotation Spectra in a Single Beam Experiment

Here we show the results of measurements performed when the backward beam was blocked to obtain single beam configuration. For crossed polarizers the measured FS signal is proportional to the square of the sine of the Faraday rotation angle. The spectra of FS signal for Doppler-broadened $S_{1/2}$ (F_g=2) - $P_{3/2}$ transition for varying of magnetic fields for light power 147.8 μW are shown in Fig.3a. The signal increases with increasing magnetic field, reaches its maximum at B = 4.6 G and decreases with the further increase of magnetic field up to 8.5 G. This leads to the M-shaped curve in φ - (± B) dependence (Fig.3b) [31], which is a result of crossed polarizers scheme and detection of FS light intensity and is similar to that obtained in [16].

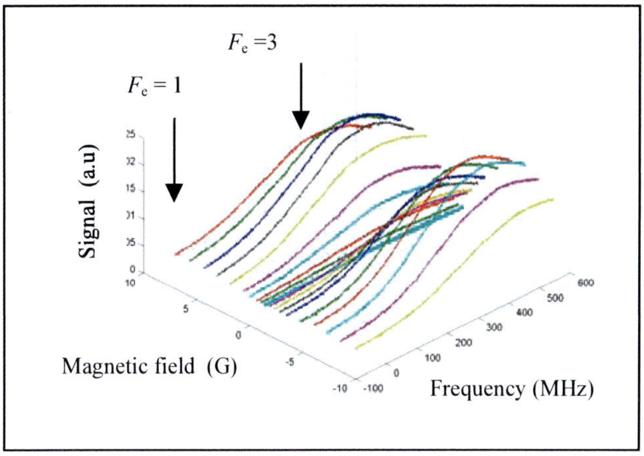

(a)

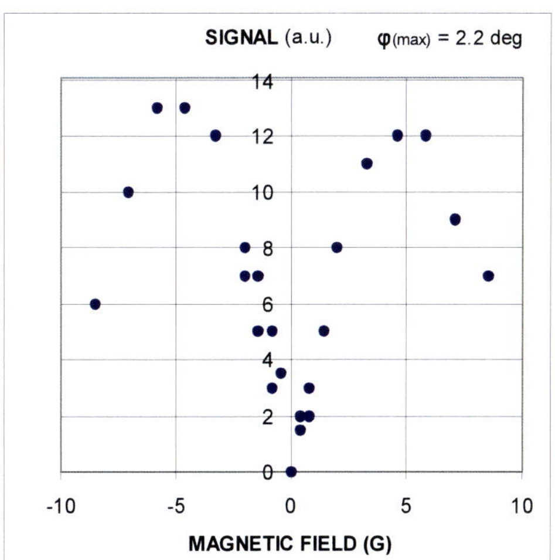

(b)

Figure 3. (a) Spectra of FS signal due to Faraday rotation for Doppler-broadened $S_{1/2}$ (F_g =2) - $P_{3/2}$ transition for varying magnetic field strength for forward beam power P_{FW} = 147.8 μW. Rotation spectra peak values correspond to $F_g = 2 - F_e = 3$ closed transition. 0-frequency corresponds to $F_g = 2 - F_e = 1$ transition. (b) Dependence of Doppler-broadened signal peak value on the magnetic field strength.

Fig. 3a shows that FS signal is maximum when the probe is resonant with the $F_g = 2$ - $F_e = 3$ hyperfine transition. Fig. 4 shows that the rotation maximum is reached at the same magnetic field strength of 4.6 G for all frequencies within Doppler broadened $S_{1/2}$ (F_g=2) - $P_{3/2}$ transition.

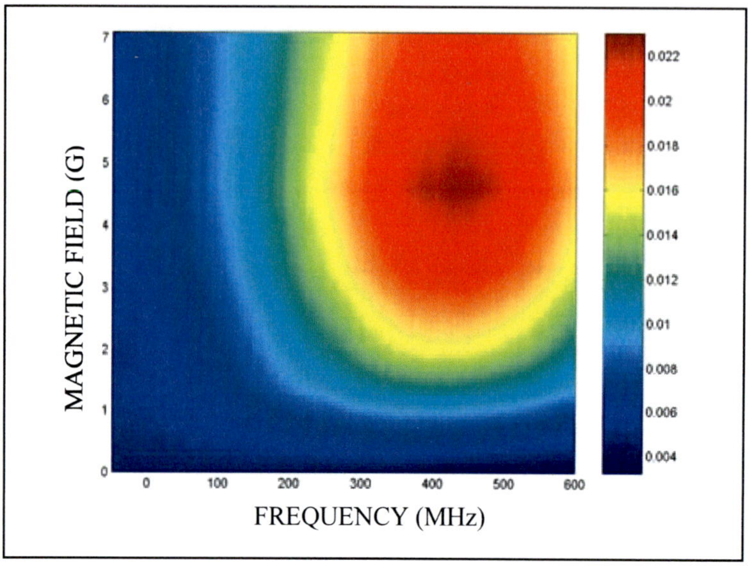

Figure 4. FS spectra – magnetic field dependence, showing that rotation maximum is reached at the same magnetic field strength of 4.6 G for all frequencies within Doppler broadened $S_{1/2}$ (F_g=2) - $P_{3/2}$ transition. 0-frequency corresponds to $F_g = 2 - F_e = 1$ transition.

The absolute value of NFR angle (φ_{max} = 2.2° for B = 4.6 G) was measured by compensating with polarizer P_2 until the transmitted signal reduced to the minimum value determined by the extinction ratio (10^{-5}) of the crossed polarizers.

The magnetic field was stepped by ~ 0.5 G and ~1 G in the ranges of 0-2 G and ~2-10 G, respectively, which was enough to reveal the FS signal – magnetic field dependent shape related to the Bennett structure, but not sufficient to detect the NFR signal, associated with the coherence. The complicated closed $F_g = 2 - F_e = 3$ V –system displays narrow bright resonances due to EIA at low laser intensities [4, 41]. For higher laser intensities more than a few mW/cm^2 the bright resonances disappear as was shown theoretically by F. Renzoni et al [41]. In that work it was also mentioned that in the experiment of G. Theobald et al [42] the absence of sharp bright resonance for the $F_g = 3 - F_e = 4$ and $F_g = 4 - F_e = 5$ transition of Cs is due to high laser intensity of a few mW/cm^2 against the saturation intensity of ~ 1.1 mW/cm^2. This analysis shows that for the conditions of our single beam experiment, when the 147 μW light power (20 mW/cm^2) was used, we could not expect to reveal narrow features attributed to the coherence effect in the FS signal vs magnetic field dependence.

Spectra of Doppler-broadened FS signals for different powers of forward beam are shown in Fig. 5a. The dependence of the FS signal peak value on light power at magnetic field strength H=4.6 G is shown in Fig. 5b. The FS signal increases nonlinearly with the FW beam power showing the nonlinear nature of Faraday rotation.

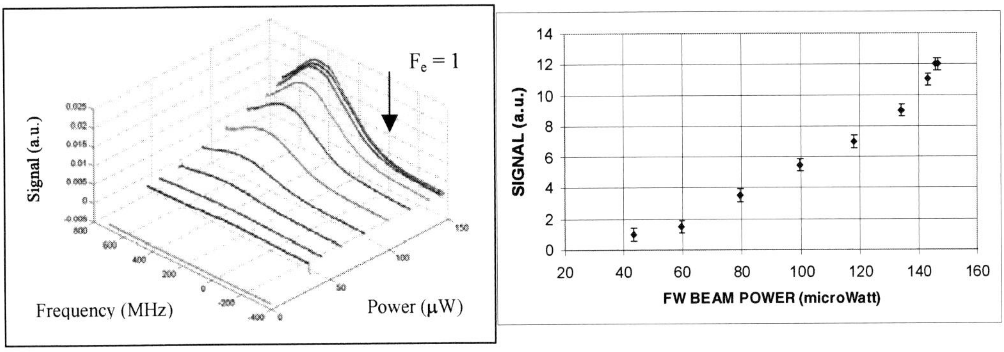

(a) (b)

Figure 5. (a) Spectra of Doppler-broadened FS signals for different powers of forward beam and B = 4.6 G. 0-frequency corresponds to $F_g = 2 - F_e = 1$ transition. Peak values of rotation signals correspond to $F_g = 2 - F_e = 3$ transition (424 MHz on frequency scale). (b) Dependence of Doppler-broadened FS signal peak value on the power of forward beam at magnetic field strength 4.6 G .

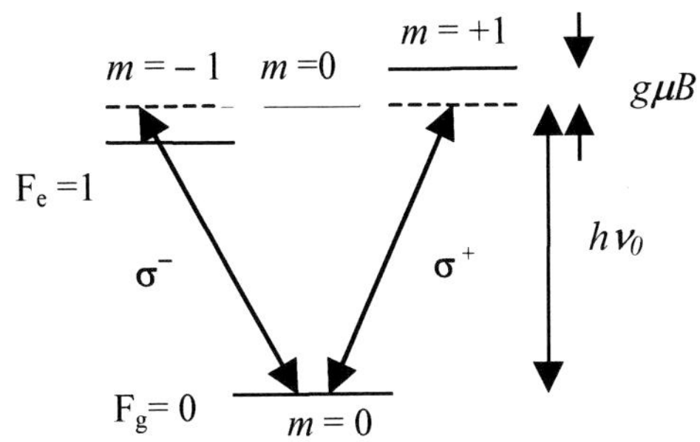

Figure 6. The Zeeman split energy levels for $F_g = 0 - F_e = 1$ transition.

Consider the influence of Bennett structure on Faraday rotation using the energy level scheme for $F_g = 0 - F_e = 1$ transition, shown in Fig. 6.

Application of magnetic field leads to the Zeeman shift of levels. In the presence of a magnetic field, the left and right circularly polarized light components interact with different velocity groups of atoms $w_{1,2}$. The resulting Bennett structures corresponding to the velocity groups resonant to the right and left circular polarizations of light are shifted in opposite directions (Fig. 7a). The values $w_{1,2}$ are determined from relation $\nu(1 - w_{1,2}/c) = (\nu_0 \pm \nu_Z)$, where ν is the light frequency, ν_0 is the resonant frequency of the transition and c is the velocity of light. This causes a difference in refractive indices for $\sigma^{\pm}$ circular components (Fig.7b) and leads to Faraday rotation (Fig. 7c). The peak in the magnetic field dependence of the rotation corresponds to the $B = h\Gamma/2g\mu$ determined by the width of Bennett structure, i.e. the homogeneous width Γ of the transition.

The NFR spectra in Fig.3 shows that the rotation has a maximum at the frequency corresponding to the $F_g = 2 - F_e = 3$ closed transition. This is the strongest transition and has no decay to the other ground state (Fig.2). For B = 4.6 G, when the rotation reaches its maximum (φ_{max} = 2.2°), $2\nu_Z \approx 8$ MHz (for example, for m = 0, F_g =2 – m = +1, F_e =3 transition and g = 2/3 for F_e = 3 state) which is consistent with $\Gamma \sim 6$ MHz natural linewidth of

$5S_{1/2}$-$5P_{3/2}$ transition. Thus the Bennett structure is responsible for NFR observed in a single beam experiment.

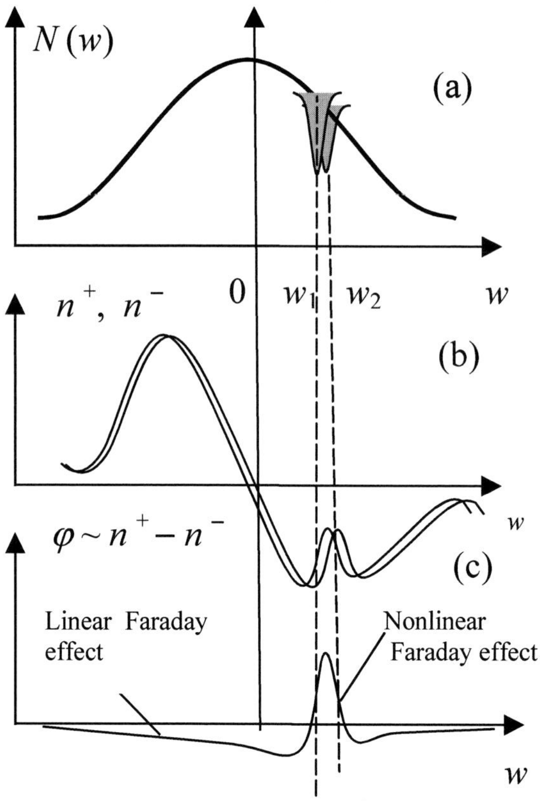

Figure 7. (a) Bennett – structure effect for $F = 0$ state in V-system with longitudinal magnetic field shown in Fig.6. N is the number of atoms, w the atomic velocity projections on the light propagation direction, (b) refractive indices n^- and n^+ in magnetic field, (c) polarization rotation of the laser light, proportional to the difference ($n^- - n^+$). Since the Bennett structure is due to the velocity selective optical pumping of Doppler distributed atoms, the Faraday rotation of atoms with such distribution can be considered as rotation produced by the Doppler-distributed atoms without Bennett "hole" minus the rotation produced by the pumped out atoms. Because the hole width is much narrower than the Doppler width, the small field rotation due to the Bennett structure should be correspondingly larger and opposite in sign compared to the linear effect [6].

The bell-shaped Doppler broadened NFR signals are a consequence of nearly crossed positioning of polarizers P_1 and P_2 [16].

The NFR angle can be estimated by [7]:

$$\varphi = (1/2)\pi^{3/2}\, \alpha\, l\, (P/P_s)\, \{2g\mu B/h\, \Gamma / [(1 + (2g\mu B/h\Gamma)^2\,]\}\, A(\nu) \quad (3)$$

where α is absorption coefficient, l is the medium length, P is the beam power, P_s is the saturation power, $A(\nu)$ is the frequency spectrum of the effect, determined by the frequency detuning (ν - ν_0), and probabilities of optical transitions between states with angular momentum J, nuclear spin I, total angular momentum F and magnetic quantum number m. The saturation power is given by [7]:

$$P_s = \{(2I+1)(2J_g+1)\, c\, h^2 S\, (\Gamma/2)\, \gamma_g\} / 8\pi\, |d|^2 \tag{4}$$

where S is the beam cross section, γ_g is the linewidth of the ground state, determined by the time of flight of atoms through the laser beam. For $I = 3/2$, $J_g = 1/2$, $S \sim 10^{-2}$ cm^2, $|d| = 8 \times 10^{-18}$ ESU (for ^{87}Rb $S_{1/2}$ - $P_{3/2}$ transition) and $\gamma_g \approx 2.5\times10^5$ Hz (for beam size 1 mm and atomic velocity 2.5×10^4 cm/sec), P_s is estimated ~ 20 μW. αl was measured to be ~ 0.6 and $P = 120$ μW. The maximum value of rotation corresponds to $(\nu_Z/\Gamma) = 1/2$. For $F_g = 2 - F_e = 3$ closed transition and exact resonance $(\nu_{ge} - \nu) = 0$ the parameter $A(\nu)$ was estimated to be ~ 0.01. For these parameters φ_{max} was estimated to be ~45 mrad ≈ 2.5° , which is consistent with the measured value 2.2°. The measured results are in good agreement with those obtained in [9, 13] for NFR related to the Bennett structure.

IV. Nonlinear Faraday Rotation in Counter-Propagating Beam Geometry

It is usual in saturation absorption experiments for the probe beam to be significantly weaker in intensity than the pump. However, the results, presented in this paper for CPB NFR relate to the case when pump power was less than the probe power to avoid the polarization rotation due to the anisotropy induced by backward circularly polarized "pump" beam in the absence of a magnetic field.

The Faraday rotation angle φ is given by $\varphi \sim (n^- - n^+)\, l$, where n^- and n^+ are the refractive indices for left and right circularly polarized waves. Fig. 8 shows the Zeeman split absorption and dispersion components, corresponding to the Zeeman split energy levels of the $F_g = 0 - F_e = 1$ transition shown in Fig.6, and the rotation curve $(n^- - n^+)$ (Verdet factor). The Verdet factor as shown corresponds to the case when transition probabilities for σ^+ and σ^- light components are equal. However, when one of the transitions dominates, then the frequency dependence of the Verdet factor will have a simple dispersive shape.

For counter-propagating beams with the same frequencies and zero detuning from resonance, only one atomic velocity group interacts with both beams (zero-frequency resonance) giving rise to Doppler-free spectra. When the probe beam passes through the crossed polarizers the use of counter-propagating beam geometry reveals the Faraday rotation spectra for Doppler-free genuine and crossover resonances. Faraday rotation in this case can be considered as arising from two effects. The first is the rotation caused by the forward beam in the absence of the backward beam. This rotation leads to a Doppler-broadened background signal, with the *value* of φ_{bgr} determined by the Bennett dip of linewidth Γ in the atomic velocity distribution, when the light intensity is high enough (nonlinear Faraday rotation). The second is determined by the Doppler-free rotation φ_{hpf} at the frequencies of each hyperfine and crossover resonance due to the presence of the backward beam. The total rotation angle is $\varphi(\nu) = \varphi_{bgr}(\nu) + \varphi_{hpf}(\nu)$. With crossed polarizers the forward probe beam intensity $I_F \sim \sin^2(\varphi_{bgr} + \varphi_{hpf})$ is monitored. For small rotation angles

$$I_F \sim \sin^2(\varphi_{bgr} + \varphi_{hpf}) \sim (\varphi_{bgr} + \varphi_{hpf})^2 = (\varphi_{bgr})^2 + 2\varphi_{bgr}\,\varphi_{hpf} + (\varphi_{hpf})^2 \tag{4}$$

The first term describes the Doppler-broadened rotation signal. The Doppler-free rotation components are described by terms $2\varphi_{bgr}\,\varphi_{hpf}$ and $(\varphi_{hpf})^2$. Because the φ_{bgr} is constant (for fixed value of magnetic field and forward beam intensity) the second term in (4) determines the signal, which is proportional to the rotation angle φ_{hpf} . For the case of equal transition probabilities for σ^+ and σ^- light components, the spectral profile of φ_{hpf} is determined by (n^- - n^+), shown in Fig.8. When one of the transitions (σ^+ or σ^-) dominates, then the frequency dependence of the Verdet factor will have a simple dispersive shape. $(\varphi_{hpf})^2$ should have a spectral profile with reversed negative part of (n^- - n^+) curve.

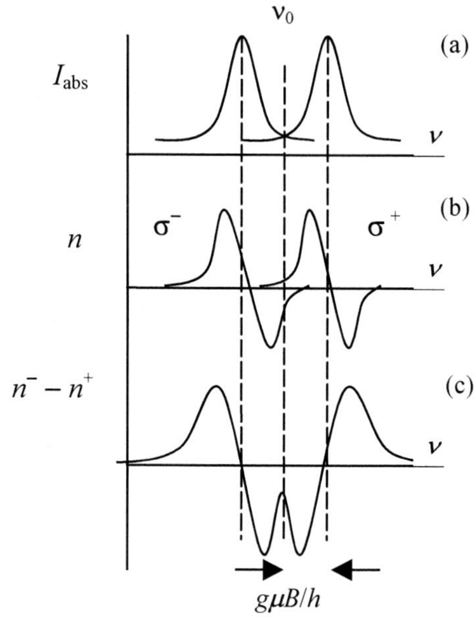

Figure 8. The Zeeman split absorption (a), and dispersion (b) spectral components, and rotation curve ($n^- - n^+$) (Verdet factor) (c) for $F_g = 0 - F_e = 1$ transition shown in Fig.6 and for the case of $g\mu B > h\Gamma$. The homogeneous line broadening Γ is assumed. The use of the simplified energy level diagram is consistent with our discussion, which supposes that the backward pump beam leads to velocity selective modification of the ground state population, particularly by optical pumping of atoms to magnetic sublevels depending on the pump polarization.

Thus Doppler-free rotation signals will appear on the Doppler broadened background determined by $(\varphi_{bgr})^2$. The resulting signal will have either a frequency profile determined by the term $2\varphi_{bgr}\varphi_{hpf}$ and shown in Fig.8, or frequency profile determined by the term $(\varphi_{hpf})^2$, depending on which term dominates. In our experiment the first case is realized, because rotation angles are small ($\varphi_{bgr} \sim 2^\circ$) and the forward beam power 140 μW is more than the backward power 30 μW.

To understand these physical processes with an applied magnetic field, it is convenient to analyze the system in a Σ frame with z quantization axis parallel to the magnetic field. The application of a pump beam leads to redistribution of atomic population over the ground state magnetic sublevels, as well as to optical pumping of atoms via the excited state.

A. Doppler-Free NFR Spectra

Circular Polarization of Backward Pump Beam

As was mentioned in the Introduction NFR in CPB geometry was performed for both linear and circular polarizations of backward beam. The results for linearly polarized backward beam are reported in [31]. Here we present a more detailed description and discussion of experimental results for circular polarization of backward pump beam. In general, this scheme corresponds to polarization spectroscopy in an external magnetic field. However, in the present experiment the backward pump beam was sufficiently low in power compared to the forward probe beam, to avoiding the polarization rotation due to the anisotropy induced by backward circularly polarized "pump" beam at B=0, although the BW beam power was large enough for velocity selective modification of the ground state population.

The spectra obtained for $\sigma^{\pm}$ circular polarizations of backward pump beam, with B = 2 G, and probe and pump powers 120 and 30 μW are shown in Fig. 9. The Doppler-free signals for both genuine and crossover CO23, CO13 resonances have a nearly dispersive shape with opposite signs for σ^+ and σ^- pump beam polarizations. Fig.10 shows the evolution of NFR spectra in magnetic field for σ^+ pump beam. The Doppler-free signals are absent when B=0. It means that the ellipticity and rotation due to anisotropy are small for the conditions of present experiment. This allows us to state that we deal predominantly with Faraday rotation signals.

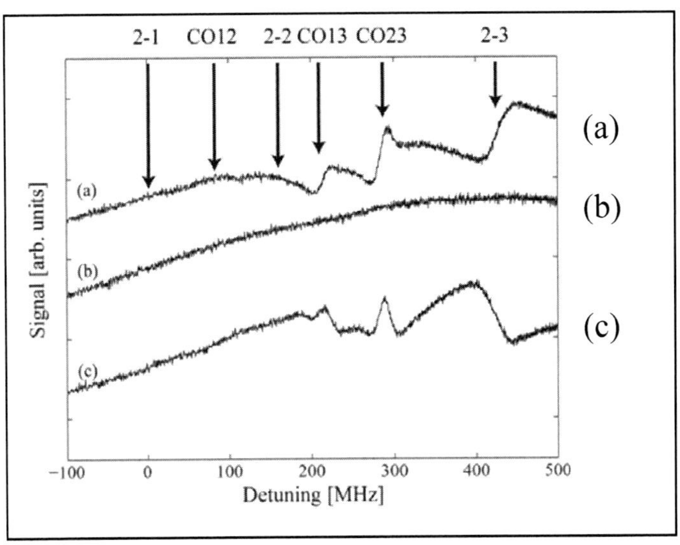

Figure 9. Doppler-free NFR spectra for σ^+ (a) and σ^- (c) circular polarizations of BW pump beam, B = 2 G, and probe and pump powers 120 and 30 μW, respectively. The spectrum (b) corresponds to the blocked BW pump beam.

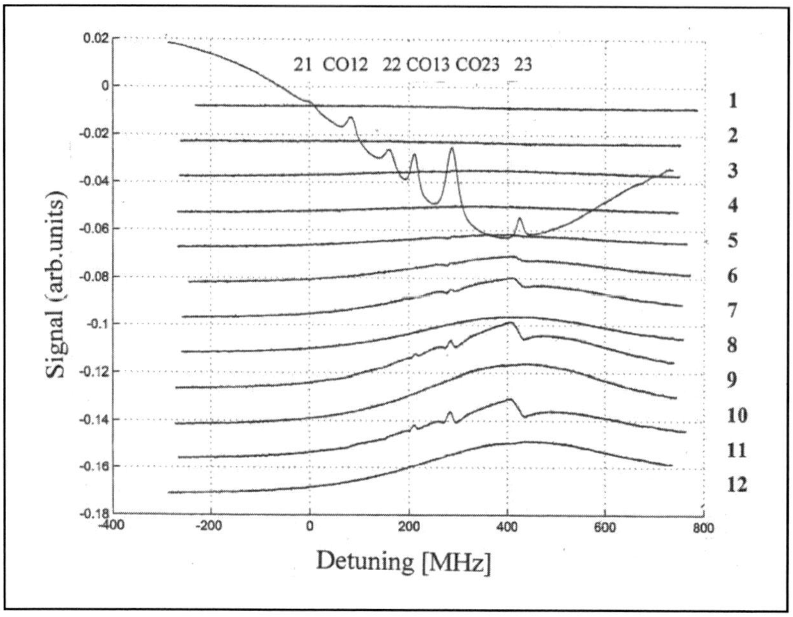

Figure 10. Evolution of NFR spectra in magnetic field for σ^+ pump beam, P_{FW} = 130 μW and P_{BW} = 30 μW. The spectra numbered as 1, 3, 5, 6, 7, 9, 11 correspond to the magnetic field strength B = 0, 0.4, 0.8, 1.4, 2, 4.6, 7.1 G, respectively. Other spectra correspond to the case when BW beam was blocked. The reference saturated absorption spectrum also is shown in the figure.

B. Discussion

1. Closed Transition $F_g = 2 - F_e = 3$.

Circular polarization of backward pump beam.

Considering first the direct resonance $F_g = 2 - F_e = 3$, we note that atoms moving transversely to the beam axis are optically pumped to the outermost $m = +2$, $F_g = 2$ or $m = -2$, $F_g = 2$ states by the σ^+ or σ^- polarized pump, respectively.

Fig.11a shows optical pumping of atoms to the $m = +2$, $F_g = 2$ outermost state by the σ^+ polarized pump. The σ^+ probe component couples the $m = +2$, $F_g = 2$ - $m = +3$, $F_g = 3$ transition with probability 15, and σ^- probe component couples the transition $m = +2$, $F_g = 2$ - $m = +1$, $F_g = 3$ with probability 1 (Fig. 11b). Thus the n_m^- contribution is negligible compared to n_m^+ and the overall signal will have nearly dispersive shape, determined by $(-n_m^+)$.

A σ^- pump beam behaves similarly. Although in this case it is n_m^+ that is negligible compared to n_m^- for probe beam. Again one obtains a nearly dispersive NFR signal, however, opposite in sign to the case of σ^+ polarized pump beam (see spectra in Fig.9).

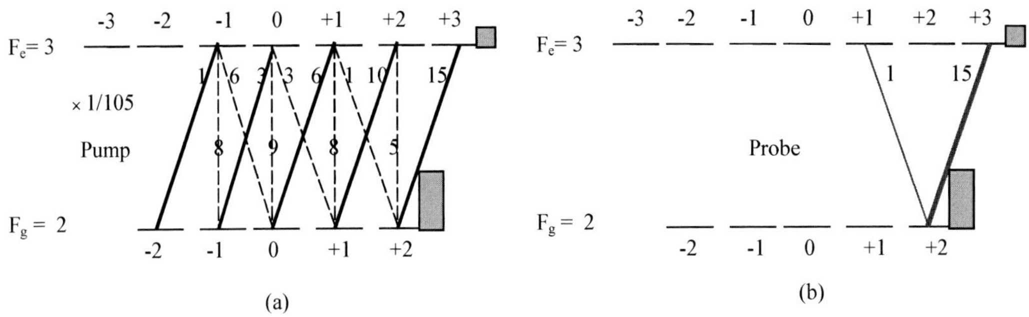

Figure 11. (a) Optical pumping of atoms to the outermost m= +2, F_g =2 state by σ^+ circularly polarized *pump* beam resonant to ^{87}Rb D_2 F_g =2 - F_e = 3 closed transition. The numbers near the lines showing each transition are the relative transition probabilities. (b) Faraday rotation signal shape is determined by $\varphi \sim (n^- - n^+)\, l$, where n^- and n^+ are the refractive indices for left and right circularly polarized waves. Because the probability 15 of m = +2, F_g = 2 - m = +3, F_e = 3 transition coupled by σ^+ *probe* component, is much higher than the probability 1 of the transition m = +2, F_g = 2 - m = +1, F_e = 3 coupled by σ^- probe component, the rotation signal is determined predominantly by $\varphi \sim (-n^+ l)$ and signal should have a dispersive shape.

2. Crossover Resonances

Contrasting the closed transitions seen on the genuine resonances, crossover resonances appear with the laser tuned halfway between two primary resonances. The forward probe and backward propagating beams interact with a certain velocity group via two different hyperfine transitions sharing a common lower level. A second group of atoms with opposite velocity component contributes indistinguishably to this resonance with the role of pump and probe beams exchanged (Fig.12). Crossover resonances involve three energy levels and one of the excited states will have decay to the other (uncoupled) ground state. As such, one would expect, in general, that the two velocity groups would see qualitatively different effects and thus the analysis of the Faraday rotation signal must differ qualitatively from the genuine resonances.

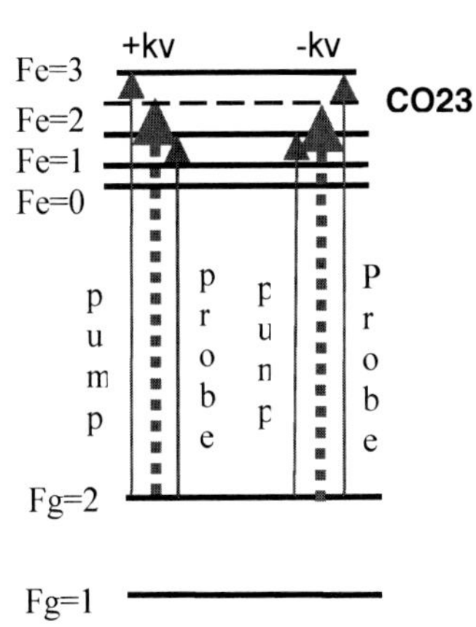

Figure 12. Atomic transitions responsible for the appearance of the crossover resonance F_g = 2 – F_e = 2,3 (CO23), shown by dashed arrows.

Circular polarization of pump beam

We first discuss the CO23 crossover resonance with a circularly polarized pump. For one velocity group, the pump beam is resonant with $F_g = 2 - F_e = 3$ transition and the probe beam resonant with $F_g = 2 - F_e = 2$ transition (Fig. 13a).The σ^+ polarized pump beam optically pumps atoms to the $m = +2$, $F_g = 2$ state. The σ^- component of probe beam is in resonance with $m = +2$, $F_g = 2$ - $m = +1$, $F_e = 2$ transition (with probability 2/30), but there are no levels in resonance with the σ^+ polarization component of probe beam. The NFR signal for this group of atoms is proportional to $(n_m^-)^{open}$. For the other velocity group the pump is in resonance with $F_g = 2 - F_e = 2$ and the probe is resonant with $F_g = 2 - F_e = 3$ (Fig. 13b). The σ^+ polarized pump beam again optically pumps atoms to the $m = +2$, $F_g = 2$ states. The σ^+ polarization component of probe beam is in resonance with $m = +2$, $F_g = 2$ - $m = +3$, $F_e = 3$ transition with probability 15/105 and σ^- component is in resonance with $m = +2$, $F_g = 2$ - $m = +1$, $F_e = 3$ transition with probability 1/105. The signal is proportional to $(n_m^- - n_m^+)^{closed}$. Because $n_m^- << n_m^+$ the NFR signal will be proportional to $(- n_m^+)^{closed}$. The total signal is the sum of the signals $\{(- n_m^+)^{closed} + (n_m^-)^{open}\}$, corresponding to the different velocity groups. However, we must take into account that the open $F_g = 2 - F_e = 2$ transition, driven by the pump beam, decays to the $F_{g'} = 1$ state with a rate $\beta = \Gamma_{Fe = 2 - Fg' = 1} / \Gamma_{Fe = 2 - Fg = 2} = 1$ (50% population loss) [43]. Thus $(n_m^-)^{open} << (- n_m^+)^{closed}$ and the NFR signal is determined mainly by $(- n_m^+)^{closed}$ with nearly dispersive shape. The deviation from an ideally dispersive shape is determined by the residual contribution of $(n_m^-)^{open}$. The case of σ^- polarized pump beam can be considered analogously. Again we observe a nearly dispersive shaped NFR signal, however opposite in sign to the case of σ^+ polarized pump beam (see Fig. 9).

The CO13 crossover resonance displays similar properties. For example, a σ^+ polarized pump beam, which is resonant with the $F_g = 2 - F_e = 3$ transition optically pumps atoms to the $m = +2$, $F_g = 2$ state. The σ^- component of the probe beam is resonant with the $m = +2$, $F_g = 2$ - $m = +1$, $F_e = 1$ transition with probability 6/30 (see Fig. 2b) (which is greater than the probability 2/30 for $m = +2$, $F_g = 2$ - $m = +1$, $F_e = 2$ transition for CO23), but the σ^+ polarization component of the probe is not resonant. The NFR signal is determined by $(n_m^-)^{open}$. For the other velocity group the pump is in resonance with $F_g = 2 - F_e = 1$ again optically pumping atoms to the $m = +2$, $F_g = 2$ state and the probe is resonant with $F_g = 2 - F_e = 3$ transition. Due to the difference of transition probabilities 15/105 for $m = +2$, $F_g = 2$ - $m = +3$, $F_e = 3$ and 1/105 for $m = +2$, $F_g = 2$ - $m = +1$, $F_e = 3$ transitions the signal value is proportional to $(- n_m^+)^{closed}$. The total signal is determined by $\{(- n_m^+)^{closed} + (n_m^-)^{open}\}$. However, in this case the optical decay β to other ground state $F_{g'} = 1$, is higher for the $F_g = 2$ - $F_e = 1$ hyperfine transition ($\beta = 5$) than the decay on the $F_g = 2$ - $F_e = 2$ transition ($\beta = 1$) involved in CO23 resonance [43]. Again the condition $(n_m^-)^{open} << (- n_m^+)^{closed}$ is satisfied and we conclude that the signal for CO13 will be nearly dispersive with approximately same value as the CO23 resonance.

It is worth mentioning the absence of NFR signals at $F_g = 2 - F_e = 1, 2$ direct and CO12 crossover resonances. The probability of the $F_g = 2 - F_e = 1$ transition is the lowest among the manifold of ^{87}Rb D_2 hyperfine transitions (Fig.2 a) and is not resolved in either NFR or reference saturated absorption spectra in our experiment (Fig. 9,10). On the other hand the $F_g = 2 - F_e = 2$ and CO12 resonances are seen in SA spectrum, but are absent in the NFR spectrum. These resonances only involve open transitions, which suffer atomic population loss due to the hyperfine optical pumping to the non-interacting ground-state $F_{g'} = 1$ level. In the SA spectrum the optical pumping between hyperfine levels leads to an increase in the probe transmission signal [44], while for NFR spectrum it leads to a decrease of the rotation signal due to atomic population loss.

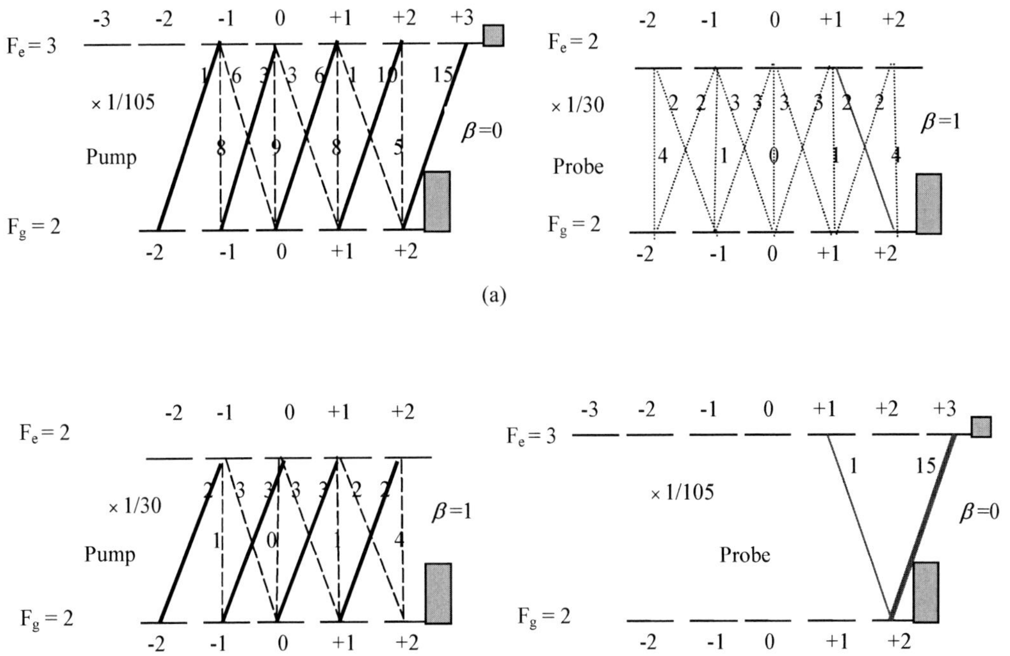

Figure 13. Atomic transitions responsible for NFR for circularly polarized pump beam. (a) pump is in resonance with $F_g=2 - F_e = 3$ transition and the probe beam is resonant with $F_g = 2 - F_e = 2$ transition. (b) the pump beam is in resonance with $F_g=2 - F_e = 2$ transition and the probe beam is resonant with $F_g = 2 - F_e = 3$ transition.

It should be noted that in polarization spectroscopy [26,45] for perfectly crossed polarizers the transmitted signal shows a Lorentzian lineshape on resonance. However, if the polarizers are not perfectly crossed, rotation produces a dispersive signal on a constant background. In our NFR pump-probe experiment the latter case was observed. The existence of background Doppler broadened Faraday rotation signal of 1-2° at $B \neq 0$ reveals the dispersive Doppler-free signals on the Doppler-broadened background even when polarizers P_1 and P_2 remained nearly crossed [27].

V. Applications

Faraday effect and particularly NFR have important applications in diverse areas such as polarization control, Faraday optical isolators and filters, sensitive magnetometry [25] and laser-frequency stabilization. Lasers with stable frequencies and narrow linewidth are essential for many fields of research. The diode laser with grating feedback has narrow linewidth < 1 MHz [46]. However, over time the central laser frequency will drift because of current, temperature and mechanical fluctuations. This drift can be reduced by stabilizing the laser to an external reference. The dispersive shape optical signals are most suitable for laser stabilization [45]. There are many ways of achieving laser frequency stability [28, 45-49]. A simple method is dichroic atomic vapor laser lock (DAVLL) [47], which uses single probe beam and therefore the error signal has Doppler broadened spectral features, but wide capture range of 0.5 GHz. Doppler-free technique can be based on saturated absorption [45], polarization spectroscopy [48], combination of saturated absorption and DAVLL [28] and selective reflection [49].

Our NFR results may find application in the control of spectral features for frequency stabilization of diode lasers [29, 30]. The magnetic field and BW pump beam polarization provide an extra control to modify the NFR dispersive signals. Diode lasers are widely used as light sources for atomic frequency standards [50] and cooling of atoms [51]. In laser-cooling experiments, usually, the laser frequency is red detuned by a few linewidth from a resonant transition and the dispersive shape signals obtained at crossover resonances can be used for frequency stabilization of diode lasers in this case.

VI. Conclusions

In conclusion, the nonlinear Faraday effect was studied for ^{87}Rb atoms. Observations across D_2 line were made. The experiment was realized in both the single and counter- propagating beam geometries. Our main results were obtained for forward probe beam power higher than the backward pump power to induce a large value of Faraday rotation and eliminate the influence of polarization rotation due to the anisotropy induced by backward circularly polarized "pump" beam in the absence of magnetic field. This enabled the observations of NFR for each hyperfine and crossover resonances.

In a single beam configuration the Doppler broadened NFR signals were observed. The peak in the magnetic field dependence of the rotation signal corresponds to the $B = h\Gamma/2g\mu$ determined by the width of Bennett structure, i.e. the homogeneous width Γ of the resonant transition.

The counter-propagating beam geometry reveals the Doppler-free signals, which have a nearly dispersive shape for genuine and crossover resonances with σ^+ and σ^- circularly polarized backward pump beam. The total rotation is being considered as the sum of two effects: rotation φ_{bgr} caused by FW beam in the absence of BW beam leading to Doppler-broadened background and Doppler-free rotation φ_{hpf} at the frequencies of each hyperfine and crossover resonance due to the presence of backward beam. The results were explained taking into account the optical pumping of atoms by the backward pump beam and different transitions probabilities for σ^+ and σ^- polarization components of probe beam, leading to the different frequency profile of Faraday rotation signal.

The results obtained may find application in the control of spectral features for frequency stabilization of diode lasers, which are widely used as light sources for atomic frequency standards and cooling of atoms.

Acknowledgments

The authors express their gratitude to the Dr. D. Sarkisyan from IPR for the loan of the sapphire cell used in this experiment and fruitful discussions, to Prof. J. Marangos from Imperial College (UK) for discussions of the experimental results and R. Bence and M. Percy for technical assistance. Dr. R. Drampyan thanks the Royal Society (UK) for financial support of study visit and Dr. A.D. Greentree is supported by the Australian Research Council, the Australian Government and by the US National Security Agency (NSA), Advanced Research and Development Activity (ARDA) and the Army Research Office (ARO) under contract number DAAD19-01-0653

References

[1] Macaluso, M.; Corbino, O.M. *Nuovo Chimento*, 1898, 8, 257-258.

[2] Arimondo, E. "*Coherent Population Trapping in Laser Spectroscopy*" in Progress in Optics, edited by E.Wolf (Elsevier Science B.V. New-York, 1996), XXXV, 257-354.

[3] Boller, K.-J.; Imamoglu, A.; Harris, S.E. *Phys. Rev. Lett.* 1991, 66, 2593-2596; see for review Marangos, J.P. "Electromagnetically induced transparency", *Journal of Modern Optics*, 1998, 45, 471-503; Fleischhauer, M.; Imamoglu, A.; Marangos, J.P. "Electromagnetically induced transparency: Optics in coherent media", *Reviews of Modern Physics*, 2005, 77, 633-673.

[4] Akulshin, A.M.; Barreiro, S.; Lezama, A. *Phys. Rev. A,* 1998, 57, 2996-3002.

[5] Born, M. *Optik*, Springer, Heidelberg, 1972.

[6] Budker, D.; Orlando, D.J.; Yashchuk, V. *Am. J. Phys.* 1999, 67, 584-592

[7] Kanorsky, S.I.; Weis, A.; Wurster, J.; Hansch, T.W. *Phys. Rev. A,* 1993, 47, 1220-1226.

[8] Letokhov, V.S; Chebotaev, V.P. *Nonlinear Laser Spectroscopy*, Springer Series in Optical Sciences, Springer Verlag, Berlin, 1977.

[9] Barkov, L.M.; Melik-Pashayev, D.A.; Zolotorev, M.S. *Opt. Comm.* 1989, 70, 467- 472.

[10] Bennet, W.R. *Phys. Rev.* 1962, 126, 580-593.

[11] Schmidt, O.; Knaak, K.-M.; Wynnads, R.; Meshede, D. *Appl. Phys. B*, 1994, 59, 167-178.

[12] Drampyan, R.Kh.; Greentree, A.D.; Durrant, A.V. "*Pump-probe spectroscopy of Rb atoms in an external magnetic field*" in "Optics Research Trends" (Editor P.V. Gallico), Nova Science Publishers, New York, 2007, Chapter 7, pp. 271-303.

[13] Budker, D.; .Kimball, D.F.; Rochester, S.M.; Yashchuk, V.V. *Phys. Rev. A*, 2002, 65, 033401-1-7.

[14] Drampyan, R. Kh.; Greentree, A. D.; Durrant, A.V. *Journal of Contemporary Physics, Armenian Academy of Sciences*, 2004, 39, 17-22.

[15] Gawlik, W.; Kowalski, J.; Neumann, R.; Trager, F. *Opt. Comm.* 1974, 12, 400-404.

[16] Drake, K.H.; Lange, W.; Mlynek, J. *Opt. Comm.* 1988, 66, 315-320 .

[17] Baird, P.E.G.; Irie, M.; Wolfenden, T.D. *J. Phys. B: At. Mol. Opt. Phys.* 1989, 22, 1733-1742.

[18] Chen, X.; Telegdi, V.L.; Weis, A. *Opt. Comm.* 1990, 74, 301-305.

[19] Franke-Arnold, S.; Arndt, M.; Zeilinger, A. *J. Phys. B: At. Mol. Opt. Phys.* 2001, 34, 2527-2536.

[20] Budker, D.; Kimball, D.F.; Rochester, S.M.; Yashchuk,V.V. *Phys. Rev. Lett.* 2000, 85, 2088-2091.

[21] Sautenkov, V.A.; Lukin, M.D.; Bednar, C.J.; Novikova, I.; Mikhailov, E.; Fleischhauer, M.; Velichansky,V.L.; Welch, G.R.; Scully, M.O. *Phys. Rev. A*, 2000, 62, 023810-1-4.

[22] Novikova, I.; Matsko, A.B.; Welch, G.R. *Optics Letters*, 2001, 26, 1016-1018.

[23] Akhmedjanov, R.A.; Zelensky, I.V. *JETP Lett.* 2002, 76, 419 -423.

[24] Budker, D.; Yashchuk, V.; Zolotorev, M. *Phys. Rev. Lett.* 1998, 81, 5788 – 5791.

[25] Budker, D.; Gawlik, W.; Kimball, D.F.; Rochester, S.M.; Yashchuk, V.V.; Weis, A. *Rev. Mod. Phys.* 2002, 74, 1163-1201.

[26] Wieman, C.; Hansch, T.W. *Phys. Rev. Lett.* 1976, 36, 1170-1173.

[27] Nakayama, S.; Series, G.W.; Gawlik,W. *Opt. Comm.* 1980, 34, 382-388.

[28] Wasik, G.; Gawlik, W.; Zachorowski, J.; Zawadski, W. *Appl. Phys. B*, 2002, 75, 613-619.

[29] Shevy,Y.; Ianelli, J.; Kitching, J.; Yariv, A. *Optics Letters*, 1992, 17, 661-663.

[30] Lee, W. D.; Campbell, J. C. *Appl. Phys. Lett.* 1991, 58, 995-997.

[31] Drampyan, R. Kh.; Greentree, A.D.; Durrant, A.V. *Opt. Comm.* 2007, 276, 251-260.

[32] Budker, D.; Kimball, D. F.; Rochester, S. M.; Yashchuk, V. V.; Zolotorev, M. *Phys. Rev. A*, 2000, 62, 043403.

[33] Fleischhauer, M.; Matsko, A.; Scully, M. O. *Phys. Rev. A*, 2002, 62, 013808.

[34] Barkov, L. M.; Zolotorev, M. S.; Melik-Pashaev, D. *Sov. JETP Pis'ma*, 1988, 48, 144-146; Hunter, L.R. *Science*, 1991, 252, 73-79; Schuh, B.; Kanorsky, S.I.; Weis, A.; Hansh,T.W. *Opt. Comm.* 1993, 100, 451-455.

[35] Petrosyan, D.; Malakyan,Yu.P. *Phys. Rev. A*, 2004, 70, 023822.

[36] Budker, D.; Kimball, D.F.; Yashchuk, V.V.; Zolotorev, M. *Phys. Rev. A*, 2002, 65, 055403.

[37] Malakyan, Yu.P.; Rochester, S.M.; Budker, D.; Kimball, D.F.; Yashchuk, V.V. *Phys.Rev.A*, 2004, 69, 013817.

[38] Gawlik, W.; Krzemien, L.; Pustelny, S.; Sangla, D.; Zachorowski, J.; Graf, M.; Sushkov, A.O.; Budker, D. *Appl. Phys.Lett.* 2006, 88, 131108-1-3.

[39] Novikova, I.; Matsko, A.B.; Sautenkov, V.A.; Velichansky, V. L.; Welch, G.R.; Scully, M.O. *Optics Letters*, 2000, 25, 1651-1653.

[40] Drampyan, R.; Greentree, A.D.; Durrant, A. *Proceedings of National Conference "Laser Physics 2002"*, Gitutjun Publ. 2003, pp. 16-19.

[41] Renzoni, F.; Zimmermann, C.; Verkerk, P.; Arimondo, E. *J. Opt. B: Quantum Semiclass. Opt.* 2001, 3, S7-S14.

[42] Theobald , G.; Dimarcq, N.; Giordano,V.; Cerez, P. *Opt. Comm.* 1989, 76, 256-262.

[43] 43. Alzetta, G.; Cartaleva, S.; Dancheva, Y.; Andreeva, Ch.; Gozzini, S.; Botti, L.; Rossi, A. *J. Opt. B: Quantum Semiclass. Opt.* 2001, 3, 181-188.

[44] Smith, D.A.; Hughes, I.G. *Am. J. Phys.* 2004, 72, 631- 637.

[45] Demtroder, W. *Laser Spectroscopy, Basic Concepts and Instrumentation,* 2nd edn. (Berlin, Springer), 1998.

[46] MacAdaam, K.B.; Steinbach, A.; Wieman, C. *Am. J. Phys.* 1992, 60, 1098-1111.

[47] Corwin, K.L.; Lu, Z.-T.; Hand, C.F.; Epstein, R.J.; Wieman, C.E. *Appl. Opt.* 1998, 37, 3295-3298.

[48] Pearman, C. P.; Adams, C. S.; Cox, S. G.; Griffin, P. F.; Smith, D. A.; Hughes, I. G. *J. Phys. B: At. Mol. Opt. Phys.* 2002, 35, 5141-5151.

[49] Muller, R.; Weis, A. *Appl. Phys. B*, 1998, 66, 323-326.
[50] Hashimoto, M.; Ohtsu, M. *IEEE, J. of Quant. Electron.* 1987, QE-23, 446-451; Knappe, S.; Shah, V.; Schwindt, P.D.D.; Hollberg, L.; Kitching, J. *Appl. Phys. Lett.* 2004, 85, 1460-1462.
[51] Metcalf, H.J.; Van der Straten, P. *Laser Cooling and Trapping* (Berlin: Springer), 1999.

In: New Trends in Quantum Coherence...
Editor: Rafael Drampyan, pp. 243-261
ISBN: 978-1-60741-025-6

Chapter 9

PHOTOEXCITATION AND PHOTOREGISTRATION OF ATOMIC MOTION ON THE SURFACES OF SOLID MATERIALS

***T.A. Vartanyan*[1]*, S.G. Przhibelskii*[2] *and V.V. Khromov*[3]**
Center for Information Optical Technologies, St. Petersburg State University of Information Technologies, Mechanics and Optics, Kronverkskiy pr. 49, St. Petersburg 197101, Russian Federation

Abstract

This review is devoted to the photostimulated phenomena in the model systems consisting of individual atoms interacting with surfaces of solid materials. Original results concerning photoinduced desorption and photostimulated surface diffusion of alkali atoms on the surface of wide band gap dielectric materials are presented. The energy is deposited first into electronic degrees of freedom via optical excitation. Then, the routes of the energy conversion are tracked by the optical means. The energy partition between the foreign atom and the surface as well as the accompanied mass transport from and through the surface is described. Unparalleled sensitivity of optical tools is essential in getting detailed information on atomic motion on surfaces.

I. Introduction

Atom-surface interactions are in the focus of current research. They equally attract theorists, experimenters and practitioners. From the theoretical point of view this kind of interactions are challenging because there are no rigorously solvable model that is able to represent even the basic features of an atom interacting with a solid surface. Due to the lack of conventional model every theoretical work starts with selection of properties of the solid surface that are to be considered. Experimenters, on the other hand, are encouraged by vast opportunities to

[1] E-mail address: tigran@vartanyan.com
[2] E-mail address: przhi@mail.ru
[3] E-mail address: khvv@gtn.ru

explore odd effects that are easily observable in this field. Finally, modern technologies, in particular, nanotechnology rely heavily on the available knowledge of atom-surface interaction. Even more, practitioners need tools to modify and control the surface processes. This is the point where light, or broadly speaking, electromagnetic fields can contribute considerably. Irradiation of the surface not only sheds light on the tiny changes that take place on the surface without illumination but it is also able to invert them or stimulate completely new behavior.

This chapter is based mainly to the results obtained in the group of the late professor A. M. Bonch-Bruevich at the S. I. Vavilov State Optical Institute and later on at the St. Petersburg State University of Information Technologies, Mechanics and Optics. Earlier accounts of the results of this group were published in [Bonch 1995, Bonch 1998b]

II. Spectroscopy of the Adsorbed Atoms

Let us start with a basic question of the electronic structure of an atom adsorbed on a solid surface. Such atoms are commonly called adatoms. In some cases the structure of the electronic levels changes so dramatically that the adatom looses its similarity to the free atom at all. It is known since the beginning of the last century [Taylor 1933] that the alkali atoms being adsorbed on the surface of refractory metals are deprived of their valence electrons and reside on the surface as positive ions. Hence, optical properties of adatoms at this extreme have no connection to the optical properties of isolated atoms.

The opposite case of very weak interactions is characteristic for alkali metal/cryogenic solid rare-gas surface. Optical absorption spectra consist of several bands with rich structure resembling that of few-atom molecules or impurities in solids [Boatz 1994].

In this chapter we concentrate on the intermediate case of rather strong interactions between adatoms and solid surfaces that nevertheless preserve adatoms identity [Fröhlich 1946]. Let us consider an alkali atom, in particular, sodium atom on the surface of sapphire. Optical absorption of this system is dominated by a broad structureless band (Fig.1) somewhat shifted to the red from the position of the characteristic yellow doublet of free sodium at 589 nm [Bonch 1985]. The absorption cross section of the atom at the maximum was found to be equal to 3×10^{-16} cm^2 while the integral of the absorption cross section over the wavelength corresponds nicely to the oscillator strength of the corresponding transition in the free atom.

These results were obtained with the use of standard spectrometers "Perkin - Elmer" and "Carry -17" and a specially designed spectroscopic cell. To enhance the adatom absorption 17 polished sapphire plates were introduced in the cell that was evacuated and filled with sodium. The cell design allows for independent control of the sapphire plates temperature and the sodium vapor density. The last was achieved through the variation of the temperature of the metal drop placed in an appendix attached to the cell. The standard experimental run consists of absorption spectra registration at the fixed cell temperature and variable temperature of the appendix. Absorption due to the free sodium vapor was measured in a separate cell with the same optical length and subtracted.

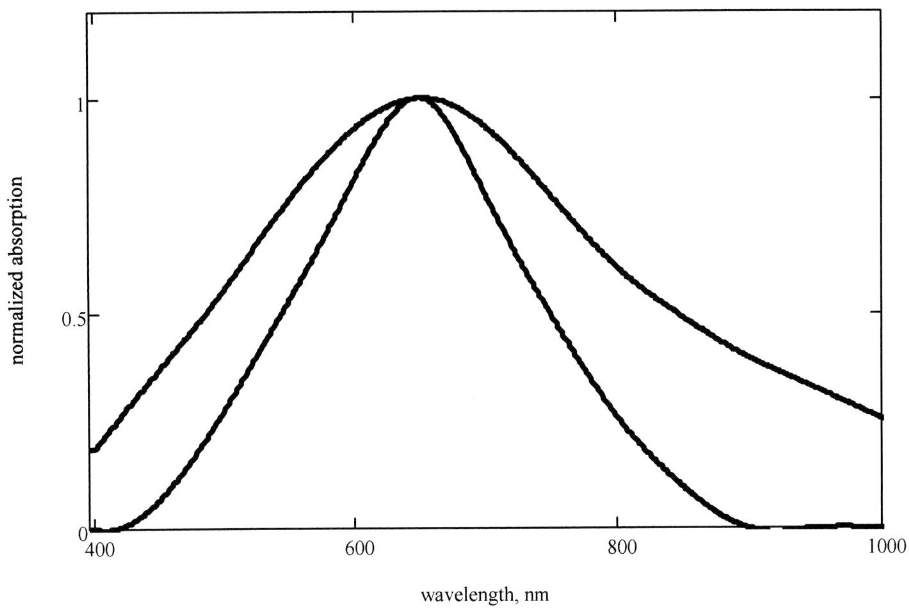

Figure 1. Absorption spectra of sodium adatoms on the surface of sapphire kept at 470 K (narrow curve) and 670 K (broader curve). The spectra were normalized to facilitate the comparison of the broadening due to the temperature rise.

These measurements proved that sodium atoms are adsorbed on the sapphire surface in the atomic form and preserve their identity.

In spite of importance of such measurements they are still very rare. Different indirect techniques were proposed to overcome the difficulties of adatom optical spectroscopy at low surface number densities. Silver ions Ag^+ were deposited on the (110) plane of a single crystalline AgCl surface with almost zero kinetic energy [Latyshev 2003]. Upon soft landing on the surface the ions were neutralized. Absorption spectrum of silver adatoms on silver chloride single crystal was found to be a broad structureless band centered at 652 nm with a full width at half maximum (FWHM) of about 1800 cm^{-1}. This band cannot be ascribed to the shifted and broadened transition in free silver atom as the lowest energy transition from the ground state to the first excited corresponds to the wavelength of 338 nm. Instead, in accordance with the complicated registration procedure, it is ascribed to the transition of the electron from its ground state in the silver adatom to the conduction band of the substrate.

Cavity ringdown spectroscopy was used to study optical absorption spectra of gold atoms adsorbed on amorphous silica [Antonietti 2005]. The samples were prepared by depositing 5×10^{12} cm^{-2} gold atoms onto a SiO_2 surface. Atoms cations were, first, mass selected, then, neutralized and directed onto the surface. Ringdown times of a high Q-factor optical cavity before and after the gold deposition were used to deduce the adatom absorption. Four rather narrow (FWHM less when 400 cm^{-1}) absorption bands at 652, 636, 500 and 450 nm were observed. Again, these bands are too far away from the first allowed transition in free gold atom at 242.8 nm to be directly traced to it. According to the density functional calculations, observed optical transitions can be attributed to the single gold atoms bound to the defect sites on the silica surface. Theory also suggests that the defect sites concentration is large enough to accommodate 5×10^{12} cm^{-2} of gold atoms with binding energy in the range of 0.9 to 3.2 eV.

The adsorption energy on the regular surface is much smaller, thus enabling rapid diffusing of adsorbed species.

III. Photodesorption Process

Photodesorption process is a very promising tool for direct laser writing of complicated patterns on surfaces. Although an ordinary heating of the substrate by absorbed radiation can cause thermal desorption of adsorbed species from the illuminated area, nonthermal mechanisms of desorption attracts much more attention. The reasons for that are higher spatial resolution, higher selectivity as well as lower inertia of nonthermal process as compared to thermal ones.

Desorption induced by electronic transitions was a very active research field at the end of the last century. Eleven international conferences and workshops were organized since 1982 [DIET 1983, DIET 2005], the last being held in Berlin in 2007. They encompass the results of investigations of many different mechanisms of desorption. Leaving apart the possibility to excite electrons by electron impact, let us consider electronic transitions induced by absorption of photons. Photons can be absorbed either by the substrate or by the adsorbed species. In the last case electronic excitation of an adatom is transformed into its motion away from the surface.

One particular example of such a process is photodesorption of sodium atoms from sapphire. Sapphire is transparent in visible, while the adsorbed sodium atoms, as it was detailed in the previous section, possess a broad absorption band in this spectral range. Nonthermal nature of sodium photodesorption from sapphire was proved in [Abramova 1984]. More details of the process were reported in [Bonch 1990a]. Figure 2 shows the idea of the desorption experiment and the registration scheme. Optical methods provide unequal sensitivity that allows for sure registration of the atoms desorbed with different velocities in one laser shot of 10 ns duration.

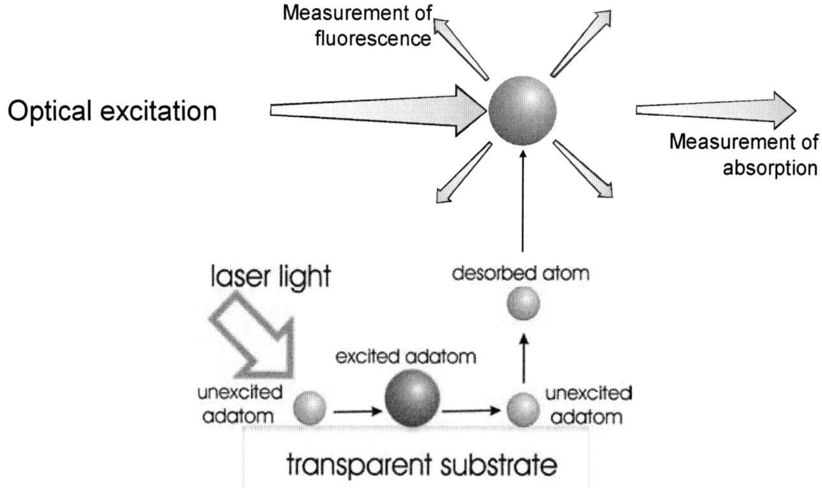

Figure 2. Registration of photodesorption via absorption or fluorescence measurements.

Fig. 3 plots the photodesorption spectrum and contrasts it with the absorption spectrum already given in Fig.1. Although the maxima of the two spectra almost coincide, quantum efficiency of desorption diminishes for larger wavelengths and drops to zero at 800 nm. This threshold corresponds to the photon energy of 1.55 eV. It is in excess of the binding energy of adatoms E_{ad}=0.7 eV that was deduced from the temperature dependence of absorption spectrum.

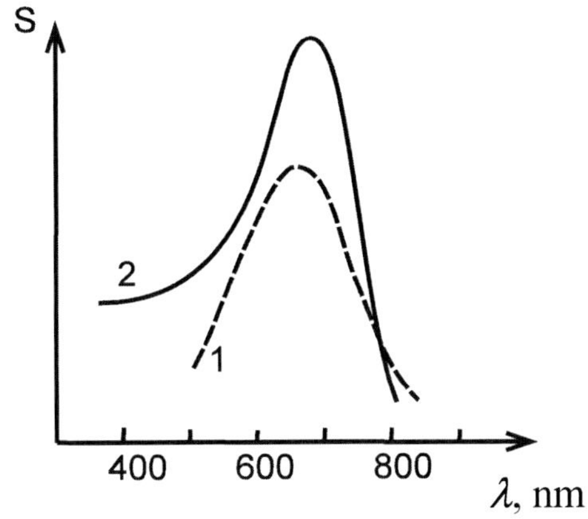

Figure 3. Absorption (1) and photodesorption (2) spectra of individual sodium atoms adsorbed on the sapphire surface.

Time-of-flight measurements provide additional important information about photodesorption process. Fig.4 plots time-of-flight spectra obtained under the action of two different wavelengths of excitation light.

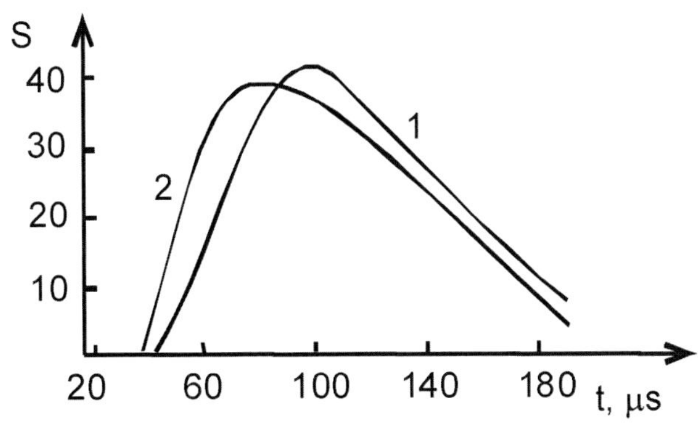

Figure 4. Time-of-flight spectra under the action of laser pulses at 530 nm (1) and 353 nm (2).

Although both spectra are broad and admit Maxwellian fits, as it is documented in Fig.5, nonthermal nature of the photodesorption process becomes obvious when one takes into account that the effective temperatures of these fits are independent of the laser fluence.

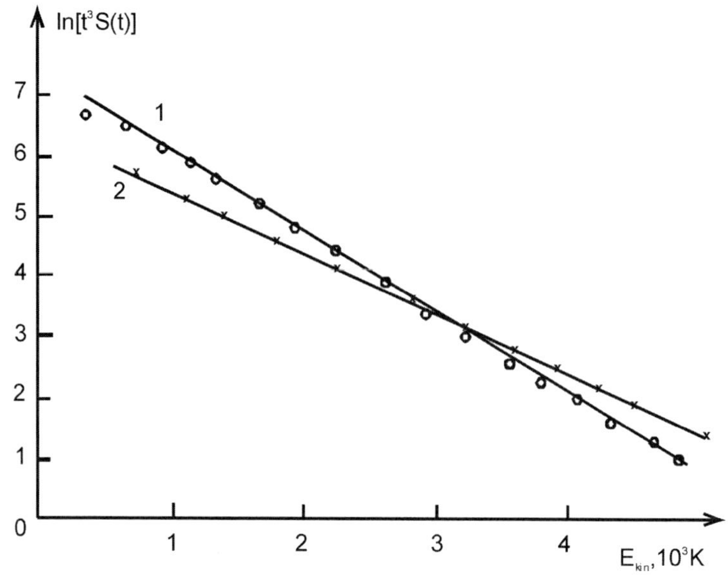

Figure 5. Same as Fig. 4 redisplayed to show the quality of Maxwellian fit.

IV. Interpretation of the Photodesorption Process

Contradictory observations presented in the preceding section may be rationalized by the following model. First, the photon is absorbed by the adatom and the adatom is transferred into an excited electronic state. Immediately after that, the multidimensional potential energy surface of the adsorption complex that comprises the adatom and the nearest nodes of the sapphire crystal lattice changes its shape. During oscillations about new equilibrium positions the adatom as well as its nearest neighbors gain the kinetic energy. After a time period of about several picoseconds the electronic excitation is quenched and the adatom is left in an unexcited electronic state. If its kinetic energy at this moment is large enough there is a chance for the atom to escape. According to the Figure 6 that is drawn approximately to the scale, the number of degrees of freedom that are involved in the process (z) is rather large. For this reason the energy partition between these degrees of freedom is likely to be stochastic.

Defining an effective temperature of the complex, T_{eff}, as the mean energy that falls to one degree of freedom share,

$$T_{eff} = T_s + \frac{2\hbar\omega}{kz}, \tag{1}$$

where T_s is the surface temperature, k is the Boltzmann constant, and $\hbar\omega$ is the energy of the photon, the probability to gain the energy E can be expressed as

$$f = \left(\frac{E}{kT_{eff}} \right)^{1/2} \exp\left(-\frac{E}{kT_{eff}} \right). \tag{2}$$

For $E > E_{ad}$ the desorption process becomes possible. Despite of many shortcomings, this model accounts for all experimental findings. It predicts Maxwellian velocity distribution of desorbed atoms with the effective temperature that depends on the energy of the photon. At the same time, the effective temperature is independent of the photon flux. According to Eq. (2), quantum efficiency of desorption for the blue light is larger than that for the red light as it was observed in the experiments.

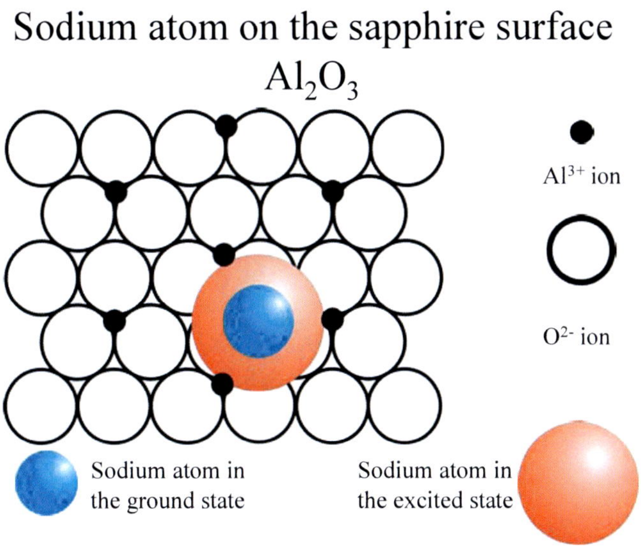

Figure 6. Sketch of sodium adatoms on sapphire surface.

V. Photodesorption from 1D Chains of Structural Defects

The cesium adsorption on sapphire provides a possibility to vary the surface number density of adatoms in a wider range as compared to the case of the sodium adsorption on sapphire. In particular, it is possible to reach so high surface concentrations that the interaction between adatoms becomes significant [Bonch 1997a, Bonch 1997b, Bonch 1998a].

An analysis of the experimental data shows that saturation is reached when the Cs atoms occupy only a small fraction of the surface of sapphire. On the other hand, lateral interactions among the atoms make a significant contribution to the energy of adsorption. This seeming contradiction is resolved by the fact that adsorption occurs at adsorption centers which form agglomerations. In general, these agglomerations can be one- or two-dimensional. The results of photodesorption experiments count in favor of the one-dimensional model [Bonch 1997a].

Photodesorption of cesium adatoms from sapphire was induced by pulses of a ruby laser. The wavelength of excitation, 694 nm, lies in an absorption band of the adatoms. To detect the desorbed particles a 2-cm-diameter probe beam resonant with the strong 894.3 nm absorption line of the free cesium atoms was directed parallel to the sapphire surface, almost touching it. In this geometry, the changes in the probe beam absorption were measured when the desorbed atoms crossed the beam. This made it possible to estimate the number of atoms and their average velocity based on their time of flight across the cross section of the detector beam. Note that the presence of a stronger, but constant in time, absorption by cesium atoms in the gas phase does not interfere with these measurements. The number of desorbed atoms was determined from the amplitude of the change in the intensity of the probe beam.

The experimentally measured dependence of the number of photodesorbed atoms on the substrate temperature for a fixed concentration of gas phase atoms is shown in Fig. 7.The photodesorption quantum yield at room temperature was 5×10^{-6}.

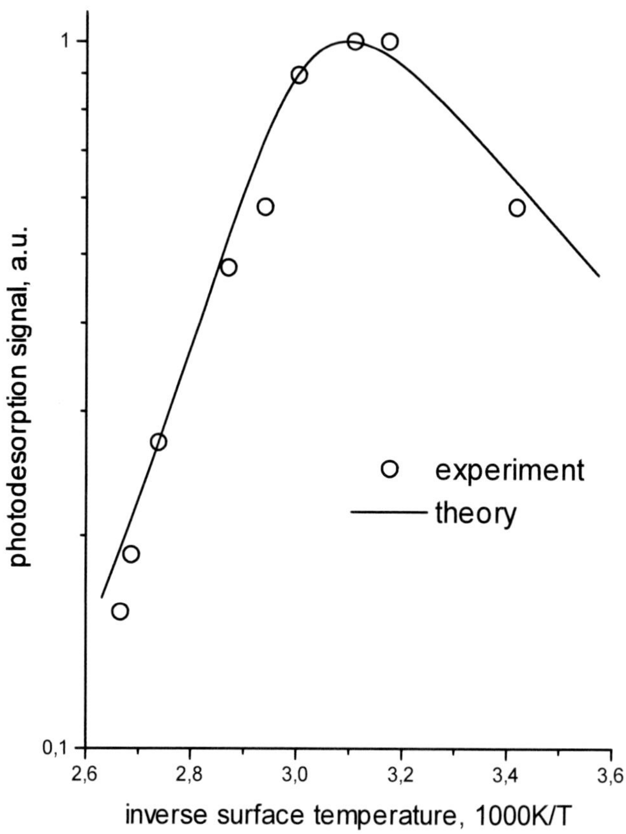

Figure 7. Temperature dependence of the photodesorption yield.

A linear dependence of the number of desorbed cesium atoms on the incident radiation power, which is typical of the photodesorption process described above, is observed for intensities below 10^4 W cm^{-2}. At higher intensities a pseudothermal desorption phenomenon is observed. More details of this process in the case of sodium are given in section VI. Pseudothermal desorption is characterized by a sudden rise in the number of desorbed atoms, which is soon replaced by saturation owing to complete desorption of adatoms during the time the laser pulse acts. The number of atoms desorbed in this regime is determined only by the surface concentration of adatoms, and may be used to measure the latter as a function of the substrate temperature. The measured dependence of the adatom surface concentration on the substrate temperature for a fixed concentration of gaseous phase atoms is shown in Fig. 8. At low temperatures, the surface concentration as a function of the temperature undergoes saturation, reaching a level on the order of 10^{13} cm^{-2}. Since this value of the surface concentration is substantially lower than that at the dense packing, 10^{15} cm^{-2}, it follows that adsorption does not take place over the entire surface, but only at adsorption centers whose density determines the maximum concentration of adatoms.

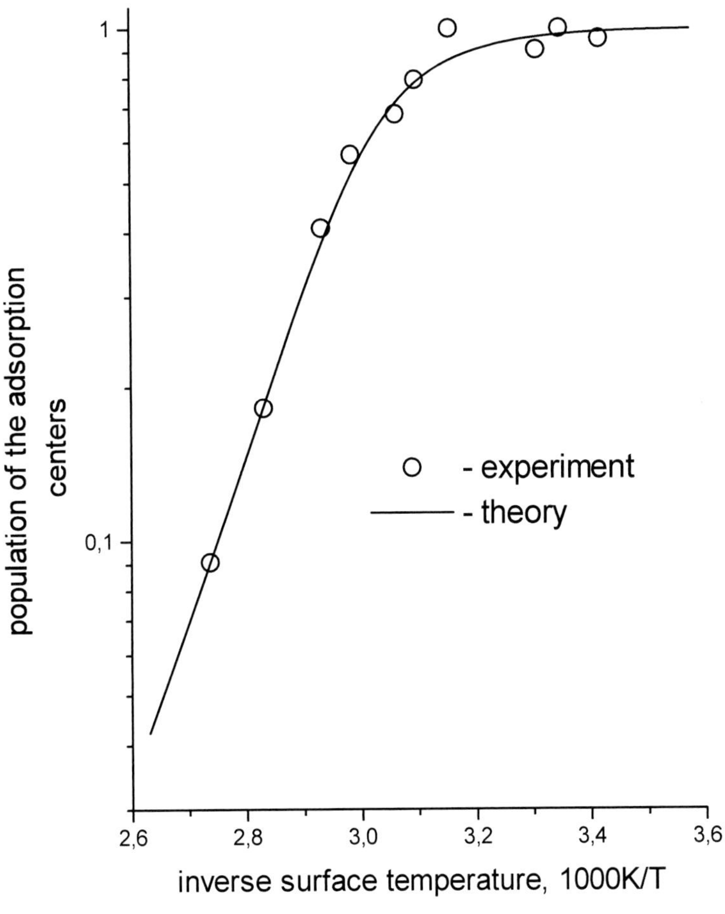

Figure 8. Temperature dependence of the pseudothermal desorption yield.

It seems natural to interpret the experimental data in the framework of Langmuir model. Let n_0 be the surface density of the adsorption centers. Then the surface density of adatoms is given by

$$n = \frac{n_0}{1 + \exp\left(\frac{E_{ad}}{kT_*} - \frac{E_{ad}}{kT}\right)}, \tag{3}$$

where T_* is determined by the density of the atoms in the gas phase. At $T = T_*$ half of the adsorption centers are filled. When temperature rises above T_*, the surface density of the adatoms falls off exponentially, while for $T < T_*$ it rapidly reaches its maximum value, n_0. The best agreement between Eq.(3) and the experimental data is obtained for T_*=336 K and E_{ad}=0.95 eV. Despite the qualitative agreement between the theoretical and experimental data on thermal desorption this estimate for E_{ad} is at variance with the data on photodesorption.

This disagreement between the point-center model and the experimental data may be solved by considering models of adsorption in reduced dimension. It is plausible to assume that one dimensional adsorption is realized in the form of chains of adatoms at the steps that always exist on a real crystalline surface. An increase of the adsorption energy at steps is well documented. Taking into account lateral interactions between adatoms delocalized within one-dimensional adsorption sites at the steps of a crystalline surface, it is possible to provide an adequate description of the experimental findings.

Adsorption on linear chains of centers can be described theoretically in terms of a one-dimensional lattice gas model. Then the lateral interaction is determined by a single parameter, the nearest-neighbor interaction energy E_i. The thermodynamic parameters of this type of model can be calculated exactly using the methods employed in the theoretical analysis of the Ising model [Huang 1963]. Two major differences between our model and the conventional one are the need to account for the material contact between the lattice system and the volume gas phase as well as the nonzero value of E_{ad}. The final result reads as

$$n = \frac{n_0}{2}\left[1 + \frac{\sinh\left(\frac{E_{ad}+E_i}{2T} - \frac{E_{ad}+E_i}{2T_*}\right)}{\sqrt{\sinh^2\left(\frac{E_{ad}+E_i}{2T} - \frac{E_{ad}+E_i}{2T_*}\right) + \exp\left(-\frac{E_i}{T}\right)}}\right]. \tag{4}$$

Obviously, for E_i=0 Eq.(4) reduces to Eq.(3). From the best fit of the data obtained from the pseudothermal desorption experiments one can get the following relation between the parameters in Eq.(4)

$$E_{ad} = E_{ad}^0 \exp\left(-\frac{E_i}{2T_*}\right) - E_i, \tag{5}$$

where E_{ad}^0 is the estimate of the adsorption energy derived from Eq.(4) by disregarding lateral interactions, i. e. for E_i=0. All the theoretical curves plotted according Eq.(4) and Eq.(5) for different E_i fall into the error bars of experimental data since they have the same slope at $T = T_*$. To resolve this uncertainty one has to consider the nonthermal photodesorption data presented in Fig. 7.

It is worth mentioning that in this model the adsorption energy depends on the position of an adatom in the chain: for single adatoms it is equal to E_a; for adatoms at the end of the chain it is equal to E_a+E_i; finally, for adatoms in the middle of the chain it is equal to E_a+2E_i. In the framework of the theoretical model the share of adatoms in different positions may be computed exactly. Nevertheless, the corresponding result will not be presented here, because in the temperature interval of our concern adatoms of the last sort dominate.

Now we are in a position to account for the measured value of the photodesorption quantum yield at the extreme low end of the temperature interval T_l=293 K. Since f=5×10^{-6}, according to Eq.(1) one finds that

$$\frac{E_{ad} + 2E_i}{T_l + \dfrac{2\hbar\omega}{z}} = 13.5. \tag{6}$$

Varying only one parameter E_i with conditions (5) and (6) held one gets a very good agreement between the theory and the experimental data on photodesorption Fig. 7. The best fit corresponds to E_i=26 meV. Hence, according to Eq.(5) and Eq.(6) one finds E_i=0.58 eV and z=165.

We emphasize that the temperature dependence of surface coverage taken in a limited temperature interval may lead to confusion if lateral interactions are important. If the condition (5) is held theoretical curves plotted in accordance with Eq.(4) fit the experimental data in a wide range of E_i, including the negative values that correspond to repulsion of neighbor atoms. It was the experimental data on nonthermal photodesorption that led us to the conclusion that adatoms in the chain attract each other and the energy of attraction is 26 meV. The adsorption energy for an isolated adatom was found to be equal to 0.58 eV and that for an adatom in the middle of the chain was found to be equal to 0.63 eV. If one disregards lateral interactions and uses the conventional Langmuir formula, an incorrect value of adsorption energy E_a=0.95 eV will be obtained.

VI. Nonlinear Photodesorption Process

With the very sensitive optical registration scheme it was possible to check the photodesorption process for linearity over five orders of magnitude of the incident light intensity from as low intensity as 1 W cm^{-2} to 10^5 W cm^{-2} (for sodium). Nevertheless, for still larger intensities the desorption rate grows exponentially (Fig.10). Simultaneously, time of flight spectra of desorbed atoms become intensity dependent, effective temperature of the Maxwellian fit being proportional to the laser intensity [Bonch 1999]. This pseudothermal photodesorption process is due to the low rate of the energy exchange between an adatom and the sapphire surface.

The exponential rise of the photodesorption flux ends up at the intensity of 10^7 W cm^{-2}. This is due to the depletion of the adatoms at the illuminated site on the surface in the course of a single laser shot.

Another kind of nonlinearity was observed for illumination with the wavelengths below the photodesorption threshold. In this case the energy one photon is inadequate to desorb the adatom. Nevertheless, it is suffices to transport the adatom from the adsorption center with large binding energy to another place on the surface where the energy of adsorption is lower. Then, the absorption of one another photon may lead to the liberation of the adatom. This process was indeed observed in the system of sodium adatoms on sapphire illuminated with the laser light at 1.06 μm.

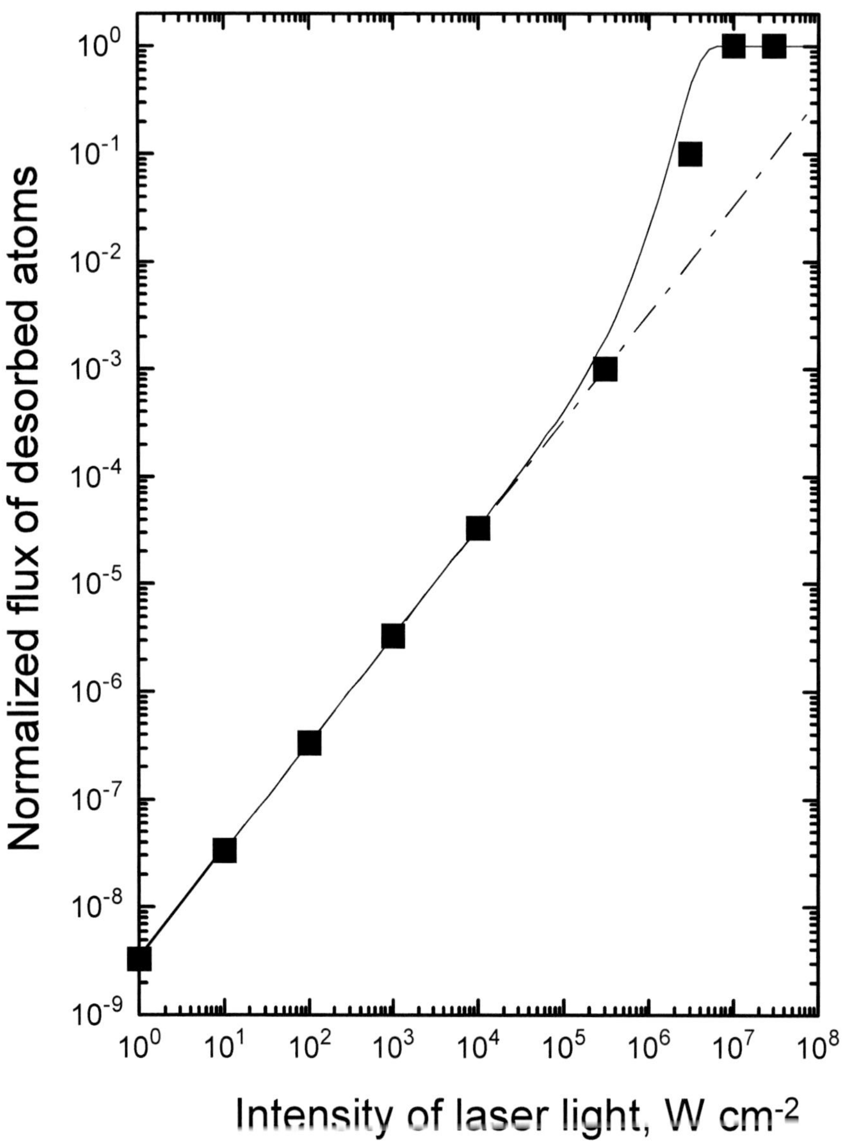

Figure 9. Squares represent the measured flux of sodium atoms desorbed from sapphire surface under the action of laser pulses of 10 ns duration at 530 nm. The solid line corresponds to the theory developed in [Bonch 1999].

In accord with the model described above, the photodesorption flux is proportional to the square of the laser intensity in the range from 10^7 W cm^{-2} to 10^8 W cm^{-2} (Fig. 10).

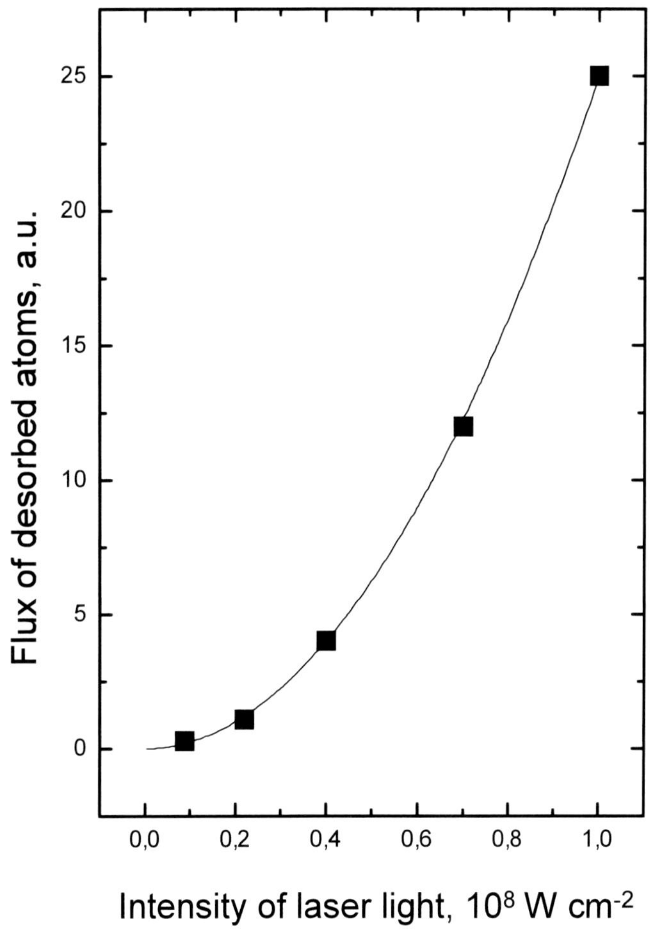

Figure 10. Squares represent the measured flux of sodium atoms desorbed from sapphire surface under the action of laser pulses of 10 ns duration at 1.06 μm. The line demonstrates the quadratic dependence of the desorption yield on the laser intensity.

VII. Photodiffusion Process

According to Eq.(2) the photodesorption probability is low. Most of the adatoms do not gain sufficient energy to leave the surface. At the same time the energy barrier for surface diffusion is 4 to 6 time less than the adsorption energy (Fig.11). Hence, surface diffusion is the major effect of illumination. When the adsorbate is inhomogeneously distributed over the surface, homogenous illumination of the surface can contribute considerably to the diffusion coefficient. On the other hand, inhomogeneous illumination of homogeneous adsorbate system can add to the diffusion process a new perspective. In this case illumination serves as a tool to redistribute adatoms according to the illumination pattern.

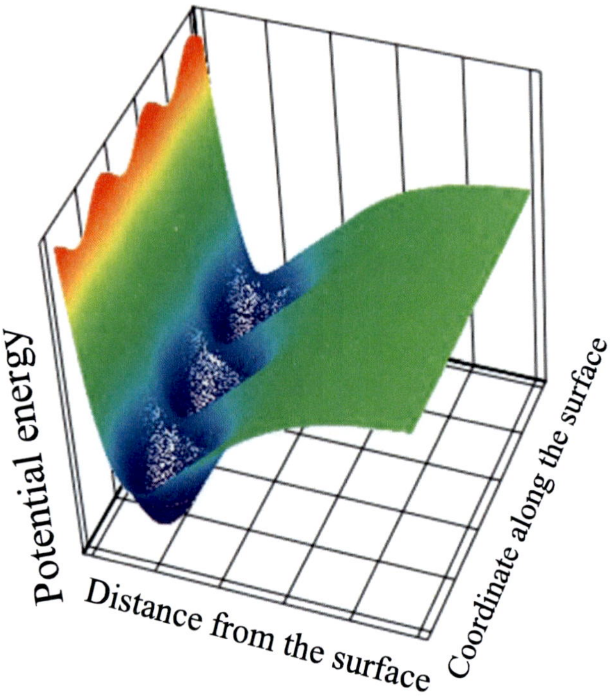

Figure 11. Sketch of the potential energy surface of an atom at the substrate surface. Only two coordinates are depicted: distance from the substrate surface and one coordinate along the substrate surface. Potential barriers for surface diffusion are much less than the adsorption energy.

As the object of study, we used a system of cesium atoms adsorbed on the surface of single-crystal sapphire. The advantages of this system are the chemical inertness of sapphire and its transparency in a wide spectral range, covering the entire visible and near-IR regions, in which the adsorbed cesium atoms exhibit strong optical absorption. This allows one, using convenient laser sources, to selectively excite and diagnose the adsorbed cesium atoms, leaving the substrate unexcited.

Saturated vapors of cesium atoms were placed into a sealed glass cell with sapphire windows. The volume density of the atoms at room temperature was 10^{10} cm^{-3}. The experiments were performed under conditions of dynamic equilibrium between the volume phase and the phase adsorbed on the surface. The energy of adsorption of cesium atoms on the sapphire surface being about 0.6 eV, the surface number density of cesium adatoms at room temperature, is n_0=10^{13}cm^{-2}.

Thus, the surface density of the adsorbed atoms appears to be fairly high, though still much lower than a monolayer. Variations of the surface density of the adsorbed atoms induced by high-power radiation were detected by measuring the transmission of a relatively weak probe light of a cw semiconductor laser. Using the value of the absorption cross section of the adsorbed atom σ=3×10^{-16}cm^2, one finds that the absorption in the adsorbed layer makes up σn_0= 3×10^{-3} of the total intensity of the light passing through the layer. Thus, absorption of this kind can be reliably measured using low-noise high-intensity diode lasers.

Among the important advantages of the chosen system is a relatively low quantum yield of the photodesorption process (10^{-5}). This allowed us to hope that the main mechanism for light-induced changes in the surface density of the atoms would be the photoinduced diffusion rather than photodesorption. As shown below, these expectations were fulfilled. Since the excitation of an adsorption system by high-power optical radiation induces both the photodesorption of atoms from the surface and their photodiffusion over the surface and, in addition, is accompanied by dark desorption and diffusion, we performed experiments of two types [Bonch 2003, Bonch 2004]

In the experiments of the first type, we studied the processes of photodesorption and dark desorption. For this purpose, the cesium atoms adsorbed on the sapphire surface were desorbed by a single pulse of a ruby laser at the wavelength of 694 nm. The diameter of the irradiated spot varied from 1 to 4 mm. After the end of the strong desorbing pulse, the surface density of the adsorbed atoms gradually recovered. The kinetics of the recovery of the surface density was detected by measuring the intensity of a diode laser beam transmitted through the irradiated area of the sapphire window of the cell. The intensity of the transmitted light at the wavelength of 840 nm was measured by a photodiode and was monitored by an oscilloscope. At room temperature, the characteristic time of recovery of the surface density was 25 s. It was verified that this time did not depend on the radius of the irradiated spot within the limits indicated above. This allowed us to assign it exclusively to the process of deposition of atoms from the gas phase, rather than to their surface diffusion from dark regions, because in the latter case the recovery time would be dependent on the radius of the irradiated spot. From these experimental data, we determined the rate of thermal desorption to be 0.04 s^{-1}.

The second type of experiments was devoted to the photostimulated diffusion itself. A cw beam of an Ar^{+} laser at the wavelength 514.5 nm and a power of 1 W served as a pump. The probe beam of a diode laser at the wavelength of 804 nm focused to the spot size 20×30 μm^2 was scanned through the sapphire surface in such a way that it could repeatedly pass in sequence the spot illuminated by the pump laser and the dark regions of the surface. The scanning was implemented using a lightweight mirror glued to the diaphragm of a loudspeaker powered by an audio-frequency generator. The scanning range of the probe beam exceeded by approximately an order of magnitude the diameter of the pump beam, which varied within the range 30 to 130 μm. After passing through the cell, the probe beam was detected by a photodiode and a lock-in amplifier. The sensitivity of the described setup was 3×10^{-4} as regard to the intensity of the probe laser. The time resolution was 0.1 s.

Because of the different values of absorption in the regions with high and low surface densities of adsorbed atoms, the intensity of the transmitted beam was modulated at the probe beam scanning frequency. Under the conditions described above, when the changes in the transmittance are concentrated within a region that is small compared to the whole scanning length, the amplitude of the first modulation harmonic in the transmitted beam, detected in the experiment, is proportional to the integral of the variation of the surface density of the adsorbed atoms along the beam scanning path. In the experiment, we measured the kinetics of the changes in the surface density of atoms after the argon-ion laser beam was rapidly switched on and off by a mechanical shutter. It was found that, as in the first experiments, for large diameter of the pump beam spot, 500 μm, this kinetics was exponential, with a rise and decay time of 25 s. It was verified that the amplitudes of the observed signals vary linearly with the pump beam power, while the adatom surface density was reduced by 20% of the equilibrium value at most. For smaller diameters of the pump beam, great difference in kinetics was observed. In particular, the steady-state surface density of adatoms was established much faster than in the case of larger spot sizes. Since the characteristic time of the steady state settling did not depend on the beam power but decreased quadratically with a decrease in the diameter of the pump beam from 130 to 30 μm, we concluded that the time of transition to the steady state was governed by the rate of dark diffusion of the adatoms over the surface. In addition, it can be concluded that the main contribution to the decrease in the surface density of atoms is made by photodiffusion rather than by photodesorption. Indeed, if the decrease in the surface density were related to the photodesorption, such a decrease of the density would occur until the surface was totally depleted. Moreover, recovery of the steady-state value of the surface density after the pump radiation has been turned off could occur only from the gas phase with a characteristic time of 25 s.

VIII. Superelastic Surface Scattering

It is very well known that an excited atom, upon a collision with a solid surface, is transferred into its ground state with the probability very near to unity. On the other hand, little is known about partitioning of the electronic excitation energy lost in this process. In an the experiments with cesium atoms colliding with sapphire surface it was shown that an appreciable part of this energy goes into kinetic energy of a deexcited atom leaving the surface [Bonch 2000, Bonch 2005].

IX. Influence of Surface Vibrations on the Optical Properties of Nearby Atoms

With the current interest in the matter wave optics and demonstration of an atom chip [Folman 2002] it becomes especially important to study the modifications of electronic structure of an atom approaching a solid surface. A host of information was obtained by selective reflection spectroscopy [Bloch 2004].

In the course of investigations of spectral line broadening due to atom-wall collisions, evidence was found for an important role playing by surface vibrations [Bonch 2006, Khromov 2007]. The shape of the luminescence spectrum of Cs atoms flying near a sapphire surface is explored experimentally and described phenomenologically for the detuning range of 5 to 50 cm^{-1} on both sides of D2 line. The experimentally determined abnormally high intensity of the antistatic wing is explained by the effect of phase relaxation of the atomic transition, caused by fluctuations of the electric field of oscillating ions of the sapphire crystal lattice.

X. Applications

Recent progress in nanotechnology is based mostly on the "upside down" paradigm. Notwithstanding with obvious achievements, this approach is inevitably connected with gross energy consumption and material waste. An attractive alternative is "bottom-up" manufacturing. In this case the quantity of raw material required could be considerably reduced.

Photoexcitation of atomic motion on the surface of solid materials described above can serve as a reliable tool for manipulation of matter at the nanoscale and forms a firm basis for the "bottom-up" paradigm.

It is common to describe the growth of islands on a dielectric substrate in the framework of Volmer-Weber model. The atoms, first, get trapped on the surface, then, migrate over it, and, finally, develop nucleus of the future island. An elementary opportunity to influence the nucleation stage of the island film deposition is to change the surface number density of the atoms adsorbed on the surface. Laser induced desorption provides a versatile tool for reduction of surface number density of adsorbed atoms at a predetermined pattern on the surface. Experimental observation of the process was reported in [Abramova 1984]. It was found that a metal film would not form on the illuminated part of the surface. The details of the subsequent study of sodium photodesorption from sapphire were presented in the part II. From the point of view of applications, it is important to mention that the metal film formation is a threshold effect that depends on both the rate of deposition and the laser intensity. Hence, a complicated illumination pattern is reproduced in the pattern of metal deposits with enhanced contrast: only the darkest parts of the surface are covered with metal. This opens up the perspective of direct laser writing techniques with subwavelength resolution. Due to large difference in the desorption probabilities from the bare substrate and metal nanoparticles [Abramova 1984, Hoheisel 1988, Bonch 1990b] multiple exposure seems also feasible, as the subsequent illumination does not harm the already formed metal structures.

The nature of the dual effect of enhanced adsorption upon illumination that was observed in [Abramova 1984] in the case of aluminum is poorly understood. A recent paper [Afanasiev 2007] reports on the observation of the laser-induced adsorption of Rb atoms on the surface of YAG crystal. Considerable part of the atoms approaching the surface was in the first excited state $5^2P_{3/2}$ due to the laser illumination at the wavelength of 780 nm. The authors believe that the loading of Rb atoms into a surface potential well occurs through the loss of their kinetic energy in the course of inelastic collisions of two excited atoms in front of the surface. Although this explanation seems questionable because of extremely small probability of the mentioned process the effect itself is so attractive for direct laser writing that it is worth farther examination. Even more important observation of [Afanasiev 2007] that needs explanation is that under illumination Rb particles do not evaporate but rather stabilize at the temperatures at which they are expected to quickly evaporate through the ordinary thermal mechanism.

Still another opportunity is to control the adsorption potential or the sticking probability of atoms impinging on the surface via illumination.

One more possibility is to employ photoinduced diffusion to redistribute the atoms over the surface avoiding any loss of material delivered at the surface.

Conclusion

In this review we tried to argue that several options in surface study and modification are still underestimated by research community. Diode laser optical spectroscopy of adatoms may be used to study surface process at very low coverage. It is free from harmful charging effect that restricts applications of established surface science techniques to the nonconducting surfaces. Nonthermal processes induced by optical radiation provide a new dimension to the emerging bottom-up techniques of creating surface micro and nano structures.

This work was supported by Federal Agency for Science and Innovations of Russian Federation, Russian Foundation for Basic Research (RFBR 08-02-00695, 07-02-00591) and the grant of the President of Russian Federation for support of the Leading Scientific Schools NSh-3398.2008.2.

References

[Abramova 1984] Abramova I.N.; Aleksandrov E.B.; Bonch-Bruevich A.M.; Khromov V.V. *JETP Letters* 1984, 39, 203-205.

[Afanasiev 2007] Afanasiev, A.E.; Melentiev, P.N.; Balykin, V.I.; *JETP Letters*, 2007, 86, 172-177.

[Antonietti 2005] Antonietti, J.-M.; Michalski, M.; Heiz, U.; Jones, H.; Lim, K.H.; Roesch, N.; Vitto, A.D.; Pacchioni G, *Phys. Rev. Lett.* 2005, 94, 213402.

[Bloch 2004]. Bloch D.; Ducloy M. (2004) *Atom-Wall Interaction* http://arxiv.org/ftp/physics/papers/0503/0503146.pdf

[Boatz 1994] Boatz, J.A.; Fajardo, M.E; *J. Chem. Phys*. 1994, 101, 3472-3487.

[Bonch 1985] Bonch-Bruevich A.M.; Maksimov Yu.N.; Khromov V.V.; *Opt. Spectr.* 58, 1985, 854-856.

[Bonch 1990a] Bonch-Bruevich, A.M.; Vartanyan, T.A.; Gorlanov, A.V.; Maksimov Yu.N.; Przhibel'skii S.G.; Khromov. V.V., Sov. Phys. JETP, 1990, 70, 604 - 608.

[Bonch 1990b] Bonch-Bruevich, A.M.; Vartanyan, T.A.; Maksimov, Yu.N.; Przhibelskii, S.G.; Khromov, V.V. *Sov. Phys. JETP*, 1990, 70, pp.993 - 996.

[Bonch 1995] Bonch-Bruevich, A.M.; Vartanyan, T.A.; Przhibelskii, S.G.; Khromov. V.V. In: *Methods of Surface Treatment and Modification: (Advanced Laser Technologies'94 International Conference)*; Pustovoy, V.; Ed., Proceedings SPIE, 1995; Vol. 2498. pp 11-18.

[Bonch 1997a] Bonch-Bruevich, A.M.; Vartanyan, T.A.; Maksimov, Yu.N.; Przhibel'skii, S.G.; Khromov, V.V. In *Nonresonant Laser-Matter Interaction'96 International Conference*; Libenson, M., Ed., Proceedings SPIE, 1997; Vol. 3093, pp. 100 – 107.

[Bonch 1997b] Bonch-Bruevich A.M.; Vartanyan, T.A.; Maksimov, Yu.N.; Przhibel'skii, S.G.; Khromov, V.V. *J. Exp. Theor. Phys*. 1997, 85, 200-204.

[Bonch 1998a] Bonch-Bruevich, A.M.; Vartanyan, T.A.; Maksimov, Yu.N.; Przhibel'skii, S.G.; Khromov, V.V. *Surf. Rev. Lett.*, 1998, 5, 331 – 335.

[Bonch 1998b] Bonch-Bruevich, A.M.; Vartanyan, T.A.; Przhibelskii, S.G.; Khromov. V.V. In: *ALT'97 International Conference on Laser Surface Processing*; Pustovoy, V.; Ed., Proceedings SPIE, 1998; Vol. 3404. pp 187-191.

[Bonch 1999] Bonch-Bruevich, A.M.; Khromov, V.V.; Przhibel'skii, S.G.; Vartanyan, T.A. *Bull. Russian Acad. Sciences, Physics,* 1999, 63, 801 - 809.

[Bonch 2000] Bonch-Bruevich, A.M.; Khromov. V.V.; Przhibelskii, S.G.; Smirnov V.N.; Vartanyan, T.A. *Opt. Spectr.* 2000, 89, pp. 487 – 489.

[Bonch 2003] Bonch-Bruevich, A.M.; Vartanyan, T.A.; Przhibelskii, S.G.; Khromov.

V.V. *Opt. Spectr.* 2003, 95, 777 – 784.

[Bonch 2004] Bonch-Bruevich, A.M.; Vartanyan, T.A.; Przhibelskii, S.G.; Khromov. V.V. *J.Opt. Tech.* 2004, 71, 366 - 369.

[Bonch 2005] Bonch-Bruevich, A.M.; Vartanyan, T.A.; Przhibelskii, S.G.; Smirnov V.N.; Khromov. V.V. *J. Exp. Theor.Phys.* 2005, 100, pp. 998 – 1004.

[Bonch 2006] Bonch-Bruevich, A.M.; Vartanyan, T.A.; Leonov N.B.; Przhibelskii, S.G.; Smirnov V.N.; Khromov. V.V. *Opt. Spectr.* 2006, 100, pp. 149–151.

[DIET 1983] *Desorption Induced by Electronic Transitions, DIET-I*, Tolk, N.H.; Traum, M.M.; Tully, J.C.; Madey, T.E. Eds. 1983. Springer, Berlin.

[DIET 2005] *Proc. Tenth Intern. Workshop on Desorption Induced by Electronic Transitions, Surf. Sci.*, 2005, 593, Iss. 1-3.

[Folman 2002] Folman, R.; Krüger, P.; Schmiedmayer, J.; Denschlag, J.; Henkel, C. *Adv. Atom. Mol. Opt. Phys.* 2002, 48, pp. 263 – 356.

[Fröhlich 1946] Fröhlich, H.; Sack, R.A. *Proc. Phys.Soc.* 1946, 59, 30 – 33.

[Hoheisel 1988] Hoheisel, W.; Jungmann, K.; Vollmer, M.; Weidenauer, R.; Träger, F. Phys. Rev. Lett. 1988, 60, pp. 1649 – 1652.

[Huang 1963] Huang K. *Statistical Mechanics*, Wiley, New York, 1963.

[Khromov 2007] Khromov. V.V.; Leonov N.B.; Przhibelskii, S.G.; Smirnov, V.N.; Vaschenko, E.V.; Vartanyan, T.A. In: *ICONO* 2007: *Nonlinear Laser Spectroscopy and High-Precision Measurements; and Fundamentals of Laser Chemistry and Biophotonics*, Tikhomirov, S; Udem, T.; Yudin, V.I.,; Pshenichnikov M.; Sarkisov O., Eds. Proceedings of the SPIE, 2007, Vol. 6727A p. 53.

[Latyshev 2003] Latyshev, A.N.; Ovchinnikov, O.V.; Okhotnikov, S.S. *J. Appl. Spectr.*, 2003, 70, 817- 820.

[Taylor 1933] Taylor, J.B.; Langmuir, I. *Phys. Rev.,* 1933, 44.423-458.

INDEX

1

A

D

E

F

G

H

I

J

K

L

M

N

O

P

Q

R

S

T

U

V

W

X

Y